D1740905

Biomapping Indigenous Peoples

Readings in Post / Colonial Literatures and Cultures in English

151

Biomapping Indigenous Peoples

Towards an Understanding of the Issues

Edited by Susanne Berthier-Foglar,
Sheila Collingwood-Whittick and Sandrine Tolazzi

Amsterdam - New York, NY 2012

Cover image:
Jean Berthier, *Manuelita* (1986; acrylic on board)

Cover design:
Inge Baeten

The paper on which this book is printed meets the requirements of "ISO 9706:1994, Information and documentation - Paper for documents - Requirements for permanence".

ISBN: 978-90-420-3591-1
E-Book ISBN: 978-94-012-0866-6

Printed in The Netherlands

Table of Contents

Part III: Surviving and Resisting

Part IV: Opposing and Reclaiming

Preface

MAPPING THE HUMAN (BIOMAPPING) – MAPPING HUMAN GENES, mapping human migrations – becomes a sensitive subject when Indigenous peoples are involved. Even when scientists profess the best of intentions, an absence of financial interests, the guaranteed anonymity of individuals tested, even when research institutions pledge assistance to the targeted Indigenous communities, by giving back scholarships or grants, even then, suspicion is ever present among those tested.

There is a widespread belief among Indigenous peoples that science still functions within the same epistemological reality as in the nineteenth century, that colonial attitudes still power research institutions, and that subalterns remain subalterns with little power over their fates.

The present study is concerned with the peoples who, in the past few centuries, have been dispossessed of their territories, their cultures, as well as most of their sovereignty, by colonizing powers. The recent (2007) Declaration on the Rights of Indigenous Peoples has not yet been ratified by all the countries that have Indigenous peoples within their borders. And in those countries where the Declaration has been ratified, it is still too soon to observe any effects on Indigenous sovereignty or land ownership.

Therefore, in a context of uncertainties concerning the rights of Indigenous peoples, whenever 'testing' is offered, the fear arises that genetic identity may be used to suppress the rights of those whose DNA does not conform to certain standards. For Indigenous communities in which the ghost of past scientific and medical experiments looms large, there is also the fear that 'testing' may pave the way for medical abuse.

Moreover, genetic testing is historically linked to past classifications of individuals. Despite contemporary debates over the notion of race, and the diverse meanings of the term in various cultures and countries, and despite the fact that the term is now widely accepted as having cultural rather than biological significance, it is difficult to forget that scientific classification of humans began in the eighteenth century in a context of white supremacy.

The debate over the origins of mankind raises another set of questions. In the nineteenth century, the answer could be found in a theological interpretation of the Old Testament with the possibility that several Adams and Eves gave rise to the various races (i.e. on the basis of skin colour) or that the black race was caused by a curse on Ham, one of Noah's sons. Archaeological finds of ancient humans – or pre-humans – always cause a great stir among the non-scientific population. Giving bones names humanizes them. "Lucy" suggests a closeness in the skeleton's relationship to us that a technical reference number could never convey. Conversely, every continent yearns for its own ancestor earning its population a place among the founding peoples of the world. Yet genomics tells us that there is only one human race on earth, originating in the Rift Valley in Africa, and migrating in small groups to every corner of the world. Most of the 'pre-humans' seem to have become extinct.

It seems difficult for some people – Indigenous or not – who are not related to Africa to imagine that continent as the home country of their distant ancestors. There are others who simply cannot bear the idea that the genes in every cell of our own bodies tell the story, not only of our human life, but also of life in general from the time the first unicellular living organisms started to exist in the primeval soup covering the surface of the earth. Researchers in the field of genomics claim that we are one people. What appears as a lofty ideal for the inhabitants of developed and mostly westernized countries is a threat for Indigenous peoples who defend their existence against what they consider an invasion by newcomers on a quest to appropriate resources. Not to mention that the scientific 'truth' presented by genomics researchers goes against most creation stories, whether they belong to Indigenous peoples or not.

Testing, we are told, can be done ethically, provided that informed consent is obtained and the interests of those tested are safeguarded. The difficulty lies in the complexity of genomics itself. Even people who have had access to schooling find it hard to understand the idea of a science that is constantly evolving, continually adding to its store of knowledge, and repeatedly issuing seemingly contradictory statements. Informed consent is thus often wishful thinking.

One might think that decoding the human genome is particularly offensive only for the dispossessed. Conversely, this would mean that non-dispossessed individuals in Western nations would not feel threatened by the analysis of their genome. Project deCode in Iceland, targeting all the inhabitants of a particularly homogeneous country in demographic matters, shows that the public outcry against personal knowledge becoming public – or even semi-public – is not specific to the dispossessed. And one is left to imagine the feeling of alienation of those who already feel slighted before any testing has taken place.

Decoding the genome enables the biomapping of individuals through their genetic traits. The migrations of their ancestors become visible. For the Indigenous and conquered peoples of the earth, DNA bears witness to the additional injury of a violent past of sexual conquest. Hard science, despite its professed objectivity and impartiality, is confronted with the 'messiness' of social science, with stories and myths that spell identity, and of course with politics.

The purpose of this book is to present the issues raised by Indigenous communities and individuals in various colonized countries and geographic areas in matters of genomics and biomapping. Most of the issues are centred on cultural and ethnic identities. Both types of identity are constructed and treasured by nations, groups, clans, families who reject outside input into what they consider to be their affairs. Indigenous peoples thus tend to consider DNA analysis as a threat to their self-identification, their existence, as well as their sovereignty. In a global context where ethnic and regional identities are re-affirmed almost everywhere, cultural identity appears as a refuge threatened by hard science.

SUSANNE BERTHIER-FOGLAR

⌘

Acknowledgements

The editors wish to thank the University of Savoie (France), and particularly the LLS Laboratory for Languages and Social Sciences[1] for their support. They are also grateful for the academic research funds of the French Rhône-Alpes region and the grant obtained (2008–2011) under Cluster 14 – Issues and Representation of Science, Technology and their Uses – to study "The Indigenous Refusal of Human Genome Studies in the Anglo-Saxon Settler Colonies."

⌘

[1] Laboratoire Langages, Litératures, Sociétés; Université de Savoie, Domaine Universitaire de Jacob–Bellecombette, BP 1104, 73011 Chambéry Cedex, France.

List of Illustrations

⌘

Part I
Overview

Human Genomics and the Indigenous

SUSANNE BERTHIER–FOGLAR

A Long History of Mistrusting Science

THIS BOOK IS ABOUT MISTRUST, MISTRUST AND AT TIMES ANGER, but most of the time mistrust. The peoples discussed – and discussing their mistrust of Western science, and more specifically population genetics – are the Indigenous peoples, a group the world became conscious of in the last decades of the twentieth century, when the voices of "those who were there first" came to be heard on the international scene. 'Western' science has long been a contentious subject between those studied – objectified – by science and the scientists whose gaze they were subjected to.

The United Nation's Declaration on the Rights of Indigenous Peoples – adopted in 2007 but still in the ratification process – creates an urgent need for an unbiased forum on contemporary Indigenous claims. These claims refer to facts and events from the point of contact between the Indigenous peoples and the mostly Western colonizers to the present, whether today is labelled colonial or postcolonial. The editors' field of expertise is discourse analysis: in this case, what is being said about hard science and the Indigenous peoples as well as what is being said by the Indigenous peoples about science. While scientists have a tendency to disclaim the relevance of non-specialist input, or to accept it only if it fits the criteria of science made understandable for the general public, we believe that discourse of every type constructs the way science is perceived by the public and thus has its importance.

In the field of population genetics, an emotionally charged subject for Indigenous peoples, the point here is not to discuss whether the findings of hard science are 'true' or not, but to investigate the way they are perceived by the public who are being investigated. In order to understand the Indigenous viewpoint, we have to place the dialectic of discourse on human genetics in the context of a history of abuse and dispossession. We thus have to deal – on the one hand – with a hard science few outsiders understand and – on the other – with burgeoning reactions seemingly disconnected from scientific reality and often dismissed by scientists as being irrelevant. It does not help that geneticists rarely, if ever, write for the general public.

It seems that the perception of genomics among the public at large, like most scientific subjects, is largely the result of partial information, the written equivalent of sound bites, or keywords. If more complex information is available to the general public, it is still highly technical[1] and is reduced to perceived keywords that, in turn, serve as the basis for an expanded discourse that comes to include the fears of dispossession and annihilation of the Indigenous peoples and thus a source of anguish. Genomics are consequently seen as a threat to Indigenous identity and as a possible tool in the political agenda if 'miscegenation,' and mixed-blood identity, should become a motive for reducing Indigenous rights. That is the starting point for the cross-disciplinary essays in this book, which aim to serve as a framework for the ongoing dialectic of Indigenous studies.

Previous Scientific Endeavours Centred on 'the Primitive'

Today's scientists do not feel bound by what others have done, in previous centuries, to the Indigenous peoples in the name of science. Despite the wrongs of the past there is no sense of original sin – and guilt – linking today's science to experiments and theories of the colonial nineteenth century when people were ranked according to racial characteristics, with white Westerners in a dominant position and the coloured Others below.

However, from the point of view of the Indigenous peoples, a discussion of their mistrust of population genetics – one of the branches of 'biomapping': i.e. mapping the biological – has to be considered as inherently linked to a long line of scientific or pseudo-scientific endeavours going back to the first contact between their ancestors and those who conquered them. Indigenous peoples consider science as a continuum, whereas today's scientists implicitly believe that the epistemological breaks that have occurred since the nineteenth century absolve them from being held accountable for past research and pseudo-science.

---

[1] Stewart Scherer, *Short Guide to the Human Genome* (Woodbury NY: Cold Spring Harbor Laboratory, 2008).

The Indigenous peoples became objects of scientific investigations almost from the beginning of conquest, when they were collected as specimens of interest, and sent back to the conquerors' homelands. Taxonomy, as practised by Carl von Linné (Linnaeus), was an endeavour to classify all living organisms, including humans, for which latter category five subspecies were defined: *afer* (African), *americanus* (American), *asiaticus* (Asian), *europaeus* (European), and – a catchall category – *montrosus*, for those imagined or thought to exist somewhere.[2] Racism intruded into science, particularly in the nineteenth century, when it became increasingly necessary to rationalize white domination over coloured peoples and various theorists attempted to establish the superiority of the white race. Anglosaxonism was the ideological tool devised to explain why 'civilization' had moved from East to West in the world and why, for the Reverend Josiah Strong, the Anglo-Saxons were God's chosen race, elected to lead the world and win "the final competition of races, for which the Anglo-Saxon is being schooled."[3] Other colonial powers had similar rationalizations for their expansion, a fact that gave rise to theories about the "irreversible differences between races" which led some scientists to eventually promote the notion of polygenesis: i.e. separate origins, separate Adams and Eves for the different races.[4]

It was thought that cultural differences among the races originated in the structure of the brain. Again, the main point was to provide evidence for the superiority of the white race. Phrenologists, such as George Combe, imagined a relation between brain areas and specific faculties and emotions. In the days before brain imaging, the approach was largely conjectural and fantasized. The size of a particular area was supposed to indicate the extent of its power.[5] In the nineteenth century, mapping brain areas was controversial almost from the start, at least among medical doctors, but proved a success with the non-scientific mainstream, probably because it seemed easy to relate skull shapes to imagined or stereotypical racial characteristics.[6]

---

[2] Spencer Wells, *Deep Ancestry* (Washington DC: National Geographic, 2007): 17.

[3] Josiah Strong, *Our Country: Its Possible Future and Its Present Crisis* (New York: The American Home Missionary Society, 1885): 175.

[4] Reginald Horsman, *Race and Manifest Destiny: The Origins of American Anglo-Saxonism* (Cambridge MA: Harvard UP, 1981): 125.

[5] Horsman, *Race and Manifest Destiny*, 56–59.

[6] Roy Porter, *The Cambridge History of Medicine* (Cambridge: Cambridge UP, 2006): 115–16.

Meanwhile, 'real' scientists preferred to deal with measurable factors such as cranial capacity. The so-called 'craniologists' experimented with various ways to accurately measure the internal capacity of a cranium. The tables produced by Samuel George Morton in *Crania Americana* (1839) have an air of scientific respectability about them.[7] However, when we consider the whole process, it is interesting to focus on how the human bodies that provided the testing material were treated by the scientists. Before they could be measured, the "specimens" had to be "harvested." In their eagerness to build collections, scientists and their helpers scoured burial sites and even recent battlefields of the Indian wars to send back freshly culled crania to Morton. The captions of the plates illustrating *Crania Americana* indicate where each specimen was found, "From a Mound" (Plate 52), or "obtained [...during...] a sojourn on the Columbia River" (Plate 43) an enumeration suggesting trivial souvenirs.[8] After collecting came the unsavoury process of cleaning out the cranial cavity, filling the space left with grain, then emptying the grain into a measuring vessel.[9] To the layman, any post-mortem medical procedure is dehumanizing per se; the fact that those under the medical gaze were the Indigenous peoples or colonized Others thus readily conjures up powerful and lasting images of domination that have now become – over a century and a half later – synonymous with scientific research in the eyes of the Indigenous peoples.

The Māori theorist Linda Tuhiwai Smith argues that the word "research itself is probably one of the dirtiest words in the indigenous world's vocabulary,"[10] and she illustrates her point by describing the "obnoxious" measurements of the cranial capacity of her ancestors. Insensitive vocabulary is often a characteristic of science. In the case of contemporary genomics, genes have to be obtained by "grinding [them] in a mortar and cooking them in a beaker";[11]

---

[7] Samuel George Morton, *Crania Americana* (Philadelphia PA: J. Dobson, 1839).

[8] Donna Spalding Andréolle & Susanne Berthier–Foglar, "Science and the American Empire: The American School of Anthropology and the Justification of Expansionism," in *Science and Empire in Nineteenth Century: A Journey of Imperial Conquest and Scientific Progress*, ed. Catherine Delmas, Christine Vandamme & Donna Spalding Andréolle (Newcastle upon Tyne: Cambridge Scholars, 2010): 83–98.

[9] Samuel Morton, *Crania Americana*, 11.

[10] Linda Tuhiwai Smith, *Decolonizing Methodologies: Research and Indigenous* Peoples (Dunedin: U of Otago P, 1999): 1.

[11] Herman J. Muller, quoted in Horace F. Judson, "A History of the Science and Technology Behind Gene Mapping and Sequencing," in *The Code of Codes*, ed.

they can be "snipped" and "spliced,"[12] a process that seems disconnected from their status as human material. These contemporary phrases and terms do not come from an ideologically charged source but from a reputable book on genomic research made understandable to the general public, which suggests that the connotation of words in scientific texts remains an unaddressed issue.

Science and the Disenchantment of the World

Misunderstanding science also stems from other causes such as the uneasy relation between science, knowledge, and belief-systems. Structuring beliefs and knowledge into a coherent world-view has been among Man's earliest endeavours. To this effect, archaic religions provided an all-encompassing rationalization of life and cosmic phenomena.[13] Advancing scientific knowledge led to the first major epistemological break of the "modern" world coinciding with the onset of the industrial age[14] incidentally occurring at a time of "confident racism"[15] as two facets of the same trend. From then on, science in the Western world diverged from all other aspects of knowledge which came to be assigned to the non-scientific spheres of the religious, the traditional, or the folkloristic. For Max Weber, this epistemological break is at the root of the "disenchantment" of the world,[16] a break that separates the world into a magical 'before' and a rational 'after' producing a sense of loss that is, for some individuals, associated with the passage from 'magic' to science. Today, the education of most Western individuals is determined by the divergence between science and that which is deemed non-scientific. Scientific knowledge is transmitted through textbooks and specialized journals and, as science has become more complex, the term 'shelf-life' has come to be applied to scientific publications, which, in turn, have lost their accessibility for the layman. Knowledge has become perishable, and in the span of a human lifetime obsolescence occurs. Moreover, the objects studied by science have become

---

Daniel J. Kevles & Leroy Hood (Cambridge MA: Harvard UP, 1992): 48.

[12] Daniel J. Kevles, "Out of Eugenics," *The Code of Codes*, 19.

[13] Åke Hultkrantz, *Native Religions of North America: The Power of Visions and Fertility* (Long Grove IL: Waveland, 1987): 21–28.

[14] Michel Foucault, *Les mots et les choses* (Paris: Gallimard, 1996): 346.

[15] Nicholas Thomas, *Colonialism's Culture* (Princeton NJ: Princeton UP, 1994): 77.

[16] Max Weber, *The Protestant Ethic and the Spirit of Capitalism* (Mineola NY: Dover, 2003): 105.

largely invisible to the naked eye. The study of the universe has expanded into remote galaxies while the structure of matter is pursued down to the subatomic scale, and research concerning biological species delves into cells and their nuclei. Scientific discoveries are reported to the general public at an accelerated pace. To the non-scientific public, science and science fiction seem, and are at times, intertwined, and the public's hunger for the extraordinary has spawned a new literary genre, the techno-thriller, which has become a conveyor of "science" to the general public. The leaders of the genre play on the public's wish to be thrilled by scientific discoveries and non-fiction. When Dan Brown states, in an author's note on the first page of *Deception Point*, that "All technologies described in this novel exist,"[17] he panders to the public's tastes for non-fiction. In the same way, Michael Crichton has made it a trademark of his novels to take scientific reality one step further. Dinosaur DNA found in fossilized samples – this is reality – is used to reconstruct a living specimen – which is science fiction.[18]

The need for 'magic' transcends Max Weber's disenchantment of the world and gives the main role to the created story. Science is fascinating and may be endowed with an aura of magic. However, fascination does not entail understanding. And despite a wealth of scientific programmes on TV channels in Western nations, despite all the efforts made to render science palatable, to give it the aura of 'entertainment' in order to alleviate the fears associated with the unknown,[19] science is at times rejected in favor of a mystical understanding of nature and creation.

For Indigenous activists, the acceptability of science is particularly problematical and any – however well-meant – attempts to present the positive side of scientific research concerning Indigenous peoples are seen as the imposition of an imperialistic viewpoint and are met with distrust. Sometimes, it seems that this anti-science sentiment is being used as a platform, and thus magnified, by "Indian tribal politicians, preying on the technological distrust of their constituents."[20] Minorities have often been mistreated by Western

---

[17] Dan Brown, *Deception Point* (London: Corgi, 2001).

[18] Michael Crichton, *Jurassic Park* (New York: Alfred A. Knopf, 1990).

[19] Daniel J. Kevles & Leroy Hood, *The Code of Codes*, vii–x; Matt Ridley, *Genome: The Autobiography of a Species in 23 Chapters* (New York: HarperCollins, 2000): 1–3.

[20] Kimberly TallBear, "Genetics, Culture and Identity in Indian Country," quoted in Eric Beckenhauer, "Redefining race: can genetic testing provide biological proof of Indian ethnicity?" *Stanford Law Review* 56.1 (October 2003): 182.

medical science, but none as much as the Indigenous peoples and the enslaved. Even enlightened contact with science may have a negative side. One of these confrontations between the Indigenous peoples and Western science occurred in the Bosque Redondo camp where the Navajo were deported between 1864 and 1868. The military MD, Dr Guyther, tried his best to understand his charges, their reluctance to use the hospital he built, their beliefs in the negative effect the spirits of the dead have on the living, the Navajo way of using herbs and ceremonies in therapeutic procedures. However, he also offhandedly observed that a Navajo could eat "twice as much as a white man," not mentioning that this extraordinary appetite was a temporary condition brought about by starvation.[21]

Scientific research has not always been kind to the Indigenous peoples, a fact that has given Western medicine such a bad reputation in the eyes of many Indigenous peoples. The forced sterilization of Indigenous women is attested in the Anglo-Saxon settler colonies,[22] as is medical research that did not state its aim: i.e. samples of human tissue collected for a specific purpose and used for another line of research.[23] That does not mean that today the Indigenous peoples necessarily reject Western science per se (such as diabetes studies). Medical projects aimed at Indigenous communities, and serving their immediate needs, have gained acceptance.

The focus in this collection of essays is on DNA and genomics, a science that is mostly obscure to the non-scientific general public. While the acronym DNA is known (at least, has been heard of), the reality behind the three letters is murky. It is generally acknowledged by the layman that DNA is used in criminology, forensics, and paternity suits. Its use in medical science is less well publicized. In fact, for the general public, the term DNA in itself has acquired the restricted meaning of identity – so much so, that commercial brands unrelated to the field of biology now claim to have their own 'DNA',

---

[21] George Gwyther, "An Indian Reservation," *Overland Monthly* (10 February (1873): 126; Susanne Berthier–Foglar, "A patient/hospital relationship in 1863–65: Mainstream Doctors and Navajo Patients in the Bosque Redondo Camp," in *The Patient*, ed. Kimberly R Myers & Harold Schweizer (Lewisburg PA: Bucknell UP, 2010): 87–90.

[22] Daniel J. Kevles, "Out of Eugenics," *The Code of Codes*, 10–11.

[23] Samer Muscati, "Indigenous Peoples Denounce DNA Conference" (1996), http://www.yvwiiusdinvnohii.net/articles/dna-con.htm (accessed 26 May 2011).

meaning the way they appear through their logo and corporate identity.[24] It seems to be only marginally understood that DNA or, more precisely, genomics, the study of individual genes, might lead to personalized therapies based on the genetic makeup of each individual, taking into account the way s/he would react to a particular medicine. One statistically efficient way to introduce personalized medicine involves drugs that have a positive effect on a specific ethnic group. In 2005, the US Food and Drug Administration approved for the first time a "race-specific" drug, "for the treatment of heart failure in self-identified black patients."[25] Racial self-definition in this case becomes a matter of life or death and not a mere statement of identity. For the obvious lack of a precise definition of 'race', race-targeting of drugs has been criticized.[26] If we posit that 'race', identity, and DNA are related terms with possibly related meanings – at least in mainstream public opinion – it is easy to understand the fear of Indigenous activists, who focus on the risk of imposed identities and on the limitation in self-definition that the widespread use of DNA analysis would entail. Incidentally, the debate surrounding race-targeted drugs is a practical illustration of the quandary of the layman who receives conflicting information from reputable sources.

The idea for this book started with a long-term involvement with public policies concerning the tribes – now nations – in the USA. It then expanded into a global study of the Indigenous peoples – the colonized – and their uneasy relationship with science, notably genomics.[27]

As for who is Indigenous, and who is not, the debate went on for over two decades in the United Nations' Working Group on Indigenous Populations, WGIP, a forum created to draft the Declaration on the Rights of Indigenous Peoples. No universally accepted answer has yet been found.[28] Unrelatedly,

---

[24] A Seattle company calls itself DNA and advertises for their "expertise in brand strategy."

[25] *FDA Approves BiDil Heart Failure Drug for Black Patients* (2005).

[26] Sergio D. Pena, "The fallacy of racial pharmacogenomics," *Brazilian Journal of Medical and Biological Research* 44.4 (April 2011): 269–72.

[27] This project received regional academic financing in France:"Cluster 14" Projects, Rhône–Alpes, France, 2008–2011. Project Leader Susanne Berthier–Foglar, University of Savoie, France; participants: Sandrine Tolazzi and Sheila Collingwood–Whittick, both from the University of Grenoble, France.

[28] Justin Kenrick & Jerome Lewis, "Indigenous Peoples' Rights and the Politics of the Term 'Indigenous',".*Anthropology Today* 20.2 (2004): 4–9.

several projects involving the study of human DNA among Indigenous peoples were started and failed, mostly because patenting and ownership of human tissues remained an ill-defined domain. However, in the Anglo-Saxon settler colonies, the issues raised became a matter of concern among Indigenous activists, who took an oppositional stance against all forms of genomic studies involving their people.[29]

Population Genetics and the Essence of Humanity

The study of DNA is first and foremost a study of life. That is how it was presented in 1953 when Watson and Crick discovered the molecular structure of DNA as a double helix in the nucleus of every cell in every living organism.[30] The typical cell in the human body contains six feet of densely coiled DNA in its nucleus, irregularly divided between 23 pairs of strands called chromosomes.[31] When human DNA was decoded in 2000, less than half a century later, it was announced to the non-scientifically trained public as probably the greatest accomplishment of all times. Depending on the commentator's point of view, the genetic material contained in DNA was seen as the "Book of Life," or, in a more down-to-earth way, the blueprint or the manual for the human species,[32] or, more flamboyantly, as a new quest akin to the one Lewis and Clark pursued in their voyage of discovery across the North American continent.[33]

The hyperbole and the recourse to metaphor are symptomatic of an arcane subject where meaning exceeds the descriptive capacity of everyday language. When DNA is made understandable to the general public, disturbing information comes to light. What is called the human genome also contains building blocks of previous evolutionary phases "from when we were single-celled creatures."[34] The genome which has collected information over some

---

[29] Debra Harry, "Acts of Self-Determination and Self-Defense: Indigenous Peoples Responses to Biocolonialism," in *Rights and Liberties in the Biotech Age*, ed. Sheldon Krimsky & Peter Shorett (Lanham MD: Rowman & Littlefield, 2005): 87–97.

[30] James D. Watson, *The Double Helix* (New York: Scribner, 1968).

[31] Robert Cook–Deegan, *The Gene Wars* (New York: W.W. Norton, 1994): 18.

[32] Nicholas Wade, "Reading The Book Of Life: Genetic Code of Human Life Is Cracked by Scientists," *New York Times* (27 June 2000): A1.

[33] Craig Venter, *A Life Decoded: My Genome: My Life* (New York: Penguin, 2007): 313.

[34] Scherer, *Short Guide to the Human Genome*, 2.

four billion years prompts questions about the essence of humanity.[35] What makes us human? What differentiates us from our closest relatives, the chimpanzees, since "we are, to a ninety-eight percent approximation, chimpanzees?"[36] And if humans are so closely related to the chimpanzees, what about the differences between humans, between human "races"? Criminology and paternity suits led to one of the uses of DNA becoming understandable to the lay public. Individuality is readable in DNA; it enables the conviction of a criminal, even years or decades after the crime, through a surviving sample of human tissue. DNA also enabled researchers to trace the (black) descendants of US President and Founding Father Thomas Jefferson over 170 years after his death.[37] Hence, powerful – albeit mysterious – markers of identity are present.

The Dark Side of Genetics: Eugenics, Mapping the Disappearing DNA

However, genetics as a science was associated in its early stages with eugenics. Eugenic policies were implemented in the 1920s and 1930s in the USA and in Europe to produce a 'better' variety of humans, "Grade A Individuals."[38] To achieve this desirable result, forced sterilization was seen as the humane way to prevent those with unwanted characteristics from reproducing. It was implied that the less desirable characteristics were prevalent among the poor and the minorities. Hence, culling the human population and defining the characteristics to be retained became an effort to reduce the birthrate of minorities. After World War Two, to distance eugenics from the Holocaust, an extreme version implying the wholesale destruction of millions of human beings, 'eugenics' became 'genetics' in a broad range of academic publications and institutions.[39] The *Annals of Eugenics* was renamed the *Annals of Human Genetics*; *Eugenics Quarterly* became *Social Biology*, the University of London's Department of Eugenics became the Department of Genetics.

---

[35] Scherer, *Short Guide to the Human Genome*, 123.

[36] *Short Guide to the Human Genome*, 28.

[37] Eugene A. Foster, "Scientific Correspondence: Jefferson Fathered Slave's Last Child," *Nature* 396 (1998): 27–28.

[38] Daniel J. Kevles, "Out of Eugenics," *The Code of Codes*, 10.

[39] Vincent Sarich & Frank Miele, *Race: The Reality of Human Differences* (Boulder CO: Westview, 2004): 91.

As for the history of genetic studies, once the initial negative connotations were out of the way, research proper started in the mid-1970s and gained momentum in 1986 with the Human Genome Initiative studying radiation-induced mutagenesis. In 1990, the project evolved into the Human Genome Project, with the participation of China, Japan, and several European nations. By then, the scope of the study had expanded to identify all human genes and to interpret human DNA.[40] However, a year later, in 1991, the project was criticized by a leading geneticist from Stanford, Luca Cavalli–Sforza, who argued that, instead of focusing on eurocentric populations, the Human Genome Project should be expanded to take into account the worldwide diversity of humans. He lamented the fact that "a number of populations of considerable interest are rapidly disappearing"[41] and suggested that the genes of "populations of special interest" be sampled rapidly.[42] Despite the goodwill behind this call for action and the sensitivity of Luca Cavalli–Sforza, his words were understood by activists to mean that he merely wanted to save the history of *Homo sapiens sapiens*, and not the Aboriginal populations themselves. They condemned the project for considering the fate of the disappearing DNA more important than the political and economic conditions that caused the Indigenous peoples to disappear physically or statistically. Moreover, his "descriptor – Isolates of Historic Interest – would be branded a moral outrage by critics who argued that researchers viewed populations as historical curiosities, and little more."[43] The objectification implied by the words used compelled a Filipino activist to declare that the funds for genome studies would be better used to promote the rights of Indigenous peoples.[44] Despite protests, work on the Human Genome Diversity Project (HGDP) started in 1993, before the human genome was fully sequenced (i.e. decoded.)

---

[40] Robert Cook–Deegan, *The Gene Wars*, 79–85.

[41] Lugi Luca Cavalli–Sforza, "The Study of Variation in the Human Genome," *Genomics* 11 (1991): 495.

[42] Lugi Luca Cavalli–Sforza, "Call for a Worldwide Survey of Human Genetic Diversity: A Vanishing Opportunity for the Human Genome Project," *Genomics* 11 (1991): 490.

[43] Samer Muscati, "Indigenous Peoples Denounce DNA Conference"; Jenny Reardon, *Race to the Finish: Identity and Governance in an Age of Genomics* (Princeton NJ: Princeton UP, 2005): 68.

[44] Human Genetics Alert, "The Human Genome Diversity Project," http://www.hgalert.org/topics/personalInfo/hgdp.htm (accessed 3 June 2011).

'Racial' Characteristics and Identity

Traditional markers of identity are often concerned with physical aspects such as skin colour and other visible physical traits – hair colour and type, eye colour, body build – traditionally used to differentiate groups of people. Some visible markers of identity, such as clothes, hairdo, and objects attached to the person, are manipulable, and deemed "media" by Marshall McLuhan.[45] DNA has, however, introduced markers of identity that reach beyond culture, beyond what individuals can control. Hence the feeling of powerlessness of a person, or group, on whom a genetic identity is imposed. Full genome sequencing, and the ensuing classification of people into categories, negates the right of individuals or groups to define their own identity, rendering void their representation of themselves. Indigenous peoples prefer self-defined identities to science-based identities, as self-definition is an important aspect of sovereignty and self-government. Moreover, the pain of imposed identities brings to mind former experiences of miscegenation, often the result of Indigenous women being raped in colonial times. For the Indigenous peoples, the ingrained remembrance in their own DNA of their ancestors' rape by conquest coalesces with a profound distrust of Western science, leading to rejection of genomics in toto.

A Brief Overview of DNA Studies: Mitochondrial DNA and the Y Chromosome

What exactly does DNA tell us?[46] Each human being carries DNA in every cell: i.e. bits of information arranged in coiled strands, linked together by rungs in the manner of twisted ladders. Technically, DNA is deoxyribonucleic acid, or nucleid acid bases on a sugar backbone constructed with four basic elements, A–C–G–T, the short form of adenine, cytosine, guanine, and thymine, arranged in pairs. There are about 30,000 genes in the human genome. However most of the genome "is a vast expanse of DNA sequence with no known function," as it is non-coding and is thus called 'junk DNA'.[47] Most of the DNA recombines for each offspring and the elements transmitted

---

[45] Marshall McLuhan, *Understanding Media: The Extensions of Man* (New York: McGraw–Hill, 1964).

[46] The information on DNA in this paragraph, unless otherwise stated, is taken from Spencer Wells, *Deep Ancestry* (Washington DC: National Geographic, 2007): 13–17.

[47] Daniel J. Kevles, "Out of Eugenics," *The Code of Codes*, 24.

by the mother and the father will be jumbled; not all will be expressed. The DNA in each cell is located in the nucleus. But there is an additional source of DNA – in fact, an altogether different type of DNA – in the mitochondria of each cell. These mitochondria are tiny elements, hundreds of which are present within each cell. They are probably of parasitic origin and are remnants of the days our very distant ancestors were unicellular beings – or close to that state – and were infected by bacteria. They never got rid of the intruders, who live on, or at least their DNA lives on, in a symbiosis within all cells. Mitochondrial DNA is not recombined during reproduction and only the mother's mitochondria are transmitted intact to the fetus. Each person thus carries a carbon copy of the maternal mitochondria, and only daughters transmit it. We can therefore assess the maternal line through mitochondria.

What is of interest here, is that lineages become traceable through a specific marker in their non-coding DNA. In the duplication process of mitochondrial DNA, random mutations: i.e. glitches, occur about once every 10,000 years. Once the mutation has happened, it is further transmitted by the daughters. Thus, the mitochondria represent a biological clock indicating a line of common ancestors. All the individuals with the same mitochondrial DNA will have a common ancestor. The notion of the common ancestor was popularized by the Oxford geneticist Brian Sykes.[48] In his bestseller about genomics and the ancestry of the female line, Sykes claims that every individual of European ancestry descends from one of the seven original mothers who have produced a line of offspring traceable to the present. *The Seven Daughters of Eve* is a rare example of genomics made understandable to the non-specialist public.

Another type of DNA, present only in males, the Y chromosome, has similar characteristics. Humans have 22 pairs of chromosomes, plus the 'sex' chromosomes, two XX for the females and XY for the males. The Y chromosome is part of a mismatched pair (X+Y) and thus is not shuffled when passed from father to son, except for random mutations which can be used as a biological clock to trace the male line.

Using mitochondrial DNA and the Y chromosome, geneticists can identify the most recent common ancestor of everyone alive who carries a given marker.[49] The different markers can be followed backwards all the way to the

---

[48] Brian Sykes, *The Seven Daughters of Eve* (New York: W.W. Norton, 2001).

[49] Solving The Impossible, "Cutting Edge Genealogy – The Use of DNA," http://solvingtheimpossible.com/creativitycompas.php (accessed 3 May 2011).

oldest common ancestor. 'Adam' thus existed 145,000 years ago and 'Eve' 190,000 years ago in Africa.[50] Population genetics do not take into account genes associated with visible characteristics such as skin or eye colour, or type of hair, and the markers have been selected only because they represent a common mutation.

Does the Concept of Race Exist?

Population genomics are about analysing the human genome and the variations between ethnic groups. However, the traditional variants used for classifying humans, such as skin colour and other physical characteristics, date back to a period when racism was widely accepted. The notion of race itself has never been debunked in the popular imagination, despite UNESCO's 1950s attempt to do so, and it is still widely ingrained in societal representations in the USA, where it is largely linked to skin colour.[51] In the early-twenty-first century, race made a comeback with Sarich and Miele's controversial book *Race: The Reality of Human Differences*, in which the authors argue that differences between humans go deeper than the skin and that they are as visible as the differences between breeds of dogs,[52] the words used being deliberately provocative.

Indigenous Activists and the Human Genome

When the Human Genome Diversity Project was started in 1993, the human genome had not yet been sequenced, meaning that the 30,000 different genes were not yet sorted out. Nevertheless, it was already known that several markers for given mutations were specific to one haplogroup (a genetic 'family'). In the sequencing phase of the human genome, when large amounts of funding were needed, pharmaceutical companies moved in, patenting genetic material on the way before knowing what the patents might be used for and often clashing publicly with the proponents of science for the public domain. The idea was to 'carpet bomb' each sequence of DNA with legalese.[53] While

[50] Luigi Luca Cavalli–Sforza, *Genes, Peoples, and Languages* (New York: North Point, 2000): 79–81.

[51] Sarich & Miele, *Race*, 92.

[52] Sarich & Miele, *Race*, 9, 198–200.

[53] Venter, *A Life Decoded*, 260.

the military term fits the patent war, it is also symptomatic of the mind-set of those who went about the business of privatizing human material.

Distancing itself from the commercial uses of genomics, and advocating a strict ethical approach, Project Genographic was launched by National Geographic as a private foundation to study human migrations, with funding for the project being provided from 2005 to 2010. The upbeat message of the project is centred on a belief in the universality of scientific knowledge and the notion that all humans are genetically related. For the geneticist Spencer Wells, heading the project, the study of human migrations had to be done as soon as possible, focusing on geographic areas not affected by the mass migrations that have occurred since the nineteenth century, and on Indigenous peoples isolated from migrations.[54] However, he meant hunter–gatherers in Africa and reindeer hunters in Siberia, not Native Americans in the USA. According to Wells, postponing large-scale analysis of human genes means facing losses of knowledge similar to those that lie behind the biodiversity crisis.[55] In 2007, Genographic had "high-quality genetic data from perhaps 10,000 indigenous people"; not truly representative of the global world population of 6.5 billion, or the indigenous population of 350 million.[56] However, Genographic follows the guidelines proposed by Luigi Luca Cavalli–Sforza in 1991.

According to Genographic's migration studies, the first group to leave the Rift Valley in Africa were the ancestors of the Australian Aborigines who started their migration around 50,000 B.P., following the coastline of Africa, and around India, and along the shores of the Indonesian archipelago into Australia. The peopling of Europe started around 40,000 B.P., with one group circling the Mediterranean and a later one taking a northern route into Central Europe. A band from this Northern group split and reached the Northern reaches of present-day Mongolia. From there a small group, probably no larger than twenty-five individuals, the typical size of a group of Siberian deer hunters, crossed the emerged territory of Beringia into North America.[57] Humans did not enter America before 20,000 B.P., and theories of older

[54] Wells, *Deep Ancestry*, 45.

[55] *Deep Ancestry*, 4.

[56] *Deep Ancestry*, 48.

[57] *Deep Ancestry*, 99; Spencer Wells, *The Journey of Man* (New York: Random House, 2002).

settlements are therefore not verified.[58] These first settlers prospered in the Americas, settling the continent from Alaska to Argentina; a later migration is present only in Western North America and never made it to South America. Moreover, it appears that the Eastern Siberian clan is related to the Navajo and both originated with a single man living around 20,000 years ago.[59] To sum up the findings of Project Genographic, Spencer Wells concludes that when "Columbus encountered Native Americans on his 1492 voyage, he was reuniting two branches of the human family tree that shared the same great-, great-… grandfather on the steppes of Asia 40,000 years before."[60]

Indigenous Protests Against All Forms of Genetic Testing

There was no Indigenous protest in 2007 when Spencer Wells described the 1492 encounter between Columbus and the Natives on Guanahani as a reunion of cousins. Protest had erupted earlier, in the USA, at the beginning of the Human Genome Diversity Project. Any form of testing, not to mention patenting of human indigenous genetic material, was deemed inappropriate by Native American activists and was reminiscent of previous violence and attempts to secretly sterilize Native women in the recent past. Most of the activism can be traced to the Indigenous Peoples Council on Biocolonialism (IPCB), an organization based in Nevada and headed by Debra Harry, a Northern Paiute and Indigenous rights activist. Tribal leaders, on the other hand, did not trust a project that might uncover miscegenation on a larger scale than previously admitted, an outcome that could put federal subsidies and recognition of the tribes at risk. Moreover, Native American genes were introduced into the European population during conquest when humans were brought back as souvenirs or servants. Miscegenation in one population or the other could undermine Indigenous claims to self-determination or tribal role ancestry. It must be said that geneticists are usually very receptive to such arguments and have been known to abandon projects when they felt the local population to be ill at ease.[61]

Conversely, in the USA, for instance, traditionalists, often speaking through politicians, argue that according to their creation story, they already

---

[58] Wells, *Deep Ancestry*, 96.

[59] *Deep Ancestry*, 94–95.

[60] *Deep Ancestry*, 102.

[61] Evelyne Heyer, Professor at the French National Museum of Natural History in Anthropology Genetics, personal communication.

know where they come from and do not want to be destabilized by scientific knowledge they do not trust. For Westerners, this position is more difficult to accept, as it comes close to the creationism professed by mainstream religious activists. While the teaching of creationism tends to be discouraged in the Western world (save for sections of the USA run by fundamentalist Republicans), the same is not necessarily true with regard to the transmission of Indigenous world-views for the sake of cultural preservation.

A Wider Rejection of Genetic Testing

Meanwhile, negative elements concerning genetic testing have come to light. They are not related to the scientific validity of genomics but to genomics as a marker of identity, and to scientific discourse as an imperialistic 'know-it-all' attitude.

Iceland is one of the best-publicized examples of a country-wide genetic-testing effort in the late-twentieth century. The Norse who settled Iceland in the ninth and tenth centuries are Indigenous, as they were the first people on the island and the Icelandic population is genetically homogeneous, as the medieval migration to Iceland was followed by the relative isolation of the population until the twentieth century. Furthermore, establishing genealogical ties has proved rather easy, since lineages were documented in sagas as well as in a mediaeval manuscript recording the settlement of the original families and their histories. The information provided by lineages, genetic testing, and medical records was supposed to provide interesting data for biomedical research and enable researchers to trace the distribution of diseases as well as the link between genes and diseases.[62] The Icelandic parliament passed a law authorizing the construction of a database and, when Project deCODE started in 1999, the company "entered into an arrangement with the pharmaceutical giant Hoffman–LaRoche" (3). It was believed that ultimately the project could lead to the development of DNA-specific drugs. Initial work on the Icelandic genealogy provided a diagram indicating the connections between individuals. Researchers could access the 'Book of Islanders', as it came to be called, without identification of the individuals mentioned while Icelanders could, with the proper privacy safeguards, access the data with the names of all the persons they were connected to (70–71). Two main problems arose

---

[62] Gísli Pálsson, *Anthropology and the New Genetics* (Cambridge: Cambridge UP, 2007): 72. Further page references in this paragraph are taken from the same book and page numbers are given in the main text.

which are not specific to Iceland and are voiced in other areas where Indigenous peoples have had their ancestry traced or tested. The first one is the scientists' belief in their right to override the family histories of those tested because their findings are 'true' and proven. Disgruntled Icelanders attempted to sue deCODE because the project disclosed gaps such as missing paternity information, or brought to light hidden family stories of teen pregnancies and inbreeding (71), or even recorded missing family members before the family knew about them (76–77). But their claims were rejected because "genealogical data [do] not constitute private property" (83). While the first set of objections came from insiders, the second objection paradoxically came from those who were left out. It is customary to present large-scale genealogies as circles, starting with the original ancestors in an inner circle and working outwards to the present-day descendants. Iceland's population has thus been represented as a ring comprising all the descendants of the original settlers and excluding all the latecomers (79). Being "outside the circle" is thus a marker of non-belonging. The project is now considered as one of the "Icelandic mistakes" and has come to a standstill (101). However, the Icelandic example concerns Indigenous peoples who do not suffer from dispossession, and one is left to imagine the pain experienced by those who *are* victims of imperialism and dispossession.

The details disclosed by genetic testing can be disturbing even if the genome in question belongs to an individual who has been dead for 4,000 years. In 2010, the well-preserved hair of an ancient dweller of Greenland was discovered in the permafrost, providing "a genome that is as good quality as a modern human genome."[63] His DNA gave information about the migrations of his ancestors who left the Siberian Far East 5,500 years ago and pointed to "a substantial and recent migration across the Bering Strait, independent of the one that gave rise to modern Inuit and Native American populations." While the description of eyes, skin, and teeth are standard elements in ethnographic writings, other types of information hint at an almost personal knowledge of the deceased and appear as too intimate to be published in a scholarly text. We learn, for instance, that the person "was an inbred male," and that he also had "an increased risk of baldness [...] dry earwax,

[63] Cassandra Brooks, "First ancient human sequenced," *The Scientist* NewsBlogs (2010), http://www.the-scientist.com/blog/display/57140/ (accessed 3 June 2011).

and a metabolism and body-mass index "adapted to a cold climate."[64] The publication about the genome of the Saqqaq man by Morten Rasmussen et al. in *Nature* shows that gene sequencing goes deeper than migration studies. It also shows us the possible causes of the already mentioned rift between modern science and the Indigenous peoples. Rasmussen envisages the next challenge as being sequencing "an ancient human genome from material outside the permafrost regions."[65]

Such a challenge might indeed prove impossible in the Anglo-Saxon settler colonies, considering the difficulties encountered by scientists who wanted to study the skeletal remains of Kennewick Man, whose 9,000-year-old bones were claimed under NAGPRA[66] by the present-day tribes of the Columbia River. Whether Kennewick Man is, or is not, an ancestor of the Umatilla tribe became secondary to the fact that "scientific property rights" clashed with human-rights arguments. And when the Pawnee lawyer Walter Echo–Hawk presented his arguments – "All we're asking for is a little common decency [...]. We are not asking for anything but to bury our dead" – the stage was set for a battle against DNA testing.[67] In the event, empathy and an emotional understanding of the plight of dispossessed peoples took precedence over scientific interests.

Dispossession can also be seen on the level of scientific discoveries. China refutes the claim that the ancestors of all modern humans in Asia have migrated out of Africa and tries to promote the theory of a parallel evolution from *homo erectus* to *homo sapiens*. For Spencer Wells, the Asian *homo erectus* is an evolutionary dead end wiped out and replaced by the fully modern humans who arrived between 40,000 B.P. and 35,000 B.P.[68] Dispossession can thus be experienced on an intellectual level.

⌘

---

[64] Morten Rasmussen, "Ancient human genome sequence of an extinct Palaeo-Eskimo," *Nature* 463 (2010): 760.

[65] Rasmussen, "Ancient human genome sequence," 761.

[66] Native American Graves Protection and Repatriation Act, Pub. L. 101-601, 25 U.S.C. 3001 et seq., 104 Stat. 3048, United States Federal Law.

[67] David H.Thomas, *Skull Wars: Kennewick Man, Archaeology, and the Battle for Native American Identity* (New York: Basic Books, 2001): 240–43.

[68] Wells, *The Journey of Man*, 118–21.

Preliminary Research of Summer 2008 in the Southwest of the USA[69]

Assessing public opinion is a difficult subject, all the more so when the subject discussed is population genomics, not a particularly mainstream theme. Activists tend to be very vocal, while the majority of people are apparently unconcerned or unwilling to discuss an arcane subject. To avoid negative publicity, Project Genographic, well under way in 2008 when the public-opinion test was carried out, affirmed strong ethnical guidelines, pledged to test willing and informed groups or individuals only, and professed a "giving back to the group attitude" in the form of educational grants.[70] However, the claim that the project was "nonmedical, nonpolitical, nongovernmental, nonprofit"[71] was rejected by its critics, who feared that the study would be yet another attempt to deprive the Indigenous peoples of one of their possessions. Arguments that amalgamated all genomic research with all medical research projects gone awry, and with pseudo-science back to the nineteenth century, were frequently put forward.

In order to alleviate Indigenous fears and to present science seen from the perspective of the 'common' Indigenous person, Project Genographic gives a few representatives of the Indigenous peoples such as Julius Indaaya, chief of a matrilocal tribe in Tanzania, and Phil Bluehouse, a Navajo, the opportunity to present their viewpoint.[72] None had any uncertainties about their identity, and the analysis of their genomes confirmed their stories. But what about the silent mass of people, some with confirmed ethnic ties, whether enrolled in a tribe or not, some who imagined they might have an "Indian" ancestor, or those who – despite a fashion for Indian ancestors – refused to admit their mixed ancestry?

The public-opinion discussions were conducted as a pilot study in the Southwest of the USA in 2008, an area chosen for its high percentage of Native American inhabitants.[73] A questionnaire-type approach, such as that commonly used to assess marketing strategies, was ruled out as being in-

---

[69] Research conducted by Susanne Berthier–Foglar.

[70] Wells, *Deep Ancestry*, 171–74.

[71] *Deep Ancestry*, 171.

[72] *Deep Ancestry*, 133–34, 141–43, 85–87.

[73] Public opinion research conducted by Susanne Berthier–Foglar.

sensitive. Instead, casual conversations were preferred,[74] the subject lending itself, obviously, to a qualitative rather than a quantitative approach. To avoid an intrusive attitude, there was no note-taking during the conversations, nor was there any sound recording, the latter being unethical without prior consent. Overt recording also places the conversation in the 'official' mode, with the discussant becoming an informant. This was to be avoided. Public opinion was to be assessed without the trappings associated with the anthropological/ sociological interview.[75]

General and preliminary discussions showed that most people, whether mainstream or minority, do not have a clear knowledge of genetics and tend to associate genome studies with health issues, either reservation-specific, such as diabetes, or mainstream, such as Kreutzfeld–Jacob (Mad Cow) disease, or HIV-AIDS. Since an understanding of science tends to be minimal in a cross-section of any population, the purpose was not to ask direct scientific questions but general questions about the main concerns of the day, be they social or political.

To assess public opinion, informal discussions were conducted on the general problems affecting the interviewee's region. Headlines in the local press at the time reflected a general awareness of the following problems (in decreasing order of importance): the rising fuel prices, with their effect on the economy and the well-being of commuters; the upcoming 2008 US presidential election; the war in Iraq; the state of the US economy. Discussions with randomly chosen individuals confirmed the validity of the above-mentioned list of issues. It was assumed that deviation from the common list of issues was significant and meant that the variant was of particular importance for the interviewee. The validity of this approach was confirmed by interviewees mentioning, as secondary issues, Native American concerns such as land and water ownership as well as tribal sovereignty. A no-prying approach was taken. In the course of the discussions (lasting five to twenty minutes), the interviewer tried to assume a non-intrusive stance, asking, for instance, if "genetic testing was an issue" or if the interviewee had "heard about genetic testing." The interviewer was conscious of the negative connotation of 'racial

---

[74] Stroma Cole, "Action Ethnography: Using Participant Observation," in *Tourism Research Methods. Integrating Theory with Practice*. ed Brent W Ritchie, Peter Burns & Catherine Palmer (Wallingford & Cambridge MA: CABI, 2005): 64–69.

[75] Shay Sayre, *Qualitative Methods for Marketplace Research* (Thousand Oaks CA: Sage, 2001): 11, 41–42, 135–36.

profiling' in choosing interviewees. They were thus chosen in locations with a high probability of their being Natives. When in the course of the interview an interviewee mentioned that s/he was not Native, the testimony became part of the pool of informants among the mainstream population.

Among the self-qualified Native Americans, three profiles were seen:[76]

—A large majority who have little concern for genetic testing, whether they have strong tribal ties or not;
—A minority of Native Americans with scientific awareness, strong tribal identity and a self-defined high blood-quantum, are interested in tests confirming their tribal identity;
—An activist minority against all forms of testing whether genetic or not.

Project Genographic was at the time updating the map of world migrations as it could be traced through genetic mutations. The lines of migrations were getting more and more detailed. While universal scientific knowledge profits from the discoveries, the question of their possible use remains open. While Genographic affirms that all tests are done without making public the names of the individuals tested, the project implicitly admits that only the San peoples in the African Rift Valley are close to their place of origin.[77] They would thus be the only ones able to claim indigenousness in its narrowest sense. All other peoples have migrated at one time or another in their past. They would still be able to claim indigenousness if they were the first to arrive in a particular region, but they would not be able to claim closeness to their place of emergence. Therein probably lies one of the reasons why Indigenous peoples refuse testing. Since they know where they come from in their own creation story, whether they emerged from a sipapu or a lake "in the north,"[78] or any other mythical place, learning that their ancestors came from the Rift Valley in Africa contradicts their original beliefs and is thus seen as an intolerable intrusion into their spiritual world and their constructed identity.

Moreover, the issue is not only about individuals, who may or may not be genetically Indigenous – the issue is political, and genetic testing conflicts

---

[76] 32 people interviewed, 12 males, 20 females, 17 to 60; self-defined tribal affiliation, 14 tribes, one unspecified "Indian."

[77] Wells, *The Journey of Man*: 129.

[78] Raymond Friday Locke, *The Book of the Navajo* (Los Angeles CA: Mankind, 1976): 8–11.

with tribal sovereignty. In countries, such as the USA, where tribes "have the power to define their membership requirements as narrowly or as broadly as they wish," descendancy from an enrolled ancestor may be a mandatory requirement.[79] This idea implies that a claimant could use genetic testing to prove his ascendancy. However, relying on genetic testing to determine membership would conflict with the sovereignty of tribes who may prefer a socially constructed definition, or tribes who have practised adoption of outsiders in the past and who might have integrated individuals lacking the genetic credentials to be tribal members. For these reasons, the lawyer Eric Beckenhauer has a critical view of "genetic essentialism" (12), arguing that, while DNA analysis can identify the ethnic ancestry of an individual, it "proves only that someone, at some point in the direct paternal or maternal lineage, descended from indigenous stock"(17). Therefore, he believes it will be of little use to authorities. As a case in point, he gives the example of the ill-fated Vermont House Bill 809 that was to set genetic criteria for defining Indianness and help the Western Mohegans, who were seeking federal recognition to prove kinship with the Mohican Indians of Wisconsin (18–19). Despite the proponent's strong wish to see "that science should have the last word" in helping the Indians, the objections to genetic testing prevailed, centred on the fact that "the 'Indian' represents a cultural identity that defies biological definition" (19).

Conclusion: Refusing the Universality of Scientific Knowledge

The main point that emerges from the ongoing debate about genetic testing to assess identity and ancestry is the Indigenous peoples' refusal to be classified according to scientific standards. Belonging to a tribe or Indigenous group is, it would seem, cultural rather than genetic. Moreover, the idea that scientific knowledge is universal is not accepted, particularly not among those whose ancestors have suffered from Western ideology.

A secondary point stems from the fact that even culturally sensitive projects such as Genographic might inevitably contribute to the proof that there is only one human species on earth, that it originated in Africa and migrated out of Africa 50,000 years ago, and that we are all "brothers after all," to para-

[79] The legal arguments in this paragraph are taken from Eric Beckenhauer, "Redefining Race: Can Genetic Testing Provide Biological Proof of Indian Ethnicity?" *Stanford Law Review* 56.1 (October 2003): 77. Further page references are in the main text.

phrase Chief Seattle's apocryphal speech. While appealing to the mainstream, the notion of universal brother- and sisterhood could stealthily undermine the Declaration on the Rights of Indigenous Peoples.

WORKS CITED

Beckenhauer, Eric. "Redefining race: can genetic testing provide biological proof of Indian ethnicity?" *Stanford Law Review* 56.1 (October 2003): 161–91.

Berthier–Foglar, Susanne. "A patient/hospital relationship in 1863–65: Mainstream Doctors and Navajo Patients in the Bosque Redondo Camp," in *The Patient*, ed. Kimberly R. Myers & Harold Schweizer (Lewisburg PA: Bucknell UP, 2010): 78–95.

Brooks, Cassandra. "First ancient human sequenced," *The Scientist* NewsBlogs (2010).

Brown, Dan, *Deception Point* (London: Corgi, 2001).

Cavalli–Sforza, Lugi Luca. "Call for a Worldwide Survey of Human Genetic Diversity: A Vanishing Opportunity for the Human Genome Project," *Genomics* 11 (1991): 490–91.

——. *Genes, Peoples, and Languages* (New York: North Point, 2000).

——. "The Study of Variation in the Human Genome," *Genomics* 11 (1991): 491–98.

Cole, Stroma. "Action Ethnography: Using Participant Observation," in *Tourism Research Methods: Integrating Theory with Practice*. ed. Brent W. Ritchie, Peter Burns & Catherine Palmer (Wallingford & Cambridge MA: CABI, 2005): 63–72.

Cook–Deegan, Robert. *The Gene Wars* (New York: W.W. Norton, 1994).

Crichton, Michael. *Jurassic Park* (New York: Alfred A. Knopf, 1990).

Food and Drug Administration (USA). *FDA Approves BiDil Heart Failure Drug for Black Patients* (press announcement 2005).

Foster, Eugene A. "Scientific Correspondence: Jefferson fathered slave's last child," *Nature* 396 (1998): 27–28.

Foucault, Michel. *Les mots et les choses* (Paris: Gallimard, 1996).

Gwyther, George. "An Indian Reservation," *Overland Monthly* (10 February 1873): 123–24.

Harry, Debra. "Acts of Self-Determination and Self-Defense: Indigenous Peoples Responses to Biocolonialism," in *Rights and Liberties in the Biotech Age*, ed. Sheldon Krimsky & Peter Shorett (Lanham MD: Rowman & Littlefield, 2005): 87–97.

Horsman, Reginald. *Race and Manifest Destiny: The Origins of American Anglo-Saxonism* (Cambridge MA: Harvard UP, 1981).

Hultkrantz, Åke. *Native Religions of North America: The Power of Visions and Fertility* (Long Grove IL: Waveland, 1987).

Human Genetics Alert. "The Human Genome Diversity Project," http://www.hgalert.org/topics/personalInfo/hgdp.htm (accessed 3 June 2011).

Judson, Horace F. "A History of the Science and Technology Behind Gene Mapping and Sequencing," in *The Code of Codes*, ed. Kevles & Hood, 37–80.

Kenrick, Justin, & Jerome Lewis. "Indigenous Peoples' Rights and the Politics of the Term 'Indigenous'," *Anthropology Today* 20.2 (2004): 4–9.

Kevles, Daniel J. "Out of Eugenics," in *The Code of Codes*, ed. Kevles & Hood, 3–36.

——, & Leroy Hood, ed. *The Code of Codes: Scientific and Social Issues in the Human Genome Project* (Cambridge MA: Harvard UP, 1992).

Locke, Raymond Friday. *The Book of the Navajo* (Los Angeles CA: Mankind, 1976).

McLuhan, Marshall. *Understanding Media: The Extensions of Man* (New York: McGraw–Hill, 1964).

Morton, Samuel George. *Crania Americana* (Philadelphia PA: J. Dobson, 1839).

Muscati, Samer. "Indigenous Peoples Denounce DNA Conference" (1996), http://www .yvwiiusdinvnohii.net/articles/dna-con.htm (accessed 26 May 2011).

Pálsson, Gísli. *Anthropology and the New Genetics* (Cambridge: Cambridge UP, 2007

Pena, Sergio D. "The fallacy of racial pharmacogenomics," *Brazilian Journal of Medical and Biological Research* 44.4 (April 2011): 268–75.

Porter, Roy. *The Cambridge History of Medicine* (Cambridge: Cambridge UP, 2006).

Rasmussen, Morten. "Ancient human genome sequence of an extinct Palaeo-Eskimo," *Nature* 463 (2010): 757–62.

Reardon, Jenny. *Race to the Finish: Identity and Governance in an Age of Genomics* (Princeton NJ: Princeton UP, 2004.

Ridley, Matt. *Genome: The Autobiography of a Species in 23 Chapters* (New York: HarperCollins, 2000).

Sarich, Vincent, & Frank Miele. *Race: The Reality of Human Differences* (Boulder CO: Westview, 2004).

Sayre, Shay. *Qualitative Methods for Marketplace Research* (Thousand Oaks CA: Sage, 2001).

Scherer, Stewart. *Short Guide to the Human Genome* (Woodbury NY: Cold Spring Harbor Laboratory, 2008).

Spalding Andréolle, Donna, & Susanne Berthier–Foglar. "Science and the American Empire: The American School of Anthropology and the Justification of Expansionism," in *Science and Empire in Nineteenth Century: A Journey of Imperial Conquest and Scientific Progress*, ed. Catherine Delmas, Christine Vandamme & Donna Spalding Andréolle (Newcastle upon Tyne: Cambridge Scholars, 2010): 83–98.

Strong, Josiah. *Our Country: Its Possible Future and Its Present Crisis* (New York: The American Home Missionary Society, 1885).

Sykes, Brian. *The Seven Daughters of Eve* (New York: W.W. Norton, 2001).

Thomas, David H. *Skull Wars: Kennewick Man, Archaeology, and the Battle for Native American Identity* (New York: Basic Books, 2001).

Thomas, Nicholas. *Colonialism's Culture* (Princeton NJ: Princeton UP, 1994).

Tuhiwai Smith, Linda. *Decolonizing Methodologies: Research and Indigenous* Peoples (Dunedin: U of Otago P, 1999).

Venter, Craig. *A Life Decoded: My Genome: My Life* (New York: Penguin, 2007).

Wade, Nicholas. "Reading The Book Of Life: Genetic Code of Human Life Is Cracked by Scientists," *New York Times* (27 June 2000): A1.

Watson, James D. *The Double Helix* (New York: Scribner, 1968).

Weber, Max. *The Protestant Ethic and the Spirit of Capitalism* (Mineola NY: Dover, 2003).

Wells, Spencer. *Deep Ancestry* (Washington DC: National Geographic, 2007).

⌘

Indigenous Peoples and Western Science

SHEILA COLLINGWOOD–WHITTICK

IT WAS THANKS PRIMARILY TO THE GROUNDBREAKING ANALYSES of Aimé Césaire, Frantz Fanon, Albert Memmi, and Edward Said that colonialism's hegemonizing discourse on non-European peoples began finally to be dismantled.[1] The deconstructive process launched by these seminal studies was complemented and extended in the 1980s when new proponents of what had by then come to be known as postcolonial theory made their appearance on the academic scene.[2] Since then, postcolonial studies have become firmly established in universities worldwide, postcolonial theoretical frameworks have been applied to research carried out in all aspects of colonial history, and a 'postcolonial approach' has been increasingly adopted in the critical overhauls regularly taking place in fields as disparate as literature and

---

[1] Aimé Césaire, *Discours sur le colonialisme* (Paris: Réclame, 1950); Frantz Fanon, *Peau noire masques blancs* (Paris: Le Seuil, 1952) and *Les Damnés de la terre*, preface by Jean–Paul Sartre (Paris: Maspéro, 1961); Albert Memmi, *Portrait du colonisé: suivi du Portrait du colonisateur* (Paris: Corrêa, 1957) and *L'Homme dominé* (Paris: Payot, 1968); Edward W. Said, *Orientalism* (London: Routledge & Kegan Paul, 1978).

[2] See, for example, Homi K. Bhabha, "Signs Taken for Wonders: Questions of Ambivalence and Authority Under a Tree Outside Delhi, May 1817," *Critical Inquiry* 12.1 (Autumn 1985): 144–65, and "The Other Question: Difference, Discrimination and the Discourse of Colonialism," in *Literature, Politics and Theory: Papers from the Essex Conference 1976–1984*, ed. Francis Barker, Peter Hulme, Margaret Iversen & Diana Loxley (London & New York: Methuen, 1986): 148–72; also Gayatri Chakravorty Spivak, *In Other Worlds: Essays in Cultural Politics* (New York: Methuen, 1987) and "Can the Subaltern Speak?" in *Marxism and the Interpretation of Culture*, ed. Cary Nelson & Lawrence Grossberg (Urbana: U of Illinois P, 1988): 271–313, to name just the first of the early influential writings of these authors.

ecocriticism, linguistics and economics. One result of this radical revisioning has been the gradual exposure of the ideological (colonial or just plain ethnocentric) ambience in which many areas of Western knowledge are steeped.

The sciences have, naturally, not escaped postcolonial scrutiny. Specialists from both within and outside a variety of scientific disciplines have identified Western science as being deeply implicated in the subjugation – the extermination even – of Indigenous peoples during the colonial era. The best of these analyses offer us illuminating insights into the political, ethical, and racial dimensions of today's genetic science and technology. They also help us recognize and understand the historical experience of scientific abuse that has made 'Third-World' peoples generally and the Indigenous peoples of former white-settler colonies in particular such resolute opponents of human genome diversity research.

Even today, there is an ingrained tendency in the West to equate science with truth, progress, intellectual independence, and enlightenment. Yet we do not have to look very far to discover multiple instances of the more value-laden, less purely disinterested positions from which scientific inquiry has often been launched. For scientists have never inhabited the socio-political vacuum in which the lay imagination locates them, nor are they endowed with any automatic resistance to the dominant ideologies, political currents, or cultural influences of their time.

In her examination of traditional views of scientific truth, Barbara Herrnstein Smith, drawing on the Polish microbiologist Ludwik Fleck, indicates the social conditioning, the collective assumptions, and 'highly contingent cultural and political factors'[3] that contribute to the production of this truth. Fleck's fundamental postulate can, she claims, be summarized thus:

> there are no pure observations, complete descriptions, or 'raw' data. All observations, including or especially those of a highly trained, extensively experienced scientist, are shaped and selected by prior belief and experience – that is, by the ideas, assumptions and practical know-how that, operating together, induce the perceptual expectations and perceptual-behavioural dispositions that, duly mutually adjusted among the members of a collective, yield what we call (scientific) knowledge.[4]

---

[3] Barbara Herrnstein Smith, *Scandalous Knowledge: Science, Truth and the Human* (Edinburgh: Edinburgh UP, 2005): 62.

[4] Herrnstein Smith, *Scandalous Knowledge*, 60.

As social beings, pioneers of nineteenth-century racial science in Britain were far from being "isolated […] in a scientific 'republic' of their own."[5] On the contrary, they were typical products of the rigid, class-bound Victorian society to which they belonged; positioned subjects whose own elite social status predisposed them to a profoundly hierarchical view of the world – a model that was seamlessly incorporated into their schema of how the different 'races of mankind' were related.[6] "For the members of a collective who share a given thought style, certain entities, categories, and connections will be especially salient and ready-to-hand and others less noticeable or invisible."[7]

In terms of ideological immunity, Victorian scientists in Britain had no greater resistance than any other section of the population to the imperialist propaganda that permeated British culture during the era of colonial expansion.[8] Consequently, much of the scientific theorizing on the Indigenous peoples of the world bears the imprint of a distinctly colonialist perspective.[9] If negative values – or lack of value – were the principal characteristics with which non-white peoples came to be almost exclusively associated in the nineteenth and early-twentieth centuries, it was because the European men of science who studied their bodies and their behaviours were ideologically predisposed to view them in that way.

Rather than privileging empirical evidence of a shared humanity as the basis for further enquiry into previously unknown populations, nineteenth- and early-twentieth-century theorizing on race was grounded in the interested a-priori judgment that human variation could only be explained in hierarchical terms. It then proceeded, as Stephen Jay Gould has magisterially demon-

---

[5] Nancy Stepan, *The Idea of Race in Science: Great Britain 1800–1960* (London: Macmillan, 1982): xiv.

[6] Geoffrey Leinhardt, *Social Anthropology* (Oxford: Oxford UP, 1964): 7, quoted in Brian V. Street, *The Savage in Literature: Representations of "Primitive" Society in English Fiction, 1858–1920* (London: Routledge & Kegan Paul, 1975): 80.

[7] Herrnstein Smith, *Scandalous Knowledge*, 59.

[8] Stepan, *The Idea of Race*, xiv–xv.

[9] The imbrication of anthropology and colonialism throughout the era of imperial expansion is succinctly summed up by Talal Asad's assertion that "social anthropology […] devoted [its efforts] to a description and analysis – carried out by Europeans, for a European audience – of non European societies dominated by European power." *Anthropology and the Colonial Encounter* (Amherst NY: Humanity, 1973): 14–15. See also, on this point, Eric Hobsbawm, *The Age of Empire 1875–1914* (London: Abacus, 1994): 252–61.

strated, to use every means at its disposal to 'prove' an inferiority that had been assumed from the outset.[10] When the figures failed to produce the anticipated proof, it was never the premise on which scientists' research was based that was called into question. Rather, it was errors in sampling, flaws in calculation techniques or mistakes in the choice of the anatomical parts to be compared etc. that were blamed.[11]

Moreover, though racial science was by no means born in the nineteenth century,[12] it was nonetheless then that "science-authorized and -legitimated notions about 'race' […] took root" throughout British society as a whole.[13] For, while earlier theories on race had not had widespread currency outside the academic milieux in which they circulated, the convergence of certain key factors from the mid-nineteenth century onwards led to the democratization of scientific knowledge about non-European peoples.

First, the Anthropological Society, founded in 1863, was able, by means of the open meetings and political debates it organized, to pass on to a much larger public ideas on race that had hitherto only been aired within the exclusive confines of the scientific community.[14] Secondly, technical innovations in the printing industry resulting in cheaper printed matter meant that a vastly greater and more socially diverse public immediately gained access to a literature (both fiction and non-fiction) that was impregnated with the latest racial theories.[15]

Two further crucial factors contributing to the vulgarization of racial science were i) the ethnological spectacles that constituted such a massively popular feature of international exhibitions in the final decades of the nineteenth century and ii) the evolutionary displays used as didactic tools by the

---

[10] Stephen Jay Gould, *The Mismeasure of Man* (New York: W.W. Norton, 1981).

[11] Richard Glover, "Scientific Racism and the Australian Aboriginal (1865–1915)," in *Maps, Dreams, History: Race and Representation in Australia*, ed. Jan Kociumbas (Sydney: Braxus, 1998): 99–101.

[12] The classification of humanity according to 'racial' characteristics had been launched in the first quarter of the eighteenth century with the publication of Carl von Linné's landmark opus *Systema Naturae* (1725).

[13] Lucius Outlaw, "Toward a Critical Theory of Race," in *Anatomy of Racism*, ed. David Theo Goldberg (Minneapolis: U of Minnesota P, 1990): 64.

[14] See Brian Street, *The Savage in Literature*.

[15] John M. Mackenzie, *Propaganda and Empire: The Manipulation of British Public Opinion 1880–1960* (Manchester: Manchester UP, 1984): 17–18.

curators of the 'modern museum' – a new nineteenth-century institution with an avowedly educative vocation.

The birth of the ethnological spectacle – the public staging of 'primitive' peoples as living evidence of 'racial types' belonging to a stage of human evolution far below that of the civilized European – was the result of an unlikely coupling between anthropology and the leisure/entertainment industry.

As early as the first decades of the nineteenth century, there already existed a considerable range of venues at which European publics could view 'savage' peoples in a recreational context. In London, for instance, "a metropolitan resident seeking amusement" could find a wide gamut of racial specimens on display in "theatres, museums, pleasure gardens, panoramas, circuses, menageries, freak shows and fairs."[16] According to Zoë Strother, however, it was the exhibition of Sartjie Baartman[17] that marked the turning point between the mere vulgar curiosity appealed to by the popular freak show and the (pseudo)-scientific interest that would come to be solicited by the organizers of ethnological expositions.[18] Henceforth, anthropologists, ethnographers, and anatomists would, in their turn, become both avid spectators of displays of 'primitive peoples' and willing participants in the organization of such spectacles.

The ethnological spectacle, which soon proved to be one of the main attractions of the international expositions that flourished throughout the Western world until the 1930s, not only set out to gratify European publics with 'evidence' of their inherent superiority over savage peoples, it also provided

---

[16] Sadiah Qureshi, "Displaying Sara Baartman, 'The Hottentot Venus'," *History of Science* 43 (2004): 236.

[17] Sara (Saartjie) Baartman (the subject of a recently released film, *Vénus Noire*, directed by Abdellatif Kechiche), was a twenty-year-old Khoikhoi servant taken to England in 1810 by the brother of her South African boss, Hendrik Cézar, and a British naval surgeon, William Dunlop. Billed as "the Hottentot Venus" and "The Greatest Phenomenon Ever exhibited in this Country" (Qureshi, "Displaying Sara Baartman," 237), Sara was first displayed in London's Piccadilly, where, for an admission fee of two shillings, spectators were allowed to poke, pinch, and prod her with a parasol in order to ascertain whether her 'extraordinary' anatomical features were genuine. See Zoë S. Strother, "Display of the Body Hottentot," in *Africans on Stage: Studies in Ethnological Show Business*, ed. Bernth Lindfors (Bloomington: Indiana UP, 1999): 27.

[18] Following the display of the "Hottentot Venus," entrepreneurs, argues Strother, "greatly expanded on the overt educational mission of the exhibitions by adding elaborate props, scenery, demonstrations, and learned lectures that all hammered home the typical nature of the presentation." "Display of the Body Hottentot," 35.

anthropologists with the unprecedented opportunity for studying 'racial types' at first hand.[19] At certain world's fairs, anthropological research was itself staged as a spectacle or, sometimes, as a learning activity in which the public was invited to participate.[20]

In tandem with and complementing these shows were ethnographic exhibits in museums. And in the modern museum, "tasked with the production and dissemination of authoritative knowledge,"[21] the evolutionary system of classification had, as Annie Coombes observes, become "the most prevalent means of displaying ethnographic material."[22]

Of key importance among the 'objects' displayed in ethnographic collections were the skulls and skeletons of Aboriginal peoples, presented systematically as evolutionary retards belonging to the lowest level of human evolution and, thus, closer to apes than to civilized Europeans. But it was not only through the racist notions they transmitted to tens of millions of visitors that ethnographic exhibitions may be said to have had a negative impact on Indigenous peoples. The displays themselves could not have been mounted, nor could research in anthropology, comparative anatomy or any of the related disciplines have been pursued, without a regular supply of human remains. As Helen MacDonald explains, "vast collections of bodies and body parts were

---

[19] Up until that point, anthropological 'knowledge' of Indigenous peoples had frequently been constructed out of second-hand observations taken from the highly subjective travel literature of the period. See, for example, John P. Jackson Jr. & Nadine M. Weidman, *Race, Racism, and Science: Social Impact and Interaction* (Santa Barbara CA: ABC–CLIO, 2004): 79.

[20] Raymond Corbey, "Ethnographic Showcases, 1870–1930," *Cultural Anthropology* 8.3 (August 1993): 354; Robert W. Rydell, John E. Findling & Kimberly D. Pelle, *Fair America: World's Fairs in the United States* (Washington DC & London: Smithsonian Institution Press, 2000): 54; Annie E.S. Coombes, "'For God and for England': Contributions to an Image of Africa in the First Decade of the Twentieth Century," *Art History* 8.4 (December 1985): 460; Eike Reichardt, *Health, 'Race' and Empire: Popular-Scientific Spectacles and National Identity in Imperial Germany, 1871-1914* (Lulu.com, 2008): 36.

[21] Eilean Hooper–Greenhill, *Museums and the Interpretation of Visual Culture* (London: Routledge, 2000): 126.

[22] Annie E.S. Coombes, *Reinventing Africa: Museums, Material Culture and Popular Imagination* (New Haven CT & London: Yale UP, 1994): 120.

necessary" to provide scientists and museum curators with the tools of their trade.[23]

It was as a direct result of the "mid-Victorian mania"[24] for collecting osteological specimens that Indigenous bodies, when not harvested directly from the scenes of massacres, were disinterred and stolen from burial sites, illicitly removed from hospitals, morgues, and prisons, and frequently mutilated through the severing of skulls from skeletons.[25] It has even been suggested that Indigenous Australian peoples were hunted down and killed for the specific purpose of satisfying the demands of metropolitan men of science and museum curators for Aboriginal remains.[26]

By elaborating knowledge systems which authoritatively 'demonstrated' the inferior, sub-human status of non-European races to academic and popular audiences, Victorian science may be regarded, then, as bearing a large share of responsibility for the racial stereotypes that persist even today in Western societies. As indicated by the example above, however, nineteenth-century

---

[23] Helen MacDonald, *Human Remains: Dissection and its Histories* (New Haven CT & London: Yale UP, 2006): 87.

[24] MacDonald, *Human Remains*, 96.

[25] Tristram Besterman, director of the Manchester Museum, states unequivocally that "the collections in our Western museums derive, at their most innocent, from grave robbing, and, at their worst, from wholesale slaughter." "Human Remains: Objects to Study or Ancestors to Bury," Debate on 2 May 2003, Institute of Ideas, http://www.instituteofideas.com/transcripts/human_remains.pdf (accessed 11 November 2010). See also Robert Hughes, *The Fatal Shore: A History of the Transportation of Convicts to Australia 1787–1868* (1986; London: Harvill, 1996); Moira Simpson, *Making Representations: Museums in the Postcolonial Era* (London: Routledge, 1996); Helen MacDonald, "Reading the 'Foreign Skull': An Episode in Nineteenth Century Colonial Human Dissection," *Australian Historical Studies* 125 (2005): 81–96; "Ancestral Remains Return Home," *Closing the Gap* (18 August 2010), http://www .indigenous.gov.au/Pages/56_ancestral_remains.aspx (accessed 11 Nov. 2010).

[26] See quotation from Bob Weatherall of the Foundation for Aboriginal and Islander Research in Simpson, *Making Representations*, 175–76; Alison Palmer, *Colonial Genocide* (Adelaide: Crawford House, 2000): 46; Paul Turnbull, "Ancestors not Specimens: Reflections on the Controversy over the Remains of Aboriginal People in European Scientific Collections," *The Electronic Journal of Australian and New Zealand History* (27 April 1997), http://www.jcu.edu.au/aff/history/articles/turnbull .htm (accessed 11 Nov. 2010).

racial theory also contributed to colonial abuse of non-European populations in more direct ways.

One way was through the overdetermined emphasis it placed on the risk of contamination and racial degeneration faced by white settler populations living in proximity to non-white races. Taking up and endlessly repeating this idea, colonial discourse represented the native in terms of "breeding swarms, of foulness, of spawn, of gesticulations."[27] "Native bodies," suggests Andrea Smith, were metaphorically transformed "into a pollution of which the colonial body must purify itself."[28]

As Robert Young argues, then, it was the fear of what might happen if the 'biological' boundaries between superior and inferior races were not vigilantly patrolled that "the onus of British colonial policy came [...] to be focussed on an effort to prevent mixing between the British and their subject peoples."[29]

Probably the most striking evidence of the 'effort' Young refers to was the racialization of space in European colonies. For, far from being a deviant system invented by rabid Afrikaner nationalists in 1948, apartheid was widely practised, if not always consitutionally enshrined, in the vigorously policed and mutually exclusive areas which settlers and natives inhabited throughout the colonized world. In Fanon's words, "This [colonial] world divided into compartments, this world cut in two is inhabited by two different species."[30]

According to Maynard Swanson, "disease and epidemiology became a widespread societal metaphor" in late-nineteenth- and early-twentieth-century South Africa (a period well before the Afrikaners came to power); a metaphor that would become "a major strand in the creation of urban apartheid."[31]

---

[27] Frantz Fanon, *The Wretched of the Earth*, preface by Jean–Paul Sartre, tr. Constance Farrington (*Les Damnés de la terre*, 1961; tr. 1963; Harmondsworth: Penguin, 1967): 33.

[28] Andrea Smith, "Not an Indian Tradition: The Sexual Colonization of Native Peoples," *Hypatia* 18.2 (Spring 2003): 672.

[29] Robert J.C. Young, *Colonial Desire: Hybridity in Theory, Culture, and Race* (London: Routledge, 1995): 144.

[30] Fanon, *The Wretched of the Earth*, 30.

[31] Maynard W Swanson, "The Sanitation Syndrome: Bubonic Plague and urban native policy in the Cape Colony, 1900–09," in *Segregation and Apartheid in Twentieth Century South Africa*, ed. William Beinart & Saul Dubow (London: Routledge, 1995): 26.

Thus, during the 1870s, the municipal authorities in Durban sought to remove Indians to a location outside of the town because of the "breeding haunts and nursery grounds of disease" that were believed to be embedded in the Indian population.[32] Subsequent to an outbreak of bubonic plague in Cape Town at the very end of the nineteenth century, asserts Swanson, "Sanitation and public health provided the legal means to effect quick removals of African populations; they then sustained the rationale for permanent urban segregation."[33]

Similarly, the so-called "Civil Lines" in British India functioned as a veritable *cordon sanitaire* protecting white expatriates from the diseases that were believed to infest the native quarters.[34] Throughout Africa, argues Barbara Bush, "spatial segregation in urban and rural areas physically enforced racial boundaries and minimised 'racial pollution'."[35] Thus, in colonial Nigeria a regulation physical distance of at least 440 yards had to separate native dwellings from European habitations. The Governor, Frederick Lugard, justified this division on 'hygiene grounds'; it was there to protect the whites "against disease and the noise of drumming":[36]

> White towns mapped out white space and arguably provided the model for the 'post-colonial,' racialised western city. As urbanisation accelerated, it was even more vital to segregate the expanding 'colonial slum' from the European cities. An obsession with sanitation took hold of the colonial imagination ...[37]

Yet, despite attempts to create a salubrious, sterile, racially hermetic space for the white inhabitants of colonies, and notwithstanding dire warnings issued by medical men on the dangers of contact with the dirty, disease-ridden, contaminating bodies of Indigenous peoples, the racial barriers separating colonizer from colonized were far from impermeable. This was mainly, as Young

---

[32] Swanson, "The Sanitation Syndrome," 27.

[33] "The Sanitation Syndrome," 39.

[34] For an excellent discussion of the pervasiveness of this idea in the imagery Kipling uses to describe areas in which 'natives' are confined in British India, see Gail Ching-Liang Low, *White Skins/Black Masks: Representation and Colonialism* (London: Routledge, 1996): 156–90.

[35] Barbara Bush, *Imperialism, Race and Resistance: Africa and Britain, 1919–1945* (London: Routledge, 2002): 76.

[36] Bush, *Imperialism, Race and Resistance*, 76.

[37] *Imperialism, Race and Resistance*, 78.

suggests, because white desire for black female bodies was driven by powerful fantasies about the quintessential lasciviousness of 'native' women.

Here again, we find that it is in nineteenth-century scientific discourse that such fantasies had their roots. For, as Sander Gilman among others has pointed out, comparative anatomists and medical men placed great emphasis on what they saw as the 'pathological' development of black women's genitalia.[38] Citing the case of Saartjie Baartman, Gilman contends that the reason why European males flocked to see her in such numbers was not just curiosity about the anatomical anomalies that scientific discourse had represented as defining characteristics of the 'Hottentot' female. They were drawn, above all, by lewd imaginings about the "sexually intensive" behaviour with which scientific literature had associated such extraordinary organs.[39]

Fuelled by medical speculation about the oversexed nature of non-white women, white male fantasies shaped interracial gender relations in all parts of the world colonized by European nations. While insisting on the repulsiveness of interracial sex, arguing that African women were "libidinous and shameless as monkies [sic] or baboons,"[40] the eighteenth-century British planter Edward Long simultaneously recounted that the white men of the West Indian colony of Jamaica were constantly led to trangress racial boundaries because of their "infatuated attachment" to black women.[41] Similarly, in spite of the strong taboo that existed on sexual intercourse between whites and blacks in the Southern States of the USA, it was observed by the American abolitionist Moncure D. Conway that "Miscegenation is already the irreversible fact of Southern Society in every thing but the recognition of it […] the mixture of

---

[38] Sander Gilman, "Black Bodies, White Bodies: Toward an Iconography of Female Sexuality in Late Nineteenth Century Art, Medicine and Literature," in *'Race', Writing and Difference* ed. Henry Louis Gates, Jr. (Chicago: U of Chicago P, 1986): 231–32.

[39] Gilman, "Black Bodies," 232. Thwarted by Baartman herself from examining her elongated *labia minora* while she was alive, Georges Cuvier, considered one of the greatest French scientists of his time, eventually excised the young Khoisan woman's genital organs and placed them in preserving fluid when he dissected her body immediately following her death in 1815. It is worth noting that the jar containing Saartjie Baartman's pickled genitalia remained (with her pickled brain and her skeleton) in the Musée de l'Homme in Paris until its repatriation with the rest of her remains to South Africa in 2002.

[40] Edward Long, *The History of Jamaica*, vol. 2: 383, quoted in Young, *Colonial Desire*, 150.

[41] Quoted in Young, *Colonial Desire*, 151.

blood has been very extensive."[42] In South Africa, too, Boer males' predilection for sex with African women remained constant throughout the country's post-contact history (including during the apartheid era), while the Anglo-Australian settler's taste for 'black velvet' – the familiar term used by white males in the outback to describe sexual relations with Aboriginal women – seems to be ongoing.[43]

The recommendations of men of science, for whom the intermingling of races "prompt[ed] the fear of degeneracy or cultural and physical pollution,"[44] were thus continually flouted by the transgressive, clandestine (and, more often than not, non-consensual) interracial sexual relations through which the white male's desire for black women was articulated. And this, in spite of the fact that miscegenation – signalled through the continually growing 'coloured' populations of European colonies – was "vehemently opposed" by "both expert and popular opinions."[45]

Prominent among those whose 'expert' opinion condemned miscegenation on the grounds of the biologically defective progeny it was alleged to produce were practitioners of medical science. One example cited by the Australian

---

[42] Moncure D. Conway, *Testimonies Concerning Slavery*, quoted in *Colonial Desire*, 146–47.

[43] Terry Goldie, *Fear and Temptation: The Image of the Indigene in Canadian, Australian and New Zealand Literature* (Montreal & Kingston, Ontario: McGill–Queen's UP, 1989); Nicolas Jose, *Black Sheep: Journey to Borroloola* (London: Profile, 2002).

[44] David T. Goldberg, *The Racial State* (Oxford: Blackwell, 2002): 85. It was also, incidentally, a dominant theme in the literatures of white settler colonies. See Street, *The Savage in Literature*, and Vernon February, *Mind Your Colour* (London & Boston MA: Kegan Paul, 1981).

[45] Henry Reynolds, *Nowhere People* (Camberwell, Victoria: Penguin, 2005): 6. Miscegenation was also criminalized in some ex-white-settler colonies. In South Africa, the Immorality Act, which made sexual relations between whites and Africans illegal from 1927, was reinforced in 1949 by the Prohibition of Mixed Marriages Act. In the USA anti-miscegenation laws that existed in as many as forty states at one time remained in force until 1967. According to Judy Scales–Trent, "as late as 1964, in five states, one could be sentenced to up to ten years in jail for miscegenation"; "Racial Purity Laws in the United States and Nazi Germany: The Targeting Process," *Human Rights Quarterly* 23 (2001): 283. See also Stefan Kuhl, *The Nazi Connection: Eugenics, American Racism and German National Socialism* (New York & Oxford: Oxford UP, 1994).

historian Henry Reynolds is that of "a British medical specialist [who] declared in 1906 that intermarriage between his countrymen and non-Europeans should be prevented because [...] of the 'terrible monstrosities' produced by such unions."[46] Australia, where the number of 'half-castes' had begun to increase dramatically by the end of the nineteenth century, was, says Reynolds, positively "obsessed with blood and biology."[47] In the USA, Virginia's 1924 Racial Integration Act "spoke in terms of the 'pollution' of America if miscegenation were allowed to continue."[48] In South Africa, claims Saul Dubow, "'race fusion' was portrayed in the most apocalyptic terms by [...] eugenist-inspired catastrophists."[49]

Both shaping and shaped by colonial discourse on the native, medical discourse repeatedly resorted to tropes of disease and contamination in order to hammer home the message of the disastrous biological consequences that would ensue from Europeans' mixing with non-white races. The discursive strategy of conflating native bodies with pollution and contagious disease was, however, not just a means of calling transgressive white males to order. It was also designed to rationalize the pathologization of Indigenous peoples and, by the same token, medical interventions recommending the quarantining of natives in isolated reservations, camps or (as was the case in South Africa) urban ghettoes.

As Roy Macleod reminds us,

> from the 1880s, the practice of medicine in the empire became a history of techniques and policies – of education, epidemiology, and quarantine – the discovery of pathogens and vectors and the segregation of races [...]. [T]his tropical medicine – its ideology European, its instrument the microscope, its epistemology the germ theory of disease

---

[46] R.R. Rentoul, *Race Culture or Race Suicide* (London: Walter Scott, 1906): 4, quoted in Reynolds, *Nowhere People*, 6. As Scales–Trent reminds us, "in Nazi Germany, Hitler called these children "monstrous beings, half man, half monkey"; "Racial Purity Laws," 274. The idea that the offspring of mixed-raced unions inherited the worst biological characteristics of both progenitors was a commonplace in scientific and non-scientific writing on the subject.

[47] Reynolds, *Nowhere People*, 9.

[48] Scales–Trent, "Racial Purity Laws," 272.

[49] Saul Dubow, "The Elaboration of Segregationist Ideology," in Beinart & Dubow, *Segregation and Apartheid*, 156.

> – served the interests of dominant economic groups and obscured the relationship of disease to social structure.[50]

Construed, predominantly, both as dangerous pollutants threatening the biological purity of the white race and as sources and vectors of disease, Indigenous populations received little attention as *patients*. They were therefore seldom the beneficiaries of colonial medicine, which until the second half of the twentieth century remained exclusively concerned with the health of white people. Citing the example of Queensland, where "the Australian Institute of Tropical Medicine set out to discover means by which to promote the health of whites in the tropics," MacLeod notes that, by contrast, "the Australian government exhibited a deep absence of concern for the health of Aborigines until the 1960s."[51]

When colonial health authorities *did* pay attention to their native populations, it was usually, Mary Ellen Kelm suggests, "as the subjects of epidemiological surveys."[52] As the tuberculosis specialist Dr Herbert A. Burns wrote apropos of Native Americans in 1932:

> The Indian, without his knowledge or consent, offers us a human experience in immunology as well as epidemiology which we can ill afford to ignore. The material is conveniently located, the data are or should easily be made available for record and study, and the results applied to regulative measures of control.[53]

In Australia, "regulative measures of control" involved the exile and internment of Indigenous syphilis and leprosy sufferers far from their homes and far from centres of white population. Forcibly removed from their communities, they were transported to places where they would be kept in absolute isolation from the rest of the world and condemned to live in frequently appalling con-

[50] Roy MacLeod, "Introduction" to *Disease, Medicine and Empire*, ed. Roy MacLeod & Milton Lewis (London & New York: Routledge, 1988): 7. See also Warwick Anderson, "Excremental Colonialism: Public Health and the Poetics of Pollution," *Critical Inquiry* 21 (Spring 1995): 641.

[51] MacLeod, "Introduction," 9. Even today, health statistics for Indigenous Australians reveal catastrophic levels of preventable diseases, high infant mortality, and low life expectancy.

[52] Mary Ellen Kelm, "Diagnosing the Discursive Indian: Medicine, Gender, and the 'Dying Race'," *Ethnohistory* 52.2 (Spring 2005): 377.

[53] Herbert A. Burns, "Tuberculosis in the Indian," *American Review of Tuberculosis* 26 (1932), quoted in Kelm, "Diagnosing the Discursive Indian," 371.

ditions.[54] Commenting on Australia's policy of compulsorily isolating Aboriginal leprosy sufferers long after international research in leprosy prophylaxis had proscribed such treatment, Suzanne Saunders observes:

> For almost thirty years Australia, which was internationally recognized for its progressive approach to medical research and treatment in many fields, denied leprosy patients adequate treatment and trenchantly enforced a negtive and inhumane policy.[55]

Even worse than the lack of treatment or inhumane treatment from which non-European peoples suffered as a result of the deeply entrenched racial outlook that pervaded medical science until well into the second half of the twentieth century has been their use as guinea pigs in medical experiments. Little by little, the grim details of how past scientific praxis has included the testing of new and sometimes dangerous drugs and treatments on non-European and Indigenous peoples are beginning to emerge. It is now known, for example, that African Americans have, since the slavery era, been used as guinea pigs in every conceivable type of experiment which medical science has sought to conduct; whether it be to deliberately expose blacks to infectious diseases or to test the toxicity of drugs, the efficacity of vaccines, the difficulties of surgical procedures, or the effects of radiation.[56] It is also known that American researchers tested the contraceptive pill on Puerto Rican women;[57] that trials for multiple vaccines were carried out on Aboriginal Australian children in institutional care;[58] and, thanks to the testimony of many Native Americans, that Canada's infamous residential schools conducted diverse medical and psychological experiments on the children interned there. As recently as October 2010, Secretary of State Hillary Clinton and US Health Secretary Kathleen Sibelius apologized to the people of Guatemala for "'outrageous and

---

[54] Ernest Hunter, "Stains on the Caring Mantle," *Medical Journal of Australia* 155 (2–16 December 1991): 780–82.

[55] Suzanne Saunders, "Isolation: The Development of Leprosy Prophylaxis in Australia," *Aboriginal History* 14.2 (1990): 181.

[56] See Harriet A. Washington, *Medical Apartheid: The Dark History of Medical Experimentation on Black Americans From Colonial Times to the Present* (Garden City NY: Doubleday, 2007), for a meticulously detailed exploration of these issues.

[57] Iris Ofelia López, *Matters of Choice: Puerto Rican Women's Struggle for Reproductive Freedom* (New Brunswick NJ: Rutgers UP, 2008).

[58] Sheila Collingwood–Whittick, "Indigenous Opposition to Genetics Research: Views from Aboriginal Australia" in Part IV of this volume.

abhorrent' experiments in Guatemala by American doctors who infected hundreds of prisoners, soldiers and mental patients with syphilis in the 1940s."[59]

Probably the best known of the callous experiments carried out in the USA itself was the Tuskegee study, a forty-year investigation conducted by white American medical researchers into the spontaneous evolution of untreated syphilis in the "Negro male." In the Tuskegee case it was not a question of trialling drugs but of witholding drugs that might have saved lives.[60] Significantly, in 1997, when President Bill Clinton apologized to the Tuskegee survivors, it was revealed that, in the opinion of doctors and medical researchers, "the Tuskegee study left such a legacy of Government distrust among black Americans that it has hindered their ability to treat blacks for AIDS or HIV, the virus that causes AIDS."[61]

Professor Jay Katz has argued that, far from being off the scale, the experiments carried out by Nazi doctors in concentration camps are simply located at "one end of a continuum." Part of that same continuum, he suggests, are, the innumerable instances of medical research *antedating* the Holocaust, in which it was the unquestioned practice of scientists to carry out medical trials on "the disadvantaged, the downtrodden."[62] Indeed, human experimentation on ethnic minority populations was such a widespread phenomenon in the USA that "Nazi doctors at Nuremberg defended *their* wartime medical 'experiments' partly by pointing to those done in the U.S. or done elsewhere by Americans."[63]

---

[59] Chris McGeal, "US says sorry for 'outrageous and abhorrent' Guatemalan syphilis tests," *The Guardian* (1 October 2010).

[60] James H. Jones, *Bad Blood: The Tuskegee Experiment: A Tragedy of Race and Medicine* (New York: Free Press, 1981).

[61] Alison Mitchell, "Survivors of Tuskegee Study Get Apology from Clinton," *New York Times* (17 May 1997): 10. Harriet Washington confirms this view, claiming that medical research "has left blacks with an ugly legacy of distrust for research and even treatment, and that it is a lingering stain on the history of medicine"; quoted in Denise Grady, "White Doctors, Black Subjects: Abuse Disguised as Research," *New York Times* (23 January 2007).

[62] Jay Katz, "Human Sacrifice and Human Experimentation: Reflections at Nuremberg," *Occasional Papers,* Paper 5 (25 October 1996), http://digitalcommons.law.yale .edu /ylsop_papers/5 (accessed 11 November 2010).

[63] Lawrence Hammar, "The Dark Side to Donovanosis: Color, Climate, Race and Racism in American South Venereology," *Journal of Medical Humanities* 18.1 (1997): 41.

Another area of scientific activity– often misconstrued in the popular imagination as originating in the racial extremism that drove Nazi ideologues under the third Reich – is that of eugenics. As Frank Dikötter reminds us, however, "eugenics belonged to the political vocabulary of virtually every significant modernizing force between the two world wars."[64] Not only was it "a 'modern' way of talking about social problems in biologizing terms,"[65] it constituted a formidable discursive tool which:

> gave scientific authority to social fears and moral panics, lent respectability to racial doctrines and provided legitimacy to sterilization acts and immigration laws. Powered by the prestige of science, it allowed modernizing elites to represent their prescriptive claims about social order as objective statements irrevocably grounded in the laws of nature; Eugenics provided a biologizing vision of society ...[66]

In white-settler nations, where genocidal logic persisted as the default position on 'race problems' long after the killing times of colonization themselves had ended, eugenics offered a providential justification for the policies enlisted to arrest the 'proliferation' of non-white races. As Scales–Trent argues, it was the Nazis who learned from the theory and practice of eugenics in the USA rather than the other way round. The "Model Sterilization Law" published in 1922 by one of the leading advocates of eugenics in the USA, Harry Laughlin, is, she points out, widely believed to have served as a prototype for the sterilization law elaborated by the Nazis.[67]

But while America's eugenicist policies initially targeted members of racially indeterminate groups (those who, in Katz's phrase, comprised "the

---

[64] Frank Dikötter, "Race Culture: Recent Perspectives on the History of Eugenics," *American Historical Review* 103.2 (April 1998): 467.

[65] Dikötter, "Race Culture," 467.

[66] "Race Culture," 468.

[67] In 1936, Laughlin was awarded an honorary degree by the University of Heidelberg for his contribution to the "science of racial cleansing." Alex Wellerstein, "Harry Laughlin's 'Model Eugenical Sterilization Law'," *History of Science*, Harvard University, http://www.people.fas.harvard.edu/~wellerst/laughlin/. Scales–Trent suggests, in fact, that eugenicists from the two countries formed a kind of mutual admiration society, with many American eugenicists thinking that "the Nazi sterilization campaign would serve as a boost for that same effort in the USA. Leading philanthropic organizations in the USA gave money to support Nazi research in this area." "Racial Purity Laws," 292.

disadvantaged, the downtrodden," sections of the population), the neo-eugenicists who established themselves in the 1950s and 1960s had altogether more racially specific populations in their sights. Because their condemnation of racial discrimination and their demands for the constitutionalization of equal rights "challenged white power structures," the ethnic groups in which civil rights, Black Power, Chicano/a, and Native American rights activism flourished became "the targets of neo-eugenicists."[68]

To begin with, reports Rebecca Kluchin, it was

> black women in the South [who] became targets of forced sterilization via hysterectomy, commonly referred to by women as "Mississipi appendectomies," because women entered hospitals to have abdominal surgery and left, unknowingly without their uteruses. The practice became widespread in communities entrenched in civil rights struggles.[69]

From the early 1970s, however, a volatile mixture of immigration fears, exasperation at the economic cost of providing public assistance to ethnic minorities, and social anxieties about growing native militancy threatened the peace of mind of the dominant white society. It was then that Mexican, Mexican American, Puerto Rican, and Native American women all became victims of forced sterilization.[70] Between twenty-five and forty-two percent of Native American women of childbearing age were sterilized in the period from 1970 to 1976.[71]

---

[68] Rebecca Marie Kluchin, "Introduction" to *Fit to be Tied: Sterilization and Reproductive Rights in America, 1950–1980* (New Brunswick NJ: Rutgers UP, 2009, 3.

[69] Kluchin, "Introduction," 6.

[70] Kluchin, "Introduction," 7. David Morgan points out that forced sterilization was legal in eighteen American states and quotes the authors of a newly published Yale study comparing US and Nazi Eugenics as saying that "The comparative histories of the eugenical sterilization campaigns in the United States and Nazi Germany reveal important similarities of motivation, intent and strategy." "Study Says U.S. Eugenics Paralleled Nazi Germany," *Chicago Tribune* (15 February 2000).

[71] Kluchin, "Introduction," 7. Sterilization programmes aimed at controlling Indigenous populations were also carried out by US medical men in neighbouring Latin American countries. Quoting from a report in *Akwesasne Notes* (1977), Charles R. England points out that "Between 1963 and 1965 more than 400,000 Colombian women were sterilized in a program funded by the Rockefeller Foundation. In Bolivia, a U.S. imposed population control program administered by the Peace Corps sterilized Quechua Indian women without their knowledge or consent." "A Look at the Indian

Although, as can be seen in "Indigenous Opposition to Genetics Research" in Part 3 of this volume, there have been insistent reports of coercive sterilization/contraception being used in Australia to control the Aboriginal population, there is as yet no authoritative study that has established sterilization as having ever existed on the kind of scale that occurred in the Americas. If, indeed, sterilization was *not* a widespread practice in Australia, it is undoubtedly because eugenics was, to use Henry Reynolds' phrase, "carried out informally" through other types of reproductive management.

Russell McGregor has pointed out, for instance, that "biological absorption" was the favoured solution of a number of medical men in prominent positions in the Aboriginal Protection Boards to the 'half-caste problem' generated by miscegenation. What this "perverse proposition" entailed was "the deliberate promotion of reproductive unions between persons of part-Aboriginal descent and white people such that after a few generations [...] no discernible trace of Aboriginal ancestry would remain."[72]

It is not possible in an introductory essay such as this to give more than a brief glimpse of the innumerable egregious ways in which Western science has deeply influenced the lives, the living bodies, and the dead remains of Indigenous peoples since the beginning of the eighteenth century. Nevertheless, even from the few examples that have been cited here, it should be obvious how scientific theories dehumanizing 'lower races' inevitably resulted in the kind of inhuman scientific practices to which the Indigenous peoples of European colonies were continually subjected from their first encounter with men of science. It is this long, traumatic history of racial persecution that needs to be remembered and reflected on before considering why certain Indigenous populations today are refusing to collaborate with Western geneticists seeking to map human diversity and the migration history of the human species.

⌘

Health Service Policy of Sterilization, 1972–1976," *Red Ink* 3.1 (Spring 1994), http://www.dickshovel.com/IHSSterPol.html (accessed 11 Nov. 2010).

[72] Russell McGregor, "'Breed Out the Colour': Reproductive Management for White Australia," in *"A Race for a Place": Eugenics, Darwinism and Social Thought and Practice in Australia*, ed. Martin Crotty et al. (Callaghan, NSW: U of Newcastle P, 2000): 61.

WORKS CITED

Anderson, Warwick. "Excremental Colonialism: Public Health and the Poetics of Pollution," *Critical Inquiry* 21 (Spring 1995): 640–69.

Anon. "Ancestral Remains Return Home," *Closing the Gap* (18 August 2010), http://www.indigenous.gov.au/Pages/56_ancestral_remains.aspx (accessed 11 November 2010).

Asad, Talal. *Anthropology and the Colonial Encounter* (Amherst NY: Humanity, 1973).

Besterman, Tristram. "Human Remains: Objects to Study or Ancestors to Bury," Debate on 2 May 2003, Institute of Ideas, http://www.instituteofideas.com/transcripts/human_remains.pdf (accessed 11 November 2010).

Bhabha, Homi K. "The Other Question: Difference, Discrimination and the Discourse of Colonialism," in *Literature, Politics and Theory: Papers from the Essex Conference 1976–1984*, ed. Francis Barker, Peter Hulme, Margaret Iversen & Diana Loxley (London & New York: Methuen, 1986): 148–72.

——. "Signs Taken for Wonders: Questions of Ambivalence and Authority Under a Tree Outside Delhi, May 1817," *Critical Inquiry* 12.1 (Autumn 1985): 144–165.

Bush, Barbara. *Imperialism, Race and Resistance: Africa and Britain, 1919–1945* (London: Routledge, 2002).

Césaire, Aimé, *Discours sur le colonialisme* (Paris: Réclame, 1950).

Coombes, Annie E.S. "'For God and for England': Contributions to an Image of Africa in the First Decade of the Twentieth Century," *Art History* 8.4 (December 1985): 453–66.

——. *Reinventing Africa: Museums, Material Culture and Popular Imagination* (New Haven CT & London: Yale UP, 1994).

Corbey, Raymond. "Ethnographic Showcases, 1870–1930," *Cultural Anthropology* 8.3 (August 1993): 338–69.

Dikötter, Frank. "Race Culture: Recent Perspectives on the History of Eugenics," *American Historical Review* 103.2 (April 1998): 467–78.

Dubow, Saul. "The Elaboration of Segregationist Ideology," in William Beinart & Saul Dubow, *Segregation and Apartheid in Twentieth Century South Africa* (London: Routledge, 1995): 145–75.

England, Charles R. "A Look at the Indian Health Service Policy of Sterilization, 1972–1976," *Red Ink* 3.1 (Spring 1994), http://www.dickshovel.com/IHSSterPol.html (accessed 11 November 2010).

Fanon, Frantz. *Les Damnés de la terre*, preface by Jean–Paul Sartre (Paris: Maspéro, 1961).

——. *Peau noire, masques blancs* (Paris: Seuil, 1952).

——. *The Wretched of the Earth*, preface by Jean–Paul Sartre, tr. Constance Farrington (*Les Damnés de la terre*, 1961; tr. 1963; Harmondsworth: Penguin, 1967).

February, Vernon. *Mind Your Colour* (London & Boston MA: Kegan Paul, 1981).

Gilman, Sander. "Black Bodies, White Bodies: Toward an Iconography of Female Sexuality in Late Nineteenth Century Art, Medicine and Literature," in *'Race', Writing and Difference*, ed. Henry Louis Gates, Jr. (Chicago: U of Chicago P, 1986).

Glover, Richard. "Scientific Racism and the Australian Aboriginal (1865–1915)," in *Maps, Dreams, History: Race and Representation in Australia*, ed. Jan Kociumbas (Sydney: Braxus, 1998): 67–130.

Goldberg, David T. *The Racial State* (Oxford: Blackwell, 2002).

Goldie, Terry. *Fear and Temptation: The Image of the Indigene in Canadian, Australian and New Zealand Literature* (Montreal & Kingston, Ontario: McGill–Queen's UP, 1989).

Gould, Stephen Jay. *The Mismeasure of Man* (New York: W.W. Norton, 1981).

Grady, Denise. "White Doctors, Black Subjects: Abuse Disguised as Research," *New York Times* (23 January 2007).

Hammar, Lawrence. "The Dark Side to Donovanosis: Color, Climate, Race and Racism in American South Venereology," *Journal of Medical Humanities* 18.1 (1997): 29–57.

Herrnstein Smith, Barbara. *Scandalous Knowledge: Science, Truth and the Human* (Edinburgh UP, 2005).

Hobsbawm, Eric. *The Age of Empire 1875–1914* (London: Abacus, 1994).

Hooper–Greenhill, Eilean. *Museums and the Interpretation of Visual Culture* (London: Routledge, 2000)

Hughes, Robert. *The Fatal Shore: A History of the Transportation of Convicts to Australia 1787–1868* (1986; London: Harvill, 1996).

Hunter, Ernest. "Stains on the Caring Mantle," *Medical Journal of Australia* 155 (2–16 December 1991): 779–83.

Jackson, John P., Jr., & Nadine M. Weidman. *Race, Racism, and Science: Social Impact and Interaction* (Santa Barbara CA: ABC–CLIO, 2004).

Jones, James H. *Bad Blood: The Tuskegee Experiment: A Tragedy of Race and Medicine* (New York: Free Press, 1981).

Jose, Nicolas. *Black Sheep: Journey to Borroloola* (London: Profile, 2002).

Katz, Jay. "Human Sacrifice and Human Experimentation: Reflections at Nuremberg," *Occasional Papers* 5 (25 October1 996), http://digitalcommons.law.yale.edu/ylsop_papers/5 (accessed 11 Nov. 2010)

Kelm, Mary Ellen. "Diagnosing the Discursive Indian: Medicine, Gender, and the 'Dying Race','' *Ethnohistory* 52.2 (Spring 2005): 371–406.

Kluchin, Rebecca Marie. *Fit to be Tied: Sterilization and Reproductive Rights in America, 1950–1980* (New Brunswick NJ: Rutgers UP, 2009).

Kuhl, Stefan. *The Nazi Connection: Eugenics, American Racism and German National Socialism* (New York & Oxford: Oxford UP, 1994).

Lopez, Iris Ofelia. *Matters of Choice: Puerto Rican Women's Struggle for Reproductive Freedom* (New Brunswick NJ: Rutgers UP, 2008).

Low, Gail Ching-Liang. *White Skins/Black Masks: Representation and Colonialism* (London: Routledge, 1996).

MacDonald, Helen. *Human Remains: Dissection and its Histories* (New Haven CT & London: Yale UP, 2006).

——. "Reading the 'Foreign Skull': An Episode in Nineteenth Century Colonial Human Dissection," *Australian Historical Studies* 125 (2005): 81–96.

McGeal, Chris. "US says sorry for 'outrageous and abhorrent' Guatemalan syphilis tests," *The Guardian* (1 October 2010).

McGregor, Russell. "'Breed Out the Colour': Reproductive Management for White Australia," in *"A Race for a Place": Eugenics, Darwinism and Social Thought and Practice in Australia*, ed. Martin Crotty et al. (Callaghan, NSW: U of Newcastle P, 2000): 61–70.

Mackenzie, John M. *Propaganda and Empire: The Manipulation of British Public Opinion 1880–1960* (Manchester UP, 1984).

MacLeod, Roy M., & Milton Lewis ed., *Disease, Medicine and Empire* (London & New York: Routledge, 1988).

Mitchell, Alison. "Survivors of Tuskegee Study Get Apology from Clinton," *New York Times* (17 May 1997).

Morgan, David. "Study Says U.S. Eugenics Paralleled Nazi Germany," *Chicago Tribune* (15 February 2000).

Outlaw, Lucius. "Toward a Critical Theory of Race," in *Anatomy of Racism*, ed. David Theo Goldberg (Minneapolis: U of Minnesota P, 1990): 58–82.

Palmer, Alison. *Colonial Genocide* (Adelaide: Crawford House, 2000).

Qureshi, Sadiah. "Displaying Sara Baartman, 'The Hottentot Venus'," *History of Science* 43 (2004): 233–57.

Reichardt, Eike. *Health, 'Race' and Empire: Popular-Scientific Spectacles and National Identity in Imperial Germany, 1871–1914* (Lulu.com, 2008).

Reynolds, Henry. *Nowhere People* (Camberwell, Victoria: Penguin, 2005).

Rydell, Robert W., John E. Findling & Kimberly D. Pelle, *Fair America: World's Fairs in the United States* (Washington DC & London: Smithsonian Institution Press, 2000).

Saunders, Suzanne. "Isolation: The Development of Leprosy Prophylaxis in Australia," *Aboriginal History* 14.2 (1990): 168–81.

Scales–Trent, Judy. "Racial Purity Laws in the United States and Nazi Germany: The Targeting Process," *Human Rights Quarterly* 23 (2001): 259–307.

Simpson, Moira. *Making Representations: Museums in the Postcolonial Era* (London: Routledge, 1996).

Smith, Andrea. "Not an Indian Tradition: The Sexual Colonization of Native Peoples," *Hypatia* 18.2 (Spring 2003): 70–85.

Spivak, Gayatri Chakravorty. "Can the Subaltern Speak?" in *Marxism and the Interpretation of Culture*, ed.Cary Nelson & Lawrence Grossberg (Urbana: U of Illinois P, 1988): 271–313.

——*In Other Worlds: Essays in Cultural Politics* (New York: Methuen, 1987).

Stepan, Nancy. *The Idea of Race in Science: Great Britain 1800–1960* (London: Macmillan, 1982).

Street, Brian V. *The Savage in Literature: Representations of "Primitive" Society in English Fiction 1858–1920* (London: Routledge & Kegan Paul, 1975).

Strother, Zoë S. "Display of the Body Hottentot," in *Africans on Stage: Studies in Ethnological Show Business*, ed. Bernth Lindfors (Bloomington: Indiana UP, 1999): 1–61.

Swanson, Maynard W. "The Sanitation Syndrome: Bubonic Plague and urban native policy in the Cape Colony, 1900–09," in *Segregation and Apartheid*, ed. Beinart & Dubow, 25–42.

Turnbull, Paul. "Ancestors not Specimens: Reflections on the Controversy over the Remains of Aboriginal People in European Scientific Collections," *Electronic Journal of Australian and New Zealand History* (27 April 1997), http://www.jcu.edu.au/aff/history/articles/turnbull.htm (accessed 11 November 2010).

Washington, Harriet A. *Medical Apartheid: The Dark History of Medical Experimentation on Black Americans from Colonial Times to the Present* (Garden City NY: Doubleday, 2007).

Wellerstein, Alex. "Harry Laughlin's 'Model Eugenical Sterilization Law'," *History of Science*, Harvard University, http://www.people.fas.harvard.edu/~wellerst/laughlin/ (accessed 11 November 2010).

Young, Robert J.C. *Colonial Desire: Hybridity in Theory, Culture, and Race* (London: Routledge, 1995).

⌘

Reconstruction of Indigenous Identities in the Twentieth Century

SANDRINE TOLAZZI

THE OBJECTIVATION OF INDIGENOUS PEOPLES BY WESTERN SCIENCE – and more particularly physical anthropology – and the use of religious arguments to justify government-planned marginalization and dispossession throughout the nineteenth century have inflicted deep scars on the Indigenous psyche. Although scientific discourse on race changed after the Second World War and decolonization, followed by attempts at reconciliation, transformed the relationship between governments and minorities, Indigenous peoples today are still defined in opposition to the dominant group. This, to Jean–Jacques Simard, lies at the heart of their ongoing "reduction," whether geographical (through reserves), economic (through government subsidies), political (through the contraction of their public space to the State's administrative structures), or judicial (through Indian status).[1] In the reversed dichotomy that ensues, many researchers now associate 'whites' with everything that seems to be going wrong with modernity – increased consumerism, environmental destruction, individualism, unchecked progress – while Indigenous peoples, who are seen as having maintained links with a specific culture and set of traditions, seem to hold the key to successful sustainable development and respectful relationships with nature and the community. Admittedly, these changes have raised awareness of the specific demands of Indigenous peoples and increased their participation in decision-making, leading to achievements in terms of recognition of land rights or self-government that constitute a breakthrough in the relationship with govern-

---

[1] Jean–Jacques Simard, *La Réduction: L'Autochtone inventé et les Amérindiens d'aujourd'hui* (Sillery, Quebec: Septentrion, 2003): 21–47.

ments. But it has also set them rigidly in outdated structures, ossifying their identity and reinforcing the distinction between white and Indigenous communities. This might explain ongoing Indigenous opposition to Western science in general, and to genetic research on Indigenous populations by non-Indigenous people in particular.

The End of Colonialism and the Transformation of Scientific Discourse on Race

The end of the Second World War coincided with the demise of colonialism as a theory justifying the subjugation of one people by another. As decolonization, supported by the newly created United Nations, progressively changed the world map, the arguments that were formerly used to account for imperialism and expansionism were challenged. Social Darwinism no longer served to rationalize domination, and, following the horrors of the Holocaust, Jules Ferry's justification of colonial expansion, that "higher races ha[d] a right over lower races" and that "they ha[d] a duty to civilize the inferior races,"[2] carried terribly bleak undertones.

The postwar context led to the questioning of many previously held assumptions, as evidence from volumes published at that time attests. In *The Science of Man and the World Crisis* (1945), the professor of anthropology Ralph Linton underlined the importance of taking into account recent knowledge on race for "intelligent planning of the new world order."[3] In the same volume, an article by the sociology professor Raymond Kennedy, entitled "The Colonial Crisis and the Future," deconstructed the five universal traits of colonialism: (1) the colour line, or the distinction between racial groups which was used to establish relations of superordination and subordination; (2) political control by the ruling power; (3) economic dependence upon and control by the mother country; (4) low stage of development of social services for native populations; and (5) lack of social contact between natives and the ruling caste.[4] After showing that the supposedly rational and scientific arguments used to justify such practices were untenable, Kennedy argued that the era of colonial rule, during which "the colonial system became institution-

---

[2] Jules Ferry, Discourse Before the Chamber of Deputies (28 July 1885).

[3] Ralph Linton, "Preface," in *The Science of Man in the World Crisis*, ed. Ralph Linton (New York: Columbia UP, 1945): vii.

[4] Raymond Kennedy, "Colonial Crisis and the Future," in *The Science of Man in the World Crisis*, ed. Ralph Linton (New York: Columbia UP, 1945): 308–11.

alized, evolving a general code of rules and customs, supported by an elaborate philosophy of rationalization based largely upon theories of the fundamental racial and cultural inferiority of the dependent peoples,'[5] was coming to an end. This, to Kennedy, was partly the result of a wave of public opinion that was now fully committed to applying the standards of democracy to colonial areas.[6]

Following this trend, the United Nations, whose 1945 Charter had affirmed the principle of self-determination of peoples, made a formal declaration on the granting of independence to colonial countries and peoples (1960), followed by the creation of a special committee on decolonization in charge of its implementation. Today, as the Second International Decade for the Eradication of Racism is drawing to an end, more than eighty former colonies have gained access to some form of independence, while only sixteen non-autonomous countries remain in the world. What it is important to underline here is the role played by anticolonial discourse in conveying a perception of Indigenous peoples as experiencing the same kind of colonization as depicted by Kennedy. Hence, in former white settler colonies public opinion has gradually realized that racism and domination towards these peoples will have to cease.

A second important element in this change of perception was the emergence of new schools of thought in disciplines such as physical or cultural anthropology which discredited old arguments used to account for differences between peoples. Hence, in the field of physical anthropology, Carleton S. Coon's theories on the racial progression of "Mediterraneans," which justified their expansion into the New World, were largely rejected two decades after the success of his volume *The Races of Europe*.[7] Although he was still serving as the president of the American Association of Physical Anthropologists in 1962, when he published *The Origin of Races*,[8] the discipline had already moved away from racial typology, and, in the context of the American civil-rights movement, his arguments could no longer be endorsed to condone segregation. As evidence from biologists trained in genetics discredited eugenics, physical anthropologists moved away from the classification of peoples into racial 'types' and instead followed Sherwood Washburn's lead in trying to explain human evolution and human variation using the new scienti-

---

[5] Kennedy, "Colonial Crisis and the Future," 345.

[6] "Colonial Crisis and the Future," 346.

[7] Carleton S. Coon, *The Races of Europe* (New York: Macmillan, 1939).

[8] Carleton S. Coon, *The Origin of Races* (New York: Random House, 1962).

fic tool of population genetics instead of the unreliable technique of morphometrics, and also taking cultural systems into account.[9] Though race was still used as a concept, its perception was transformed. To Washburn,

> the concept of race is fundamentally changed if we actually look for selection, migration, and study people as they are (who they are, where they are, how many they are); and the majority of anthropologist textbooks need substantial revision along these lines.[10]

A similar change can be seen in cultural anthropology, which gradually abandoned the cultural evolutionism that had characterized the discipline in the nineteenth century, dividing cultures into 'primitive' and 'civilized'. New schools of thought emerged, which made this distinction irrelevant. Structuralists, such as Claude Lévi–Strauss, pointed to the similarities between peoples regarding the structures of the mind.[11] Cultural relativists, such as Franz Boas, highlighted the importance of history in what made peoples evolve differently.[12] And functionalists, such as Bronislaw Malinowski, called attention to the function that each element of a culture fulfilled.[13] These new schools contributed to conveying a more positive representation of Indigenous peoples, though they did not put an end to the rationalization/classification that inevitably accompanied the study of these populations.

Finally, as the 1949 General Conference of UNESCO instructed the Director-General to "study and collect scientific materials concerning questions of race," "give wider diffusion to the scientific information collected," and "prepare an educational campaign based on the information,"[14] scientific racism was discredited by scientists themselves. Biologists who specialized in developmental genetics confronted the views of scientific racists at confe-

---

[9] Washburn's 1951 paper on "The New Physical Anthropology" is considered as a turning point in the evolution of physical anthropology as a discipline. Sherwood Larned Washburn, "The New Physical Anthropology," *Transactions of the New York Academy of Sciences* 2.13 (1951): 298–304.

[10] Washburn, "The Study of Race," *American Anthropologist* 65.3 (1963): 523.

[11] Claude Lévi–Strauss, *Anthropologie structurale* (Paris: Plon, 1958).

[12] Franz Boas, *Race, Language, and Culture* (New York: Macmillan, 1948).

[13] Bronislaw Malinowski, *Argonauts of the Western Pacific* (1922; New York: E.P. Dutton, 1961).

[14] *Records of the General Conference of the United Nations Educational, Scientific, and Cultural Organization. Fourth Session* (Paris: UNESCO, 1949): 22.

rences, and published articles arguing that the former classification of races was misleading. To Leslie Clarence Dunn, a prominent geneticist, differences between humans were mainly due to biological heredity and cultural heritage.[15] His Russian colleague, Nikolay Petrovich Dubinin, underlined the fact that while there could be such a process as natural selection among animals, historical, cultural, and environmental factors were mostly accountable for changes observed among *Homo sapiens*.[16] Most of all:

> Recent genetic knowledge suggests that all people have the ability to reason, that they are born equal, and that there is no hierarchy of superior and inferior races. Racial variations do not affect man as a social animal; in proclaiming the contrary, racialism perverts scientific data. Racialist theories are pseudo-scientific, with no basis in biological fact. Racial prejudice will disappear when colonialism and imperialism also disappear.[17]

Various statements by UNESCO (e.g., the 1967 Statement on Race and Racial Prejudice) had anticipated these views.[18] By that time, then, science could no longer justify the subjugation of Indigenous peoples.

Indigenous Empowerment

Despite decolonization and the transformation of scientific discourse on race, assimilationist theories did not disappear overnight; they were simply framed in another type of discourse which aimed at treating Indigenous peoples as citizens with the same rights and duties as other citizens – hence, for example, the citizenship and voting rights given to Canadian First Nations in the 1960s (prior to this, First Nations living on reserves were considered as wards of the government) – without taking into account their specificities. Meanwhile, these peoples started to organize nationally and internationally in a context of new movements questioning the existing social order. The 'red power' that started to emerge in the form of the American Indian Movement was echoed

[15] Leslie Clarence Dunn, "Race and Biology," in *Race, Science and Society*, ed. Leo Kuper (New York: Columbia UP, 1975): 32.

[16] Nikolay Petrovich Dubinin, "Race and Contemporary Genetics," in *Race, Science and Society*, ed. Leo Kuper (New York: Columbia UP, 1975): 69.

[17] Dubinin, "Race and Contemporary Genetics," 87.

[18] UNESCO, *Statement on Race and Racial Prejudice* (Paris: UNESCO, Meeting of Experts on Race and Racial Prejudice, 1967).

in the rise of 'black power' in Australia and the creation of the National Indian Brotherhood (now Assembly of First Nations) in Canada. These organizations were no longer willing to bow their heads before government policies that ignored their demands. Consequently, when the 1969 White Paper on Indian policy came out in Canada,[19] suggesting treating Indians like other Canadians and therefore repealing the Indian Act and the Indian Affairs branch of the Department of Indian Affairs and Northern Development, and transferring the services to government agencies that provided for all other Canadians, there was widespread opposition from First Nations organizations which prevented the implementation of this scheme. Soon, indigenous organizations became active in trying to reclaim land rights in their own countries, and gathered on the international level to raise awareness about indigenous issues. The creation of the American – then International – Treaty Council in the USA (1974) and the World Council of Indigenous Peoples in Canada (1977) was decisive, leading to the coordination by these two bodies in 1977 of the first international conference of NGOs on Indigenous issues.[20]

Following this conference, the United Nations set up the first Working Group on Indigenous Peoples in 1982 as part of the Sub-Commission on Prevention of Discrimination and Protection of Minorities. This group was to elaborate a declaration on the rights of indigenous peoples which could be used as an international standard by members of the United Nations.[21] Although the declaration itself was not adopted until 2007, and though Australia, Canada, New Zealand, and the USA immediately opposed it (New Zealand, Australia, Canada, and the USA have only recently changed their minds), it was a first step towards recognition by international institutions of the specific concerns of Indigenous peoples. 1995 then marked the beginning of the United Nations' International Decade of the World's Indigenous Peoples, during which a Permanent Forum on Indigenous Issues was set up. This Forum was to advise the United Nations Economic and Social Council on Indigenous

---

[19] Canada, Department of Indian Affairs and Northern Development, *Statement of the Government of Canada on Indian Policy (White Paper), 1969* (Ottawa: Department of Indian Affairs and Northern Development, 1969).

[20] Lotte Hughes, *The No-Nonsense Guide to Indigenous Peoples* (Toronto: New Internationalist, 2003): 87.

[21] René Boudreault, *Du mépris au respect mutuel: clefs d'interprétation des enjeux autochtones au Québec et au Canada* (Montreal: Écosociété, 2003): 52. Hughes, *The No-Nonsense Guide to Indigenous Peoples*, 22.

questions related to "economic and social development, culture, the environment, education, health, and human rights."[22]

As international attention increased, Indigenous peoples not only started reclaiming land and rights that had been previously taken away, they also began voicing their concerns about specific issues from within international institutions or outside them. Among these issues, scientific research on their communities and intellectual property rights were two sources of worry, not only because of the legacy of the past but also because of the dangers of unethical practices and even the appropriation of genetic material. The World Council of Indigenous Peoples' Declaration of Principles, ratified in September 1984, did try to introduce some safeguards by stating that "indigenous peoples and their designated authorities have the right to be consulted and to authorize the implementation of technological and scientific research conducted within their territories and the right to be informed about the results of such activities."[23] But today, as Western science pursues genetic research and patents findings, the genetic resources of indigenous peoples remain unprotected by intellectual property rights (IPR). This is why these peoples are currently trying to fight back through, for example, the Intergovernmental Committee on Intellectual Property and Genetic Resources, Traditional Knowledge and Folklore[24] or through their participation in international events such as the 2002 World Summit on Sustainable Development, where they developed a 100-point plan of implementation underlining the following:

> We will work against any IPR regime that attempts to assert patents, copyrights, or trademark monopolies for products, data, or processes derived or originating from our knowledge. Genetic material, isolated genes, life forms or other natural processes must be excluded from IPR regimes (Point 28).
>
> We will oppose biopiracy and the patenting of all life forms (Point 36).[25]

---

[22] United Nations Permanent Forum on Indigenous Issues, http://www.un.org/esa/socdev/unpfii (accessed 15 November 2010). Also Hughes, *The No-Nonsense Guide to Indigenous Peoples*, 23.

[23] Hughes, *The No-Nonsense Guide to Indigenous Peoples*, 28.

[24] *The No-Nonsense Guide to Indigenous Peoples*, 105.

[25] *Indigenous People's Plan of Implementation on Sustainable Development*, presented to the United Nations' World Summit on Sustainable Development (Johannesburg, 2 September 2002).

The empowerment of indigenous peoples also had an influence on the national level, relations with governments being considerably transformed in the second half of the twentieth century through both concrete and symbolic actions. There was a growing realization that indigenous peoples' individual rights had to be protected, but also that some of their demands concerning land, culture or self-government could only be answered through collective rights. While land rights in Canada were recognized through the Calder case in 1973 and subsequently in the 1982 Constitution Act, it took longer for Australia to overturn the doctrine of *terra nullius*[26] through the 1992 Mabo decision and enable aboriginal peoples to claim rights to the land through the 1993 Native Title Act. As government commissions were established to determine priorities,[27] as action was taken so that Indigenous peoples in Canada could have more responsibility in the field of education,[28] or as ancestral remains previously held by British museums were gradually repatriated to Australia,[29] there was an impression that institutions were finally taking into account the collective demands of indigenous communities and showing more respect for their cultures.

This impression was further reinforced by the attempts at reconciliation and apologies recently formulated by the Canadian and Australian governments. Following the 1998 Statement of Reconciliation,[30] Canada has now established a Truth and Reconciliation Commission in charge of "establishing new relationships embedded in mutual recognition and respect that will forge

[26] This notion that Australia was an empty, uninhabited land was used by the British to justify both their colonization of the continent and the resulting attitudes and policies the settlers adopted towards the Indigenous population.

[27] For example, the Royal Commission on Aboriginal Peoples created in 1991, which published its report in 1996. Canada, Royal Commission on Aboriginal Peoples, *Report of the Royal Commission on Aboriginal Peoples* (Ottawa: Minister of Supply and Services Canada, 1996).

[28] In 1973, the Department of Indian Affairs and Northern Canada gave responsibility to the First Nations in the field of education.

[29] Paul Turnbull, "Scientific Theft of Remains in Colonial Australia," *Australian Indigenous Law Review* 11.1 (2007): 92–104.

[30] Canada, Minister for Indian Affairs and Northern Development, *Gathering Strength. Canada's Aboriginal Action Plan* (Ottawa: Minister of Public Works and Government Services, 2000).

a brighter future,"[31] and in 2008 the then Prime Minister Stephen Harper apologized to the First Nations for the residential schools system set up by the Canadian government at the turn of the nineteenth century. In Australia, a Council for Aboriginal Reconciliation was established in 1991, and although it was later replaced with a private body, Reconciliation Australia, due to the lack of commitment on the part of the Howard government (1996–2007), then Prime Minister Kevin Rudd's apology to the 'Stolen Generations' in 2008 offers a glimpse of hope for Aboriginal peoples in Australia.

A Still Unbridgeable Divide

Given the new relationship which seems to be developing between Indigenous peoples, the non-Indigenous population, and institutions at the national and international levels, we could perhaps have expected a change in Indigenous attitudes to Western science and the beginnings of cooperation with scientists, especially in cases where their research is focused on diseases that affect Indigenous communities. Although there have been such cases of cooperation, Indigenous discourse has, overall, remained hostile to genetic research. Unfortunately, there are examples of research leading to a similar form of betrayal to that which Indigenous peoples experienced during colonization, when they were told that their lives would improve if they abided by Western principles or values, or that their children would benefit from going away to residential schools. One notorious example of such betrayal is the research that was conducted between 1982 and 1985, when almost 900 blood samples were taken from the Nuu-chah-nulth people, a community living on the Western coast of Canada. This was the largest-ever study of a First Nations community, involving almost 40% of its members. The Nuu-chah-nulth people were told by Richard Ward, who led the study, that this might help find a cure for arthritis, although, when questioned by a BBC reporter for a documentary called *In Search of the First Americans*, Ward stated that he was actually tracing the evolutionary history of First Nations by studying their DNA. Soon after, the scientist moved to the University of Utah, where he got funding to do further study on the blood samples, though the Nuu-chah-nulth were never informed of such a study. Finally, he was offered a position at Ox-

[31] Truth and Reconciliation Commission of Canada. Our Mandate, http://www.trc.cvr.ca/overview.html (accessed 16 November 2010).

ford, where the samples were used again for a variety of studies, including a project on the spread of lymphotropic viruses by intravenous drug users.[32]

Although this is now cited as an example of a worst-case scenario by the ethics committees which have been created since then to control such research, manipulation can still exist in other ways. It may involve getting public support for research which is presented as paving the way for finding a cure to a specific disease, so that the community that refuses to participate is blamed for its attitude. In other cases, scientists may be within the bounds of the regulations set up by the ethics committees but still withhold some information on the form of consent Indigenous peoples have to sign, or they may present well-rounded arguments dressed up in scientific jargon which disclaim indigenous world-views.

Even if ethics committees managed to find a way to ensure respect for the communities that are being studied, suspicion would still prevail simply because of the opposition that still persists, albeit less overtly, between Indigenous and non-Indigenous peoples. Although this opposition is no longer used to validate and substantiate the domination of the latter by the former, it still conveys the idea that the two groups hold irreconcilable opinions. While trying to erase former beliefs that Indigenous peoples were uncivilized, primitive peoples, the new discourse on First Nations in Canada, for example, reinstates such opposition by presenting them as being the guardians of the land who shared their knowledge with the first settlers and saved them from famine and illnesses, and now still hold 'traditional' conceptions of the world that are the key to "solving conflicts related to great moral, political, and cultural questions."[33] Nowhere is this more palpable than in the guide to understanding Aboriginal cultures published by Indian and Northern Affairs Canada to train its own staff. The guide divides the values held by the "traditional Aboriginal society" and the "non-traditional dominant society" in two columns, thus highlighting the opposition between those for whom "harmony with nature is privileged" and "the group is important," and those for whom "control of na-

---

[32] The story of the Nuu-chah-nulth study is summed up in David Wiwchar, "Nuu-chah-nulth blood returns to West coast," *Ha-Shilth-Sa* 31.25 (16 December 2004): 1–4.

[33] Canada, Affaires indiennes et du Nord Canada, *Atelier de sensibilisation aux cultures autochtones* (Ottawa: Ministre des Travaux publics et Services gouvernementaux Canada, 1999): 36. (My tr.)

ture is privileged" and "the individual is important."[34] It is interesting to note that the document also underlines the fact that problem-solving is a matter of intuition, creativity, and a holistic approach in Indigenous communities, while non-Indigenous communities would rather use rationality, logic, and a linear approach.[35] Considering this division, it is no wonder that scientific research on genes, which could be interpreted as a need to control nature and use rationality to solve problems, should be seen as opposed to Indigenous values.

Indigenous discourse is also framed in terms which emphasize this opposition, as the documentary *The Leech and the Earthworm*,[36] which considers genetic engineering from an Indigenous point of view, clearly shows. In this documentary, the debate on genetic engineering exposes once again the conflict between Indigenous and non-Indigenous values, and genetics research is described as another example of how non-Indigenous peoples – scientists, governments, and multinational corporations in particular – are trying to impose their set of values on Indigenous peoples. However, it could be argued that genetic research on Indigenous peoples offers opportunities to reverse the views propounded by racial theorists of another era and show how valuable Indigenous world-views may be. Capitalism and the emphasis on individual property leave so many people by the wayside that they can hardly be considered as progress. The type of governance that Western countries have adopted is proving incapable of dealing with long-term problems because it functions within the short time-span of elections. The greed of multinational companies, in conjunction with the consumerism of individuals, has damaged the planet beyond repair. The benefits science seemed to offer in the form of biochemicals or genetic modifications meant to help mass-produce fruit and vegetables that are resistant to pests, or increase milk yield in cows, are now leading to environmental problems. The process of imposing one value system on another still seems to be going on, yet it is becoming increasingly difficult to justify.

In the end, the hostility shown by Indigenous peoples towards genetic engineering can partly be explained by the story of colonization. But at the same time, this history has reinforced Indigenous identities, so that we are left with, on the one hand, a dominant system whose values are seen as completely

---

[34] Canada, Affaires indiennes et du Nord Canada, *Atelier de sensibilisation aux cultures autochtones*, 34–35.

[35] *Atelier de sensibilisation aux cultures autochtones*, 34.

[36] Marc Silver & Max Pugh, *The Leech and the Earthworm* (Debra Harry, 2003).

opposed to indigenous values and threatening to corrupt the latter and, on the other, another system with its own mind-set whose participants have gained in confidence and are no longer willing to accept arguments from the main-stream as indisputable. According to this representation, ethics committees can be of no use, because they cannot reconcile the scientists' points of view with Indigenous ones: while scientists may think in terms of biomapping, Indigenous peoples inevitably think in terms of biocolonizing.

Yet Simard reminds us that reality is more complex than the polarization which opposes "the modern, individualistic, technological, alienated, and artificial civilization – evil," to indigenous cultures presented as the other side – "holistic, ecological, authentic, 'natural'" – of modernity.[37] What is more, in some cases, this oversimplification which can be found in many analyses of Indigenous peoples' condition, tends to hide the very structures of alienation and oppression that they are meant to address.[38] That is not to say that the specificities of indigenous populations should not be taken into account, but that researchers should be aware that "the Amerindian and Inuit ghettos are subjected to the same forces that structure *all* advanced capitalist societies"[39] – those forces that, precisely, leave so many people destitute. As long as the dichotomy prevails – a dichotomy which is used by Indigenous peoples to support their claims and by non-Indigenous peoples in their response to such claims – the opposition to Western science on the part of Indigenous peoples will inevitably persist.

WORKS CITED

Boas, Franz. *Race, Language, and Culture* (New York: Macmillan, 1948).

Boudreault, René. *Du mépris au respect mutuel: clefs d'interprétation des enjeux autochtones au Québec et au Canada* (Montreal: Écosociété, 2003).

Canada. Affaires indiennes et du Nord Canada. *Atelier de sensibilisation aux cultures autochtones* (Ottawa: Ministre des Travaux publics et Services gouvernement aux Canada, 1999).

---

[37] Simard, *La Réduction: L'Autochtone inventé et les Amérindiens d'aujourd'hui*, 391. (My tr.)

[38] *La Réduction: L'Autochtone inventé et les Amérindiens d'aujourd'hui*, 258.

[39] *La Réduction: L'Autochtone inventé et les Amérindiens d'aujourd'hui*, 258. (My tr.)

——. Department of Indian Affairs and Northern Development. *Statement of the Government of Canada on Indian Policy (White Paper), 1969* (Ottawa: Department of Indian Affairs and Northern Development, 1969).

——. Minister for Indian Affairs and Northern Development. *Gathering Strength. Canada's Aboriginal Action Plan* (Ottawa: Minister of Public Works and Government Services, 2000).

——. Royal Commission on Aboriginal Peoples. *Report of the Royal Commission on Aboriginal Peoples* (Ottawa: Minister of Supply and Services Canada, 1996).

Coon, Carleton S. *The Origin of Races* (New York: Random House, 1962).

——. *The Races of Europe* (New York: Macmillan, 1939).

Dubinin, Nikolay Petrovich. "Race and Contemporary Genetics," in *Race, Science and Society*, ed. Leo Kuper (New York: Columbia UP, 1975).

Dunn, Leslie Clarence. "Race and Biology," in *Race, Science and Society*, ed. Leo Kuper (New York: Columbia UP, 1975).

Ferry, Jules. Discourse before the Chamber of Deputies (28 July 1885).

Hughes, Lotte. *The No-Nonsense Guide to Indigenous Peoples* (Toronto: New Internationalist, 2003).

Indigenous People's Plan of Implementation on Sustainable Development. Presented to the United Nations' World Summit on Sustainable Development (Johannesburg, 2 September, 2002).

Kennedy, Raymond. "Colonial Crisis and the Future," in *The Science of Man in the World Crisis*, ed. Ralph Linton (New York: Columbia UP, 1945): 306–46.

Lévi–Strauss, Claude. *Anthropologie structurale* (Paris: Plon, 1958).

Linton, Ralph. "Preface," in *The Science of Man in the World Crisis*, ed. Ralph Linton (New York: Columbia UP, 1945): vii–viii.

Malinowski, Bronislaw. *Argonauts of the Western Pacific* (1922; New York: E.P. Dutton, 1961).

Silver, Marc, & Max Pugh. *The Leech and the Earthworm* (Debra Harry, 2003).

Simard, Jean–Jacques. *La Réduction: L'Autochtone inventé et les Amérindiens d'aujourd'hui* (Sillery, Quebec: Septentrion, 2003).

Truth and Reconciliation Commission of Canada. Our Mandate, http://www.trc.cvr.ca/overview.html (accessed 16 November 2010).

Turnbull, Paul. "Scientific Theft of Remains in Colonial Australia," *Australian Indigenous Law Review* 11.1 (2007): 92–104.

United Nations. *Charter of the United Nations* (26 June 1945).

——. *Declaration on the Granting of Independence to Colonial Countries and Peoples – Resolution 1514 (XV)* (14 December 1960).

UNESCO. *Records of the General Conference of the United Nations Educational, Scientific and Cultural Organization. Fourth Session* (Paris: UNESCO, 1949).

——. *Statement on Race and Racial Prejudice* (Paris: UNESCO, Meeting of Experts on Race and Racial Prejudice, 1967).

United Nations Permanent Forum on Indigenous Issues, http://www.un.org/esa/socdev/unpfii (accessed 15 November 2010).

Washburn, Sherwood L. "The New Physical Anthropology," *Transactions of the New York Academy of Sciences* 2.13 (1951): 298–304.

——. "The Study of Race," *American Anthropologist* 65.3 (1963): 521–31.

Wiwchar, David. "Nuu-chah-nulth blood returns to West coast," *Ha-Shilth-Sa* 31.25 (16 December 2004): 1–4.

⌘

Part II
Defining and Mapping

Genetic Blood Testing of Native Americans in the USA*

RENATE BARTL

DECADES BEFORE DNA TESTING BECAME POPULAR, genetic blood testing was very prominent in searches for typical genetic markers and gene variations for specific biological races. The method of testing blood for typical *Native American* gene frequencies started in 1923[1] and ended in the 1980s, when DNA testing was substituted in investigations of the origin and ancestry of American Indians. This earlier era of genetic blood testing (1923–1980s) will be discussed here from an ethno-historical perspective to provide an overview of the predecessor of DNA screening and the problems inherent in this method.

The Method

The blood group antigen system according to which American Indians were tested is set up on the basis of the following factors or markers:[2]

---

* This article is not written by a trained geneticist or physician – the genetic data mentioned here are evaluated in an ethnohistorical discourse by an ethnohistorian who has been doing research on mixed groups of European, African-American, and Native American ancestry in the USA for nearly thirty years. I want to thank Helen C. Rountree for proofreading and commenting on this article.

[1] Dennis H. O'Rourke, "Blood Groups, Immunoglobulins, and Genetic Variation," in *Handbook of North American Indians: Environment, Origins, and Population*, vol. 3, ed. Douglas H. Ubelacker (Washington DC: Smithsonian Institution, 2006): 762, mentions that the study of genetic variation in indigenous populations was initiated by the study of Arthur F. Coca & Olin Deibert, "A Study of the Occurrence of the Blood Groups Among the American Indians," *Journal of Immunology* 8 (1923): 487–93.

[2] O'Rourke, "Blood Groups, Immunoglobulins, and Genetic Variation," is used here as the major source for the blood group antigen system and its variations among Native

1. Red Cell Antigen System
1.1. ABo blood group system (ABo)
1.2. MN and Ss blood group system (MNSs)
1.3. RH blood group system (RH)
1.4. Duffy blood group system (FY)
1.5. Kidd blood group system (JK)
1.6. Diego blood group system (DI)
1.7. Other Systems: Hh, Secretor, Lewis, Kell, Lutheran, Dombrock, Colton, Yt, Xg
2. White Cell Antigen System:
2.1. Human Leukocyte Antigen (HLA)
3. Serum Protein System:
3.1. Albumin (ALB)
3.2. Haptoglobin (HP)
3.3. Transferrins (TF)
3.4. Group Specific Component (CG)
3.5. Protein Properdin Factor B (BF)
3.6. Other Systems: Butyrylcholinestase E1 (BCHE1), Cholinestase E2 (CHE2)
4. Red Cell Enzymes
4.1. Phosphoglucomutase (PGM1, PGM2, PGM3)
4.2. Acid Phosphatase (ACP1)
4.3. Esterase D (ESD)
4.4. Immunoglobulin (IgG) / Gamma Globulins (GM)
4.5. Other Systems: Adenylate Kinase (AK1), Uridine Monophosphate Kinase, Phosphocluconate Dehydrogenase

According to modern research, the following frequencies are classified as typical of Native North Americans (see Table overleaf).[3]

As of today, more information on the genetic variations of Native Americans can still be drawn from genetic blood testing than from modern DNA screening.

> Much of what is known about patterns of genetic variation among indigenous populations of the Americas, and the evolutionary and population dynamics that gave rise to the observed patterns, are based almost exclusively on classical marker data. Allele frequency

---

North Americans. The information on the chromosomal location of the antigen system was available to O'Rourke and is mentioned in his article. As the researchers of the era discussed (1923–1980s) had no access to this information, it is disregarded here.

[3] Table taken from "Blood Groups, Immunoglobulins, and Genetic Variation," 773. Corrections by author.

data remain the most extensive genetic data available for indigenous populations of North America, forming the basis for the biological history and origin of populations of the region.[4]

High Frequency / Presence	Low Frequency / Absence
ABO*O	ABO*A2, ABO*B
MN*M	RH*RO
RH*R1	LU*A
FY*A	KEL*K
DI*A	HLAA*1, HLAA*3, HLAA*11
HLAA*2, HLAA*9	HLAB*29, HLAB*18
HLAB*35, HLAB*27	GM*F B, GM*A,F B
GM*A G, GM*A T	BF*F
GC*CHIP, GC*IGL	Abnormal haemoglobins
ALB*Mexico, ALB*Naskapi	
TF*DCHI, TF*BO-1	
BCHE1*U, CHE2*50-	

A Critical Analysis

Unfortunately this view may be overly optimistic, as the case studies discussed here will show. The method and the data collected contain many problems.

First of all, tribal designations and membership criteria have changed in the course of time:

> In the United States only a portion of the tribal groups in which individuals may identify membership are recognized by the federal government. Criteria for membership among tribal groups are not uniform. Many, if not most of the population samples for classical marker studies were collected prior to these political distinctions. Thus, in earlier studies it may be difficult to determine sample affiliation relative to the current definitions and recognized groups. An additional complexity obtains with terminology of group identity. Because many indigenous populations prefer to be identified by a name they have chosen rather than one assigned by outside entities, group names have changed over time, complicating retrospective comparative analyses. In some regions, different group names are used for groups and communities that are closely tied by language, culture, and geography.[5]

---

[4] O'Rourke, "Blood Groups, Immunoglobulins, and Genetic Variation," 762–63.

[5] "Blood Groups, Immunoglobulins, and Genetic Variation," 762–63.

There is also a lack of standardization of classical markers through which groups could be compared on a standardized level.[6] Furthermore gene frequencies probably have been misinterpreted as non-Native American, indicating admixture with other races.[7]

Certain blood gene frequencies between two biological races do not differ – or do so only minimally – so that it is impossible to ascribe the markers to one race or the other:

> A situation analogous to that in which the U.S. White and Indian frequencies are alike, is that wherein the gene frequencies in the American Indian and the West African Negro population are nearly identical [...]. In certain cases the allelic frequencies in the West African Negroes and the North American Whites are identical or exceedingly similar.[8]

Early studies – replicated and revised by subsequent studies – sometimes produced different and contradictory results, pointing to the methodological problems of these early studies. There seem to be incidences where faulty antisera were used, resulting in a wrong determination of blood group frequencies. There are also indications of errors in laboratory typing, and finally there is the problem of sample sizes so small that they could produce wrong data.[9] "Another problem is how best to average the values for different areas when the samples are different in size."[10] Moreover the inter-Native American gene flow is sparsely examined in most studies, although it is seen as one of the primary determinants of allele frequency variations in Native North American populations.[11]

---

[6] O'Rourke, "Blood Groups, Immunoglobulins, and Genetic Variation," 763.

[7] Race in this case means biological race (Caucasian, African, Native American, etc.), which geneticists try to identify by genetic markers and loci on the DNA string. That this definition of race is not supportable will be demonstrated in this article.

[8] Bentley Glass, "On the Unlikelihood of Significant Admixture of Genes from the North American Indians in the Present Composition of the Negroes in the United States," *American Journal of Human Genetics* 7 (1955): 374–75.

[9] O'Rourke, "Blood Groups, Immunoglobulins, and Genetic Variation," 765–67.

[10] Glass, "On the Unlikelihood of Significant Admixture of Genes from the North American Indians in the Present Composition of the Negroes in the United States," 370.

[11] O'Rourke, "Blood Groups, Immunoglobulins, and Genetic Variation," 776.

Case Studies from the Eastern USA

Genetic blood testing has mainly been used to find and prove American Indian ancestry. For this endeavour it was necessary first to search for pure-blood, typical Native American genetic blood markers, to determine race on a biological level. With the same typology, gene frequencies in blood were also ascertained for other races such as Europeans (Euro-Americans) and Africans (African Americans).

Now let us take a closer look at some cases in which this method was employed to determine race, or the racial composition of a group. One complication will be that there are several cases of Native American tribes splitting along colour lines with the aim of getting rid of their darker-skinned section.

The Nanticoke of Delaware

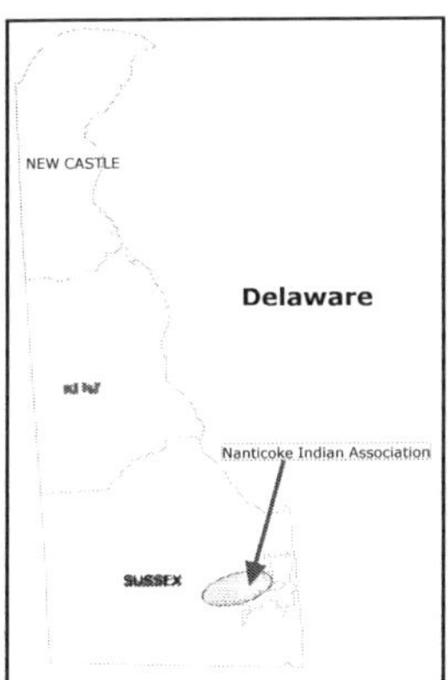

FIGURE 1. Map of Delaware (map by Renate Bartl)

The Nanticoke Indians of Delaware were recognized by the state of Delaware in 1881.[12] The same year, one part of the Nanticoke congregation started to build its own separate Indian church. The reason for this was that a new pas-

[12] Some eastern states of the USA have invented state recognition for tribes living within their territory. This has become necessary, as the USA only acknowledge tribes that had an official relationship (e.g. signing a treaty) with the US Government after 1776 (= federally acknowledged tribes). All tribes with an official relationship only to the colonial powers before 1776 are not acknowledged by the US Government as a federal Indian tribe. To compensate for this, several eastern states have established regulations to acknowledge the Indian tribes within their borders on a state level, also providing them with a state reservation in some cases.

tor was hired for the old Nanticoke Methodist Church, whom one segment of the congregation disapproved of as being too 'Negro'.[13] In consequence, the congregation split up into two groups. The opposing segment formed the Nanticoke Indian Association and constructed this new church: the Indian Mission United Methodist Church (incorporated in 1915), while the remaining group became the Harmony Group or Nanticoke Moors[14] and stayed at the old church, which is known nowadays as the Harmony African Methodist Episcopal Church (incorporated in 1875).[15]

The Halifax and Warren County Indians of North Carolina

This Indian group living in Halifax County and Warren County of North Carolina also split along colour lines – and was tested for blood gene frequencies after that schism. In 1957 the Halifax and Warren County Indians divided into the Haliwas and the Hollister Negroes, with the Haliwas classifying themselves as 'Indian' and the Hollister Negroes as 'non-Indian' or 'Negro'. In 1958–59 both groups were tested for blood frequencies.[16] The factors tested in both groups were Red Cell Antigens (see p. 74 below).

---

[13] Racial expressions like this are used here, because they were typical in this period of time. The author is fully aware that they are not politically correct now, but as this article is an historical overview, they are used in their historical context.

[14] The term *Moor* is derived from the designation for Moors from North Africa who had immigrated, or were deported to the east coast of North America since the seventeenth century. There are persons and groups in the USA today who still identify as Moors (e.g., the Federation of the Moorish Science Temple of America, Inc. [Ancient Moabites or Moors] in Maryland).

[15] William Harlen Gilbert, Jr., "Surviving Indian Groups of the Eastern United States," in *Annual Report of the Board of Regents of the Smithsonian Institution for 1948* (Washington DC: Smithsonian Institution Press, 1948): 407–38. Calvin. L. Beale, "American Triracial Isolates: Their Status and Pertinence to Genetic Research," *Eugenics Quarterly* 4.4 (1957): 187–96. Frank W. Porter III, *The Nanticoke* (Indians of North America; New York: Chelsea House, 1987): 68. Nanticoke Indian Association, "The Nanticoke Indian Tribe" (2003–2004), http://www.nanticokeindians.org/ (accessed 04 July 2010). State of Delaware, Department of State, "Harmony United Methodist Church" (30 December 2009), http://archives.delaware .gov/markers/sc /SC-187.shtml#P0_0 (accessed 5 July 2010). State of Delaware, Department of State, "Indian Mission United Methodist Church" (30 December 2009), http://archives .delaware.gov/markers/sc/SC-122.shtml#P0_0 (accessed 05 July 2010).

[16] William S. Pollitzer, R.M. Menengaz–Bock & J.C. Herion, "Factors in the Micro-

FIGURE 2. Nanticoke Indian Center, Nanticoke Indian Association, near Millsboro, Delaware (photo by Renate Bartl, 1991)

FIGURE 3. Nanticoke Indian Museum, Nanticoke Indian Association, near Millsboro, Delaware (photo by Renate Bartl, 1991)

---

evolution of a Tri-Racial Isolate," *American Journal of Human Genetics* 18 (1966): 26–38. Dr William S. Pollitzer (1923–2002) was professor in the Department of Anatomy, University of North Carolina at Chapel Hill. He and his research groups tried to define standard markers for race and analysed racially mixed groups for their 'racial composition' on the basis of these markers.

- ABo blood group system (ABo)[17]
- MN blood group system (MN, He)
- RH blood group system (RH)[18]
- Haemoglobin electrophoresis (Hb)[19]

The result of this study was:

> On the basis of blood type gene frequencies, the "Indian" community is about 18% white, 41% Negro, and 41% Indian, while the "non-Indian" community is about 35% white, 34% Negro, and 31% Indian. Members of the non-Indian group are slightly more Negroid in appearance, [...][20]

This result is highly questionable. The collection of the data uncovers several problems, including these two: there is no constant number of persons tested for all red cell antigen factors, and additional phenotype frequencies were determined only for the 'Non-Indian' group.

Before we start to discuss the method and findings of this study, the further history of the Haliwas should be mentioned. They were recognized by the State of North Carolina as a state tribe in 1965. In 1979, the year they applied for federal acknowledgement, they added a Saponi and Nansemond Indian identity to their tribal identity, and changed their name to Haliwa-Saponi Indian Tribe. The tribe is still living in Halifax County, North Carolina, and has a tribal centre there. In 2010 the Haliwa-Saponi Indian Tribe had 3,800 enrolled members – and was still waiting for a decision on its petition for federal acknowledgment.[21] No information could be obtained on what happened to the Hollister Negroes after the schism.

---

[17] A total of 396 persons were tested for ABo: 150 persons belonging to the 'Non-Indian' group, 246 persons belonging to the 'Indian' group.

[18] A total of 397 persons were tested for MN, He, and RH: 150 persons belonging to the 'Non-Indian' group, 247 persons belonging to the 'Indian' group.

[19] A total of 377 persons were tested for Hb: 145 persons belonging to the 'Non-Indian' group, 232 persons belonging to the 'Indian' group.

[20] Pollitzer, Menengaz–Bock & Herion, "Factors in the Microevolution of a Tri-Racial Isolate," 37.

[21] Haliwa-Saponi Indian Tribe, "Haliwa-Saponi Indian Tribe" (2010): www.haliwa-saponi.com (accessed 5 July 2010). North Carolina Department of Administration, Commission of Indian Affairs, "Tribes and Organizations" (2009), http://www.doa.nc.gov/cia/tribesorg.htm (accessed 26 May 2010). US Department of the Interior,

Discussion of the Sample Data For Racial Classification

The real problems occur when the results of the study on the Halifax and Warren County Indians are compared to standard data for racial identification of Europeans (Caucasians), African Americans (Africans), and Native Americans. The standard data chosen for European ancestry were English white gene frequencies; African-American ancestry was derived from Charleston or Gullah Negro data; and the standard for Native-American ancestry was taken from Cherokee Indian gene frequencies.

These methods and standard data need to be discussed more intensely here, because they contain many unproven and problematical assumptions.

European ancestry

The first assumption was that the European gene frequencies of the group examined came from an 'English white' parental population: "The white settlers of that region of North Carolina where the isolate is found today are known to be almost exclusively English."[22] The question is, whether the group really did intermix with pure-blood 'English Whites' only, or whether they also intermixed with immigrants from other European regions, or mixed-blood Europeans. Such people have lived (as a minority) in the North Carolina piedmont since colonial times.

African ancestry

The next assumption was that the African element in the group can be deduced from data on the 'Charleston Negro' or 'Gullah Negro' gene frequencies: "The Gullah Negroes of Coastal South Carolina [are] known by blood factors and morphology to be closely related to their African ancestors."[23]

First of all, the choice of data from South Carolina rather than Africa is dubious. Do data of African Americans from Charleston or the Gullah region of South Carolina really reflect the gene frequencies of African populations?

Office of Federal Acknowledgement, "List of Petitioners by State (as of September 22, 2008)" (22 September 2008): www.bia.gov/idc/groups/public/documents/text/idc-001215.pdf, 34 (accessed 6 May 2010).

[22] Pollitzer, Menengaz–Bock & Herion, "Factors in the Microevolution of a Tri-Racial Isolate," 34. By 'isolate' the authors mean the Halifax and Warren County Indians.

[23] "Factors in the Microevolution of a Tri-Racial Isolate," 34.

FIGURE 4. Map of South Carolina
(map by Renate Bartl)

Additionally, Africa is a very big continent and there are different gene frequencies for the populations from different regions. So, from what area of Africa was the African element in the Gullah Negro ancestry imported?

> It would be far more important to know the actual provenience of the slaves taken to North America, but it seems impossible to do better than to sample as wide a range of populations from West Africa as possible.[24]

As we see, these problems were already known to, and discussed by, researchers as early as 1955 – and data from Africa were already available at that time.[25] Why these data were not used by the Pollitzer research group is not known.

A further crucial point in this discussion is that all African Americans are more or less the result of intermixture with whites over many generations

---

[24] Glass, "On the Unlikelihood of Significant Admixture of Genes from the North American Indians," 370.

[25] "On the Unlikelihood of Significant Admixture of Genes from the North American Indians."

since early colonial times, which almost certainly has led to a misinterpretation of the data for the percentage of white ancestry.

A big part of the African-American ancestral group came by way of the Caribbean/West Indies with intermixture already taking place in these or other areas. Not neglected in this context should be the huge number of (free) persons of colour entering North America from early colonial time onwards. This group still is relatively sparsely investigated, although there are proofs that they intermixed with Native Americans of Eastern North America on a broad level from the beginning. Ira Berlin calls these immigrating persons of colour 'Atlantic creoles':

> Black life on mainland North America originated not in Africa or in America but in the netherworld between the two continents. Along the periphery of the Atlantic – first in Africa, then Europe, and finally in the Americas – it was the product of the momentous meeting of Africans and Europeans and then their equally fateful rendezvous with the peoples of the New World. Although the countenances of these new people of the Atlantic – "Atlantic creoles" – might bear the features of Africa, Europe, or the Americans in whole or in part, their beginning, strictly speaking, were in none of those places. [...] Atlantic creoles traced their beginnings in the historic encounter of European and Africans on the west coast of Africa. [...] Creoles of African descent were among the first Africans transported to the mainland.[26]

For these reasons, the data Pollitzer and his colleagues used for the African-American parental group are not reliable and surely caused inaccurate results.

Native American ancestry

The third assumption was that the standard data for Native American gene frequencies could be taken from the Cherokees of the Carolinas:

> During Colonial times, the Cherokees had general overlordship among Indian tribes of the Carolinas. Of the tribes of the southeastern United States, they alone have been extensively analyzed for blood factors.[27]

---

[26] Ira Berlin, *Many Thousands Gone: The First Two Centuries of Slavery in North America* (Cambridge MA: Harvard UP/Belknap Press, 1998): 17, 25.

[27] Pollitzer, Menengaz–Bock & Herion, "Factors in the Microevolution of a Tri-Racial Isolate," 34.

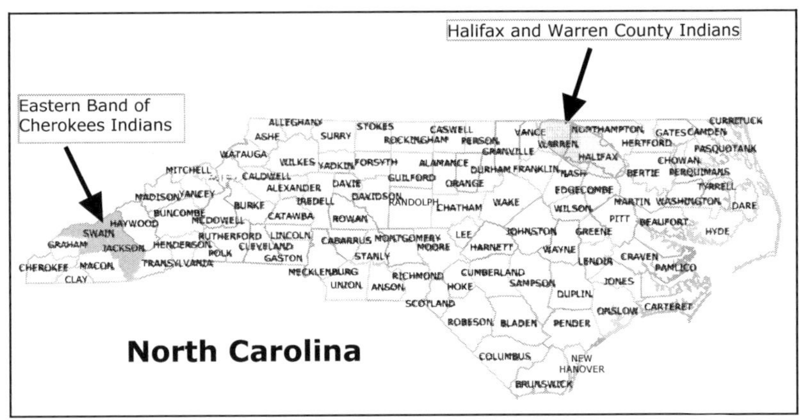

FIGURE 5. Map of North Carolina
(map by Renate Bartl)

The first problem with this assumption is that the only data on Native Americans of this region available and used for comparison were drawn from two 1957–60 surveys of the Eastern Band of Cherokees from North Carolina.[28] No data from other Indian groups and tribes of North Carolina were collected for comparison, especially data for Saponi and Nansemond were lacking – the two groups the Haliwa-Saponi Indian Tribe claims to descend from[29].

Eastern Band of Cherokee Indians data

Two surveys were made among Indians living on the Eastern Band of Cherokee Indian Reservation (also called Qualla Boundary, Jackson and Swain Counties, North Carolina). During the first survey in 1957–58, blood samples

[28] William S. Pollitzer, Robert C. Hartman, Hugh Moore et al., "Blood Types of the Cherokee Indians," *American Journal of Physical Anthropology* 20 (1962): 33–34.

[29] As of 2010, one federally recognized Indian tribe – the Eastern Band of Cherokee Indians – eleven state-recognized tribes, and twenty-six Indian groups without recognition are living in North Carolina. US Department of the Interior, Office of Federal Acknowledgement, "List of Petitioners by State (as of September 22, 2008)," 34–36. US Department of the Interior, Office of Federal Acknowledgement, "Indian Entities Recognized and Eligible To Receive Services From the United States Bureau of Indian Affairs," *Federal Register* 7.153 (11 August 2009): 40218–23, http://www.bia.gov/idc/groups/public/documents/text/idc-001736.pdf (accessed 6 May 2010). North Carolina Department of Administration, Commission of Indian Affairs. "Tribes and Organizations."

were taken from 534 school children "whose degree of Indian ancestry was known"[30] and who were genetically tested for haemoglobin patterns and ABo blood groups.

A second survey was made in 1959–60 and blood samples were taken from seventy-eight full-blood and ninety-four mixed-blood Indians. The information on the degree of Indian blood was obtained from the tribal records. No information is given on the degree of European and African-American blood quanta in the mixed-blood persons screened. In 1960, the total population of the Eastern Cherokee on the Qualla Boundary Reservation was 4,494.[31] What was done in this study was quite typical for all these studies: the researchers took a small segment of the tribal population, screened them, and declared the data to be genetic standard data for the whole tribe. In one case twelve percent, in the other case four percent of the tribal population was screened, and the results were projected onto the whole tribe and defined as standard. This time-saving but potentially faulty method of projecting the results of small-size samples onto the entire Indian tribe is not unusual for defining standard gene frequencies for Native North Americans – as will be seen later.

Using a 1928 study of the physical anthropology of the Eastern Cherokee, which stated that admixture with Scotch, Scotch-Irish, English, and Germans was reported for the group, the 1957–60 study comes to its ultimate findings: "It appears probable that hybrid Cherokees have been produced primarily by an admixture with an essentially English population."[32] This outcome of the study is somewhat dubious. The Cherokees were living in the Southeastern USA before their deportation to the Indian Territory was ordered by the Indian Removal Act of 1830. In the Old South, some of the Cherokees were planters and slaveholders, who owned – and inter-mixed with – African slaves.[33] These slaves became known as 'freedmen' after the Civil War, and

---

[30] Pollitzer, Hartman, Moore et al., "Blood Types of the Cherokee Indians," 33.

[31] There seems to be an initial assumption in this study that the non-Indian parental group of all mixed-blood Cherokees was (English) White. African-American ancestry is ignored, or even denied. The study also neglects to mention how the researchers obtained permission to take blood from the persons tested.

[32] Pollitzer, Hartman, Moore et al., "Blood Types of the Cherokee Indians," 42.

[33] For an overview of the literature on slaveholding Indians, see Renate Bartl, "Native American Tribes and Their African Slaves," in *Slave Cultures and the Cultures of Slavery*, ed. Stephan Palmié (Knoxville: U of Tennessee P, 1995): 162–75.

today the former slaveholding tribes (mainly Cherokees, Chickasaws, Choctaws, Creeks, and Seminoles) still classify their freedmen as 'freedmen with (Indian) blood' and 'freedmen without (Indian) blood'.

The Eastern Band of Cherokee Indians was part of the Cherokee Nation of the Southeast; they separated from the Nation after 1819, and after 1830 they were able to evade deportation to the Indian Territory by retreating to the mountain region of western North Carolina. They gained federal recognition in 1889.[34] In 2010, they were still the only federally recognized tribe in North Carolina. The 2000 census counted 8,092 individuals living on the reservation.[35]

The relatively high degree to which the Cherokees were intermixed with non-Indians is shown on the New Guion Miller Roll of 1909 and the New Dawes Roll of 1889–1914:[36] the degree of Indian blood ranges from "Full" to "1–128" – the latter meaning one Cherokee ancestor and 127 non-Cherokee ancestors in the seventh generation back. Additionally, "AW" is shown in the "Blood" column, the abbreviation for 'adopted white'. This points to another difficulty in genetic studies of Native Americans on the whole: how does the widespread adoption of persons from other tribes, and of non-Indian persons, affect the data? To what degree does the inter-tribal exchange of spouses change the data? Can these data, then, really be used as standard for Native Americans?

Finger, in his book on the Eastern Band of Cherokee Indians, describes the tribal (emic) perspective of the racial classifications of tribal members:

> In discussing varying degrees of Cherokee ancestry [...], I frequently refer to mixed-bloods, white Indians, and full-bloods. [...] I use these as the Cherokee themselves do, without scientific definition. A "mixed-blood" might have any degree of Indian-white (or Indian-black) ancestry, but in this book the term almost always refers to in-

---

[34] John R. Finger, *Cherokee Americans: The Eastern Band of Cherokees in the Twentieth Century* (Lincoln & London: U of Nebraska P, 1991): xi, xiii.

[35] US Department of the Interior, Office of Federal Acknowledgement, "Indian Entities Recognized and Eligible To Receive Services From the United States Bureau of Indian Affairs," Eastern Band of Cherokee, "The Eastern Band of Cherokee Indians" (2010), http://nc-cherokee.com/ (accessed 5 July 2010).

[36] Sample pages of these tribal rolls are reprinted in Bob Blankenship, "Eastern Cherokee Rolls (1817–1924)," in *Cherokee Roots*, ed. Bob Blankenship (Cherokee NC: Cherokee Roots Publications, 1992) , vol. 1: 165.

> dividuals who are not predominantly of Cherokee lineage. Contrary to its literal meaning, a "full-blood" is almost never entirely Cherokee and might more accurately be called a "fuller-blood," defined by one scholar as a person with at least three-fourths Cherokee ancestry. [...] Full-bloods, by Cherokee definition, might include individuals with considerably less Cherokee blood, depending on their behavior. Obviously there is a cultural as well as genetic component involved in describing people in these terms. Similarly, "white Indian" is a term that is partly culturally defined, and it is possible (though unlikely) that a full-blood might be called a white Indian if highly acculturated.[37]

With this in mind, we must question what the 'racial composition' of persons tested really was in the Pollitzer studies when they refer to 'full-bloods' and 'mixed-bloods'.

Data from Other American Indian Populations

The Pollitzer report on the Cherokee blood types lists standard gene frequencies for other Native American tribes, too, with the number of screened persons indicated:[38]

Cherokee: 78 persons[39]
Chippewa: 161 persons[40]
Blood: 241 persons[41]

---

[37] Finger, *Cherokee Americans: The Eastern Band of Cherokees in the Twentieth Century*, xiv.

[38] Pollitzer, Hartman, Moore et al., "Blood Types of the Cherokee Indians," 39. US census data for Indian population of tribes and reservations were not taken in the years 1940–60. To give an impression of the total size of the tribes discussed, population numbers had to be taken from other sources and time periods.

[39] Cherokee data published in 1962. 1960: 4,494 persons on the *Qualla Boundary Reservation*. Pollitzer, Hartman, Moore et al., "Blood Types of the Cherokee Indians," 39.

[40] Chippewa data published in 1954. 1650: 35,000 persons in the tribe (estimated); 1970: 41,946 persons in the US part of the tribe. Robert E. Ritzenthaler, "Southwestern Chippewa," in *Handbook of North American Indians, Northeast*, ed. Bruce E. Trigger (Washington DC: Smithsonian Institution, 1978): 743.

[41] Blood data published in 1953. 1909: 1,174 persons in the tribe; 1970: 4,262 persons in the tribe. Hugh A. Dempsey, "*Blackfoot*," in *Handbook of North American*

Navaho: 361 persons[42]
Pima: 489 persons[43]
Apache: 73 persons[44]
Pawnee: 80 persons[45]
Diegueno: 58 persons[46]

The data were taken from other publications and discussed in the footnotes of the table for typical gene frequencies. What is obvious here is that once again the samples are small, but the data are used as if they were standard data for the respective tribes. There are other problems as well. The data for the Chippewa were taken from full-bloods only. The "Blood" gene frequencies were said to represent one-sixth white admixture, while the number of persons screened for different blood group markers varied. The Navaho data were

---

Indians, Plains (vol. 13.2), ed. Raymond J. DeMallie (Washington DC: Smithsonian Institution, 2001): 622, Table 1.

[42] Navajo data published in 1949. 1930: ca. 38,800 persons in the tribe. David F. Aberle, "Navajo Economic Development," in *Handbook of North American Indians, Southwest* (vol. 9), ed. Alfonso Ortiz (Washington DC: Smithsonian Institution, 1983): 642. 1981: 166,519 persons in the tribe. Robert W. Young, "Apachean Languages," in *Handbook of North American Indians, Southwest* (vol. 10), ed. Alfonso Ortiz (Washington DC: Smithsonian Institution, 1983): 397–98; 399, Table 4.

[43] Pima data published in 1958. 1961: 11,246 persons in the tribe. Robert A. Hackenberg, "Pima and Papago Ecological Adaptations," in *Handbook of North American Indians, Southwest* (vol. 10), ed. Alfonso Ortiz (Washington DC: Smithsonian Institution, 1983): 176.

[44] Apache (Mescalero and Chiricahua) data published in 1959. 1956: 502 Chiricahua Apache. Morris E. Opler, "Chiricahua Apache," in *Handbook of North American Indians, Southwest* (vol. 10), ed. Alfonso Ortiz (Washington DC: Smithsonian Institution, 1983): 409. 1888: 431 Mescalero Apache, 1981: at least 1,000 Mescalero Apache. Morris E. Opler, "Mescalero Apache," in *Handbook of North American Indians, Southwest* (vol. 10), ed. Alfonso Ortiz (Washington DC: Smithsonian Institution, 1983): 428.

[45] Pawnee data published in 1960. 1960: 1,149 persons in the tribe. Douglas R. Parks, "Pawnee," in *Handbook of North American Indians, Plains* (vol. 13.1), ed. Raymond J. DeMallie (Washington DC: Smithsonian Institution, 2006): 543, Table 3.

[46] Diegueno (= synonym for Tipai-Ipai) data published in 1953. 1940: 957 persons in the tribe, 1968: 1,322 persons in the tribe. Katherine Luomala, "Tipai-Ipai," in *Handbook of North American Indians, California* (vol. 8), ed. Robert F. Heizer (Washington DC: Smithsonian Institution, 1978): 596, Table 1.

based on "full-bloods, except for one family with white admixture"; the degree of white admixture was not given. The Pima data were obtained from "school children, not necessarily full-bloods." The Apache data came from full-bloods. Pawnee gene frequencies were taken from persons with "tribal status," who included "full-blood Pawnees plus hybrids of Pawnee with other Indian tribes." Finally, the Diegueno blood samples were taken from full-bloods only.[47]

Thus, the 'standardization' for different tribes was produced by basing the findings on varying sample sizes and varying blood quanta – a method highly questionable for gaining standard data for comparison.

Conclusion

This discussion of genetic blood testing makes it clear that nowadays race, ethnicity, and tribal identity cannot be defined biologically or genetically – and cannot be specified by blood or DNA tests. Instead it is socially defined.

The main problem in finding genetic markers for a specific race is to find 'pure-blood' samples with which the results can be compared in order to classify a person or population racially. And 'pure-bloods' may not actually be, genetically, what they are called.

Among the Eastern Band of Cherokee Indians we have seen that the classification 'full-blood' person (usually being equal to 'pure-blood' in Western science) can also contain mixed-blood persons; the Cherokees are more interested in social reality than in biological accuracy.

The most serious problem lies with the defined standards (standard markers) for racial identification, which are based on very small sample sizes compared to the whole ethnic group or race for which the racial marker should have been defined. Unfortunately, the data are often published as if they were the basic standard data for a whole race, ethnic group, tribe, or nation. Quoting and republishing these data again and again has made them into accepted standards, and their validity is no longer questioned.

Modern DNA testing is confronted with the same problems as these genetic blood tests. How can researchers find 'pure-blood' samples for comparisons? Are there really groups and populations that have not intermixed, staying 'pure-blooded' over many generations? And what does one do with adop-

---

[47] All information and citations taken from Pollitzer, Hartman, Moore, Rosenfield, Smith, Hakim, Schmidt & Leyshon, "Blood Types of the Cherokee Indians," 39, Table 11 and fn 1–8.

ted persons, captives, or spouses from other ethnic groups and races, which has caused gene flow between American Indians and non-Indians for over four centuries? We can predict that modern DNA tests will not be able to overcome these difficulties, either.

Finally, it can be said that genetic testing – be it blood factors or DNA testing – will not be able to prove or create a realistic ethnic or racial identity for a person, because that identity is socially and historically formed, and cannot be reduced simply to genetic markers. Blood is not just blood; blood is much, much more than that.

WORKS CITED

Aberle, David F. "Navajo Economic Development," in *Handbook of North American Indians, Southwest* (vol. 9), ed. Alfonso Ortiz (Washington DC: Smithsonian Institution, 1983): 641–58.

Bartl, Renate. "Native American Tribes and Their African Slaves," in *Slave Cultures and the Cultures of Slavery*, ed. Stephan Palmié (Knoxville: U of Tennessee P, 1995): 162–75.

Beale, Calvin. L. "American Triracial Isolates: Their Status and Pertinence to Genetic Research," *Eugenics Quarterly* 4.4 (1957): 187–96.

Berlin, Ira. *Many Thousands Gone: The First Two Centuries of Slavery in North America* (Cambridge MA: Harvard UP/Belknap Press, 1998).

Blankenship, Bob. "Eastern Cherokee Rolls (1817–1924)," in *Cherokee Roots*, vol. 1, ed. Bob Blankenship (Cherokee NC: Cherokee Roots Publications, 1992).

Coca, Arthur F., & Olin Deibert. "A Study of the Occurrence of the Blood Groups Among the American Indians," *Journal of Immunology* 8 (1923): 487–93.

Dempsey, Hugh A. "Blackfoot," in *Handbook of North American Indians, Plains* (vol. 13.2), ed. Raymond J. DeMallie (Washington DC: Smithsonian Institution, 2001): 604–28.

Eastern Band of Cherokee, "The Eastern Band of Cherokee Indians" (2010), http://nc-cherokee.com/ (accessed 05 July 2010).

Finger, John R. *Cherokee Americans: The Eastern Band of Cherokees in the Twentieth Century* (Lincoln & London: U of Nebraska P, 1991).

Gilbert, William Harlen, Jr. "Surviving Indian Groups of the Eastern United States," in *Annual Report of the Board of Regents of the Smithsonian Institution for 1948*, ed. Smithsonian Institution (Washington DC: Smithsonian Institution Press, 1948): 407–38.

Glass, Bentley. "On the Unlikelihood of Significant Admixture of Genes from the North American Indians in the Present Composition of the Negroes in the United States," *American Journal of Human Genetics* 7 (1955): 368–85.

Hackenberg, Robert A. "Pima and Papago Ecological Adaptations," in *Handbook of North American Indians, Southwest* (vol. 10), ed. Alfonso Ortiz (Washington DC: Smithsonian Institution, 1983): 161–77.

Haliwa-Saponi Indian Tribe. "Haliwa-Saponi Indian Tribe" (2010): www.haliwa-saponi.com (accessed 5 July 2010).

Luomala, Katherine. "Tipai-Ipai," in *Handbook of North American Indians, California* (vol. 8), ed. Robert F. Heizer (Washington DC: Smithsonian Institution, 1978): 592–609.

Nanticoke Indian Association Inc. "The Nanticoke Indian Tribe" (2003–2004): http://www.nanticokeindians.org/ (accessed 04 July 2010).

North Carolina Department of Administration, Commission of Indian Affairs. "Tribes and Organizations" (2009): http://www.doa.nc.gov/cia/tribesorg.htm (accessed 26 May 2010).

Opler, Morris E. "Chiricahua Apache," in *Handbook of North American Indians, Southwest* (vol. 10), ed. Alfonso Ortiz (Washington DC: Smithsonian Institution, 1983): 401–16.

——. "Mescalero Apache," in *Handbook of North American Indians, Southwest* (vol. 10), ed. Alfonso Ortiz (Washington DC: Smithsonian Institution, 1983): 419–39.

O'Rourke, Dennis H. "Blood Groups, Immunoglobulins, and Genetic Variation," in *Handbook of North American Indians: Environment, Origins, and Population* (vol. 3), ed. Douglas H. Ubelaker (Washington DC: Smithsonian Institution, 2006): 762–76.

Parks, Douglas R. "Pawnee," in *Handbook of North American Indians, Plains* (vol. 13.1), ed. Raymond J. DeMallie (Washington DC: Smithsonian Institution, 2001): 515–47.

Pollitzer, William S., Robert C. Hartman, Hugh Moore et al. "Blood Types of the Cherokee Indians," *American Journal of Physical Anthropology* 20 (1962): 33–34.

——, R.M. Menengaz–Bock & J.C. Herion. "Factors in the Microevolution of a Tri-Racial Isolate," *American Journal of Human Genetics* 18 (1966): 26–38.

Porter, Frank W., III. *The Nanticoke* (Indians of North America; New York: Chelsea House, 1987).

Ritzenthaler, Robert E. "Southwestern Chippewa," in *Handbook of North American Indians, Northeast* (vol. 15), ed. Bruce E. Trigger (Washington DC: Smithsonian Institution, 1978): 743–59.

State of Delaware, Department of State. "Harmony United Methodist Church" (30 December 2009), http://archives.delaware.gov/markers/sc/SC-187.shtml#P0_0 (accessed 05 July 2010).

——. "Indian Mission United Methodist Church" (30 December 2009), http://archives.delaware.gov/markers/sc/SC-122.shtml#P0_0 (accessed 5 July 2010).

US Department of the Interior, Office of Federal Acknowledgement, "List of Petitioners by State (as of September 22, 2008)" (22 September 2008), www.bia.gov/idc/groups/public/documents/text/idc-001215.pdf (accessed 6 May 2010).

——. "Indian Entities Recognized and Eligible To Receive Services From the United States Bureau of Indian Affairs," *Federal Register* 74.153 (11 August 2009): 40218–23, http://www.bia.gov/idc/groups/public/documents/text/idc-001736.pdf (accessed 6 May 2010).

Young, Robert W. "Apachean Languages," in *Handbook of North American Indians, Southwest* (vol. 10), ed. Alfonso Ortiz (Washington DC: Smithsonian Institution, 1983): 393–400.

⌘

Indigenous Peoples: Attempts to Define[1]

ULIA POPOVA–GOSART

I. The Problem

THE LITERATURE FOCUSING ON 'INDIGENOUS PEOPLES' IS VAST and continues to grow.[2] The term commonly refers to specific cultural groups that exist in a variety of geo-political and social settings, usually as a minority.[3] Most of these groups are subjects of domestic jurisdiction;[4] they are also protected under international law. Conceptions of

---

[1] I want to express gratitude to my close friend Dan Haley for his assistance in pre-editing this article, to Professor Maurice Zeitlin for his intellectual guidance, and to Professor Anne Gilliland for her constructive comments.

[2] Godoy et al., for example, report that from 1985 to 2003 the annual growth of literature on this subject was 9.46%. Ricardo Godoy, Victoria Reyes–Garcia, Elisabeth Byron, William Leonard & Vincent Vadez, "The Effect of Market Economies on the Well-Being of Indigenous Peoples and on Their Use of Renewable Natural Resources," *Annual Review of Anthropology* 34 (2005): 121–38.

[3] I am not referring here to a legal category of 'minority' but, rather, to social conditions (and political claims arising from those conditions) of groups enjoying a status of Indigenous peoples and/or self-identifying as such. Indigenous identity implies alliance with groups whose identity as distinct peoples entails a particular life-style, threatened (and, as history testifies, in many cases destroyed) by states foreign to Indigenous economic and political structures. Indigenous peoples are often encapsulated within dominating regimes of their states; yet sometimes they constitute a majority of the population, as, for example, in Bolivia.

[4] This is despite the fact that in some cases these groups may have their own political, social, and legal institutions functioning within the boundaries of the group's territory. Such is, for example, the case of Native Americans who are currently, as Duane Champagne describes it, have an unrecognized "dual citizenship" status: i.e.

'Indigenous peoples' developed by social and legal scholars are very diverse.[5] For example, in international legal contexts this expression is rooted in the way in which colonies and dependencies of particular states were thought of and referred to, particularly beginning in the last quarter of the nineteenth century, which marks the beginning of the development of a legal idea of 'Indigenous peoples' in international law.[6] The colonial notion of indigeneity led to the way in which the concept would be defined in the first international document in which the rights of Indigenous groups and the duties of states regarding them were created – the International Labour Organization Convention 107 (1957).[7] Relatively recent literature in which legal notions of indigeneity are employed describes 'Indigenous peoples' broadly as nations/peoples that strive in different ways for political autonomy, such as the Chamorro peoples struggling for political self-determination in US-controlled Guam, or Mexican Zapatistas (Indigenous rebels in the Mexican state of Chiapas). These notions often rest on particular legal standards of treating groups recognized as indigenous peoples by the international community and by the states in which these groups are located. The way in which states' jurisdictions define indigenous groups (essentially communities that deserve special treatment as

---

simultaneously belonging to two distinct and historically controversial social contexts: that of the tribe and that of the US society of which tribes are a geographical, political, social, and cultural part. Duane Champagne, *Notes From the Center of Turtle Island* (New York, Toronto & Plymouth: AltaMira, 2010).

[5] Literature dealing with 'Indigenous peoples' also emerges from the arts and humanities. In this essay, however, I do not examine this literature, as my prime interest is the development of legal notions of indigeneity and, supporting these notions, social-science theories.

[6] Some of the most prominent examples of such groups are those inhabiting the current territories of North and South America and classified as 'Indians', an expression that has been used as a generic term for all New World inhabitants since the Columbian 'discovery' of the Americas.

[7] In the context of this Convention, indigeneity was defined as a property of groups other than colonies; it referred to populations who inhabited particular territories of the Latin American states at which the first international development programme ("Andean Indian Programme") was directed. At the same time, the Convention itself was a product of administrative knowledge of the International Labour Organization (ILO); the way in which indigeneity was codified in the text of this Convention rested upon earlier documents of the ILO, specifically the Colonial Code of the 1930s, a set of documents produced as a measure to discipline the colonial workforce.

a way to recognize and protect the rights of these groups) tend to differ from one state to the other. At times, a specific state's way of dealing with Indigenous questions may contradict international principles aimed at protecting Indigenous rights (such is the situation of the Russian Federation, for example). This diversity of legal notions of indigeneity makes it very difficult (indeed, impossible) to construct a legal notion of indigenety that could serve as a sound analytical instrument to be employed across the boundaries of states' legal and political systems. Social-science authors, whose work reinforces legal scholarship, repeatedly employ concepts of indigeneity that signify a so-called 'traditional' life-style distinguished from modernity and pursued in continuity with the traditions of concrete peoples that go back centuries (for example, nomadic peoples indigenous to the Kalahari Desert in Africa or the Penan hunter–gatherers of Malaysia in Sarawak, Borneo). This expression also targets individuals whose indigenous identity emerges from participating in human and Indigenous rights forums and political events, especially in the studies of social movements.[8]

Groups understood as 'Indigenous peoples' are estimated to comprise 300–370 million people or 5–7% of the world's population. They speak over 5,000 of the world's languages, live in areas rich in biodiversity, and are often considered among the poorest segments of the population, whose survival as distinct peoples and cultures has been endangered by the effects of what has been called modernity and globalization.[9] One of the recent factors,

[8] See *Political Theory and The Rights of Indigenous Peoples*, ed. Duncan Ivison, Paul Patton & Will Sanders (Cambridge & New York: Cambridge UP, 2000); Rachel Sieder, *Multiculturalism in Latin America: Indigenous Rights, Diversity, and Democracy* (New York: Palgrave Macmillan, 2002); Elza Stamatopoulou, "Indigenous Peoples and The United Nations: Human Rights as a Developing Dynamic," *Human Rights Quarterly* 16.1 (February 1994): 58–81; Ronald Niezen, *Origins of Indigenism* (Berkeley, Los Angeles & London: U of California P, 2003).

[9] See World Bank, "Operational Directive 4.20: Indigenous Peoples," in *Indigenous Peoples and International Organizations*, ed. Lydia Fliert (Nottingham: Russell, 1991): 83–88; United Nations, "State of The World's Indigenous Peoples," United Nations Permanent Forum on Indigenous Issues (2009), http://www.un.org/esa/socdev/unpfii /documents/SOWIP_web.pdf (accessed 29 September 2010); Patrick Thornberry, *Indigenous Peoples and Human Rights* (Manchester: Manchester UP, 2002); International Work Group for Indigenous Affairs, *The Indigenous World 2000/2001* (Copenhagen: International Work Group for Indigenous Affairs, 2001); David May-

mentioned by a number of scholars as in various ways affecting the existence of indigenous groups, is genetic research. Resulting from the current lack of comprehensive normative instruments (especially of international scope) that would ensure benefit-sharing with Indigenous groups on a fair and continuous basis, the current wave of bioprospecting and human genetic studies is seen as potentially harmful to Indigenous groups.[10] Investigations conducted with little or no ethical concern for the needs of Indigenous communities can lead to a scientific and industrial monopoly over information derived from these communities. These investigations can also serve as a means of stigmatizing indigenous scientific practices and indigenous individuals.

The multiplicity of definitions of indigeneity apparently leaves open the question of which criteria are necessary for classification as an 'Indigenous people' across areas of studies. A generally agreed-on observation is that no definition can be devised that is capable of capturing the diversity of these groups and their life-styles.[11] The lack of guiding principles for defining 'In-

---

bury–Lewis, "Indigenous Peoples," in *The Indigenous Experience: Global Perspectives,* ed. Roger Maaka & Chris Andersen (Toronto: Canadian Scholars, 2006): 17–29.

[10] This is not to say that research cannot be beneficial to both Indigenous groups and scientific communities. The point is that subjecting Indigenous individuals and their life-styles to investigation when Indigenous individuals have no control over the processes and results of research performed in their communities and/or on them is a step toward a new form of exploitation by means of science in the service of industrial and at times political interests of non-Indigenous parties.

[11] See, for example, Alan Barnard, "Kalahari Revisionism, Vienna and The 'Indigenous Peoples' Debate," *Cultural Anthropology* 14.1 (2007): 1–16; *International Law and Indigenous Peoples*, ed. James Anaya (Oxford & New York: Oxford UP, 2003); Tania Murray Li, "Articulating Indigenous Identity in Indonesia: Resource Politics and the Tribal Slot," *Comparative Studies in Society and History* 42.1 (January 2000): 149–79; Helen Quane, "The Rights of Indigenous Peoples and The Development Process," *Human Rights Quarterly* 27.2 (May 2005): 652–82; United Nations Development Programme, "UNDP and Indigenous Peoples: A Policy of Engagement," *UNDP Engagement with Civil Society* (2001), http://www.undp.org/partners/civil_society /empowering_indigenous_peoples.shtml (accessed 29 September 2010); United Nations Permanent Forum on Indigenous Issues Secretariat, "The Concept of Indigenous Peoples (Background paper)," in *Proceedings of the Workshop on Data Collection and Disaggregation for Indigenous Peoples, New York, 19–21 January 2004* (United Nations Permanent Forum on Indigenous Issues 2004, UN Doc. PFII/2004/WS.1/3).

digenous peoples' leads some authors to question the utility of the concept, and others to abandon its use altogether.[12] Nevertheless, the need to have a way to comprehensively define 'Indigenous peoples' in a somewhat coherent fashion has been persistently acknowledged by a number of authors and organizations.

While the diverse cultural and social realities of groups and individuals understood by scholars as 'Indigenous peoples' cannot be captured in a single definition, it is possible to distinguish in the thematic literature three major approaches or perspectives[13] for examining what may be called *indigenous realities* (by which I mean processes existing within the boundaries of cultures / societies termed 'indigenous peoples'):

—The social science perspective, being a mode of thought emerging from experience of analysing indigenous realities by means of scientific theories and from a social position of scientific enterprise;
—The legal perspective, resulting from conceiving indigenous realities from a position of law, and focused on solving problems emerging from social and political relations between groups with the status of 'Indigenous peoples' and the states in which they exist, by means of the law;
—The political perspective, emerging from experiences of resistance by individuals claiming Indigenous identity against those legal and social

[12] See, for example, Thornberry, *Indigenous Peoples and Human Rights,* 48; Benedict Kingsbury, "'Indigenous Peoples' in International Law: A Constructivist Approach to the Asian Controversy," *American Journal of International Law* 92.3 (July 1998): 414–57; Seán Eudaily, *The Present Politics of the Past: Indigenous Legal Activism and Resistance to (Neo)liberal Governmentality* (New York & London: Routledge, 2004); Adam Kuper, "The Return of The Native," *Current Anthropology* 44.3 (June 2003): 389–402; United Nations Economic and Social Council. Commission on Human Rights, "Indigenous Issues. Report of the Working Group. Chairperson-Rapporteur: Mr. José Urrutia (Peru)" in *Proceedings of the Fifty Third Session, December 10, 1997* (United Nations Economic and Social Council 1997, UN Doc. E/CN.4/1997/102).

[13] I base my understanding of 'perspective' on the following definition by Karl Mannheim: "the subject's whole mode of conceiving things as determined by his historical and social setting"; Mannheim, *Ideology and Utopia,* tr. Louis Wirth & Edward Shils (*Ideologie und Utopie*, 1929; London & New York: Harcourt, Brace, 1936): 266.

conceptions about indigeneity that have been imposed upon their peoples.

The realities of these knowledge-making experiences are highly interconnected, and therefore the proposed categorization must be seen only as a demonstrative device intended to achieve methodological clarity regarding the existing knowledge whose subject-matter is 'Indigenous peoples'.

II. Perspectives

The social-science perspective

This focuses on the socio-historical realities of what has been termed indigeneity.[14] This perspective can be found in a variety of disciplines, most significantly in anthropology but also in postcolonial studies, history, conservation, biology, ecology, political science (especially in studies of political resistance), sociology, and to some extent in geology, archaeology, and area studies. Among a number of proposed ways of characterizing 'Indigenous peoples', two tendencies have been overwhelmingly popular in these disciplines: an inclination to define these groups as descendants of the "original inhabitants of the region"[15] (i.e. the first settlers of a particular land); and a propensity to place the meaning of indigeneity within a larger context of a dichotomy between modernity and tradition.[16]

---

[14] Scholars use the words 'indigeneity', 'indigenism', 'indigenousness' interchangeably when referring to a specific characteristic/set of properties pertinent to groups termed Indigenous. In this essay I use the term 'indigeneity' consistently.

[15] Godoy et al., "The Effect of Market Economies on the Well-Being of Indigenous Peoples and on Their Use of Renewable Natural Resources."

[16] Thornberry, *Indigenous Peoples and Human Rights,* 33–39. I do not question that groups represented as 'Indigenous peoples' in legal and social-science literature enjoy unique life-styles threatened by post-industrial economic development and globalization; nor do I suggest that these groups' identities do not rest upon histories which genealogically and epistemologically go back to the time when their ancestors settled in new territories which these groups may or may not inhabit now. My critique, rather, is focused on *representations* of these groups that rest upon theories and methodologies of non-Indigenous origins. Such representations, which are based on worldviews that are foreign to Indigenous peoples, have historically served as a support for their subordination.

The *original inhabitants* description projects the meaning of indigeneity as native and/or autochthonous to a particular location,[17] often with an emphasis on the historical priority of inhabitation of a particular place. Apart from the power this concept has gained within political contexts, its theoretical basis rests on a set of contradictory statements. The implied possibility of a genealogical (and at times a suggested social) lineage between the first settlers of the land and the current occupants of the same territory implies that a group can inhabit the same geographical place for a number of years without moving and/or merging with other social collectives. However, it is highly unlikely that ecological and social events would not have affected a group over any considerable length of time. Even if taken as a general theoretical guiding thread, the proposition has its weaknesses. Consider a brief chronology of events in the history of the Inuit peoples living in the Nunavut Territory of Canada's Eastern Arctic.[18] The Inuit peoples' immediate ancestors, the Thule

---

[17] See, generally, Michael Dove, "Indigenous People and Environmental Politics," *Annual Review of Anthropology* 35 (2006): 191–208; Michaela Pelican, "Complexity of Indigeneity and Autochthony: An African Example," *American Ethnologist* 36.1 (February 2009): 52–65; Stefanie Wickstrom, "The Politics of Development in Indigenous Panama," *Latin American Perspectives* 30.4 (2003): 43–68. 'Rural' and 'local' are used interchangeably with 'indigenous' in works focused on rural development such as Brij Kothari, "Theoretical Streams in Marginalized Peoples' Knowledge(s): Systems, Asystems, and Subaltern Knowledge(s)," *Agriculture and Human Values* 19.3 (2002): 225–37. James Clifford provides an interesting overview of literature focused on the experiences of displaced groups who claim indigenous identity and introduces the concept of engaging diaspora theories as a way to articulate the realities of Indigenous groups. See Clifford, "Varieties of Indigenous Experience: Diasporas, Homelands, Sovereignties," in *Indigenous Experience Today,* ed. Marisol de la Cadena & Orin Starn (Oxford & New York: Berg, 2007): 197–225; Barbara Zimmerman, Carlos Peres, Jay Malcolm & Terence Turner, "Conservation and Development Alliances With the Kayapo of South-Eastern Amazonia, a Tropical Forest Indigenous People," *Environmental Conservation* 28.1 (2001): 10–22.

[18] Nunavut (meaning 'our land' in the Inuktitut language) is a relatively new territory established in 1999 under the provisions of the Nunavut Land Claims Agreement (1993), which was carved out of the Northwest Territories of Canada. Under the provisions of the Agreement, the Inuit of Nunavut received title to the 350,000 sq km of lands and a significant level of political autonomy from Canada. Currently about 82% of the Nunavut population identify themselves as Inuit. For the text of the Nunavut Land Claims Agreement, refer to *Agreement between the Inuit of the Nunavut settlement area and Her Majesty the Queen in right of Canada, 1993*. For more information,

peoples, came to Arctic Canada and Greenland six to nine centuries before the Europeans. Their occupation of these localities led to the disappearance of the Dorset culture that had existed there for about two millennia and whose peoples had themselves been the colonizers of pre-Dorset palaeoeskimos known as the first occupants of the High Arctic.[19] Thus, it can be argued that the history of the Nunavut territory begins with peoples other than Inuit, making current Nunavut occupants, technically speaking, not 'Indigenous' to these lands. At the same time, to what degree does thus constructed history (which is broadly the history of the geographical location) help us understand what determined the current social and political conditions of the population of Nunavut: namely, the emergence of Nunavut as a separate territory, as an autonomous political district and a homeland for people who identify themselves as Inuit peoples? Furthermore, how does the 'original descendancy' projection (even if it is structured in a somewhat flexible fashion, relative to the period of European ascendancy) capture the diversity of current life conditions of the individuals self-identifying as Inuit living within or outside the Nunavut borders? Finally, how does this constructed representation of indigeneity help to build theoretical parallels between the conditions of Inuit and other groups termed 'indigenous peoples' (whose land claims, for example, emerged on similar grounds yet were caused by circumstances very different from those of the Inuit)?

Interrelated with the 'original inhabitant' viewpoint is a tendency to define 'Indigenous peoples' as being *ontologically different* from those who have 'discovered' and/or studied them. This is the type of thought notorious for its history of service to political and cultural projects of domination over less powerful groups.[20] Epistemologically, this type of representation can be

---

see *Nunavut: Inuit Regain Control of Their Lands and Their Lives*, ed. Jens Dahl, Jack Hicks & Peter Jull (Copenhagen: International Working Group on Indigenous Affairs, 2000).

[19] Eric Smith & Mark Wishnie, "Conservation and Subsistence in Small-Scale Societies," *Annual Review of Anthropology* 29 (2000): 493–524.

[20] This mode of thinking about 'the Other' has been variously described, depending on the context of application. Edward W. Said, for example, characterized it as 'Orientalism', while Ter Ellingson, working on problems essentially similar to the ones developed by Said yet emerging from a very different (Rousseauvian) social context, described this mode of thought as the 'myth of the Noble Savage'. Edward Said, *Orientalism* (New York: Vintage, 1978); Ter Ellingson, *The Myth of the Noble Savage* (Berkeley & Los Angeles: U of California P, 2001).

traced back to positivist philosophy;[21] it is most prominently expressed in the theories of social evolution and modernization.[22] Despite the epistemological

[21] To see the intellectual foundations upon which these propositions rest, it would be instructive to consult the works of Auguste Comte and his follower, John Stuart Mill. Compare: "Civilization consists [...] in the development of the human mind on the one hand, and in the influence of man upon nature, which, on the other hand, is the consequence thereof. In other words, the elements of which the idea of civilization is composed, are: the sciences, fine arts and industry, the latter expression being used in its widest sense." Comte, 1822, as quoted in Wilhelm Grewe, *The Epochs of International Law*, tr. Michael Byers (*Epochen der Völkerrechtsgeschichte*, 1984; Berlin & New York: Walter de Gruyter, 2000): 449. And: "The word civilization is a word of double meaning. It sometimes stands for *human* improvement in general, and sometimes for *certain kinds* of improvement in particular. We are accustomed to call a country more civilized if we think it more improved; more eminent in the best characteristics of Man and Society; farther advanced in the road of perfection; happier, nobler, wiser. This is one sense of the word civilization. But in another sense it stands for that kind of improvement only, which distinguishes a wealthy and powerful nation from savages or barbarians." Mill, 1836, as quoted in Elspeth Young, *Third World in The First: Development and Indigenous Peoples* (New York: Routledge, 1995): 35.

[22] Among the significant social evolutionists of the nineteenth century was Herbert Spencer, an English sociologist and follower of Auguste Comte whose works were prominent not only among sociologists but also among legal theorists concerned with the construction of norms of political control over colonial (defined as 'uncivilized') populations. See James Lorimer, *Studies National and International* (Edinburgh: William Green & Sons, 1890); Herbert Spencer, *The Study of Sociology* (London: Henry S. King, 1873). Among modernization theorists there was a school of thinkers who focused on transformations of territorial states; they classified social entities using dichotomous 'ideal types' of traditional and modern societies, and postulated that transformation toward the latter, a more developed (and more desired) state, could be achieved by the introduction of technological and scientific assistance to the less rational and less industrialized groups. Among the most prominent theorists in this area was Talcott Parsons. It would take another work to see the ways in which social evolution and modernization theories ideas found implementation in policy and economic initiatives focused on Indigenous groups. For a thorough discussion, see, for example, the work of Luis Rodríguez–Piñero, *Indigenous Peoples, Postcolonialism, and International Law: The ILO Regime, 1919–1989* (New York: Oxford UP, 2005): 90, 103–12, 198–99. Of special interest is his reference to the 1953 ILO publication *Indigenous Peoples: Living and Working Conditions of Indigenous Populations in Independent Countries* (rev. 1955 and 1956), wherein Indigenous peoples are described by ILO authors as existing in undesirable living conditions and as "backward" populations in

relatedness between these two theories, each develops a somewhat different viewpoint. What relates the two (and is of interest in this study) is a particular vision of social transformations that explains differences among societies using a model of progressive stages of development. According to this vision, industrially and militarily stronger groups exemplify more advanced states of social development and are composed of morally and intellectually superior individuals as compared to conquered and/or dependent peoples. Another notion that both theories implicitly support is that progress toward 'higher levels' of civilization/development can be cultivated by means of proper education and technological assistance to the 'less advanced' groups, as societies (analogous to biological life-forms) grow toward a perceived perfection. This interpretation of social transformations has served to justify various interventions in Indigenous settings as forms of humanitarian assistance aimed at instituting and supporting progressive change. Despite the wide criticism these theories have received, their legacy can still be felt, as in studies that continue to assess Indigenous experiences as marked by 'traditionality'.[23] The authors

---

need of being integrated "into the economic life of [their respective] country" (ILO, as cited by Rodríguez–Piñero, 97, note 70). Rodríguez–Piñero presents the following statement from Jef Rens, the ILO Deputy Director–General, which displays an attitude toward the indigenous populations upon which development projects were directed: "Our Andean programme is based on. [...] [a] fundamental conception that [...] every man is worth every man. I am deeply convinced that if we give the Altiplano Indians their change, they would be able to rise to whatever level and to produce among them individuals of a high culture, comparable to the best elements in the most civilized nations [...]. We are like elder brothers whom age has allowed the way of modern and civilized life better than the younger ones. It is our duty to offer our young Indian brothers a hand" (Luis Rodríguez–Piñero, 106, note 115).

[23] A major body of literature in which such characterizations of Indigenous peoples have been developed is that concerning the protection of traditional Indigenous knowledge, described as a product of a particular life-style found across Indigenous communities, which requires and rests upon a specific mode of thinking different from and endangered by modernity. See Michael Warren, *Using Indigenous Knowledge in Agricultural Development* (Washington DC: World Bank, 1991); *The Cultural Dimension of Development: Indigenous Knowledge(s) Systems*, ed. Michael Warren, Jan Slikkerveer & David Brokensha (London: Intermediate Technology Publications, 1995); *Who Will Save the Forests? Knowledge, Power and Environmental Destruction*, ed. Tariq Banuri & Frédérique Apffel–Marglin (London: Zed, 1993). Also consider the ways in which the Indigenous/traditional vs. modern/Western knowledge-dichotomy argument had been described by the UN authors in, for example, the United Nations Con-

that employ this approach tend to structure indigeneity as an abstract property to be found across 'traditional' cultures, leading to a somewhat stereotyped representation of Indigenous groups and thus leaving the diversity of Indigenous realities obscured. While these authors as a rule celebrate Indigenous experiences instead of using them to exemplify 'backward' ways of life, their work continues to represent Indigenous communities and individuals as silent and distant objects of research (and, by extension, of legal and political action) essentially different from those who construct representations.[24]

The legal perspective

While interconnected with that of social science, the legal perspective is based on a separate history. It broadly emerges from a history of relations between various states and the specific ethnic/cultural groups living within their borders; this history has been codified in specific legal documents which make up international law (if looked upon as a field of professional knowledge) as well as in separate states' jurisdictions. These groups have been defined by the prominent legal scholar of indigenous rights, James Anaya, as "the living descendants of pre-invasion inhabitants of lands now dominated by others,"[25]

---

vention on Biological Diversity, "Knowledge, Innovations and Practices of Indigenous and Local Communities: Implementation of Article 8(j) (Note by Executive Secretary)," in *Proceedings of Conference of the Parties to the Convention on Biological Diversity, Third Meeting, Buenos Aires, Argentina, 4–15 November, 1996* (United Nations Convention on Biological Diversity 1996, UN Doc UNEP/CBD /COP/3/19); United Nations Convention to Combat Desertification, "Traditional Knowledge. Addendum. Building Linkages Between Environmental Conventions and Initiatives (Note by the Secretariat)," in *Proceedings of Conference of the Parties. Committee on Science and Technology. Third Session. Recife 16–18 November 1999* (United Nations Convention to Combat Desertification 1999, UN Doc ICCD/ COP(3)/CST/3/Add.1); World Bank, "Indigenous knowledge for development: Framework for action" (1998), http://www.worldbank.org/afr/ik/ikrept.pdf (accessed 29 September 2010).

[24] For further critiques, see, generally, Chris Tennant, "Indigenous Peoples, International Institutions, and the International Legal Literature From 1945–1993," *Human Rights Quarterly* 16 (1994): 1–57, and Arun Agrawal, "Dismantling the Divide Between Indigenous and Scientific Knowledge," *Development and Change* 26 (1995): 413–39.

[25] See James Anaya, *Indigenous Peoples in International Law*, 3. Currently, two approaches could be located as defining the concept within international law, as Benedict Kingsbury suggests: a positivist one, which treats IP as a legal category that requires a

leading to a specific conceptualization of indigeneity in international law. In contrast to the social-science perspective, the term 'Indigenous' as used in international legal contexts came from a French colonial category intended to avoid connotations associated with discrimination.[26] The term generally served, Patrick Thornberry notes, as a "euphemism" to describe non-Europeans subjected to colonial regimes.[27] In their attempts to define indigeneity, legal writers employed social-science theories that were currently prevalent, where the content of a specific legal problem determined the kind of social theory to be used.

The current approach to the Indigenous question in international law is based on two treaties – the UN ILO Convention 107 (1957) and the UN ILO Convention 169 (1989) – and one declaration, the UN Declaration on the Rights of Indigenous Peoples (2007). The history leading to this approach can be seen as developing through three distinct periods:

1. The introduction of the question of the rights of colonial populations during the last quarter of the nineteenth century, leading to the first concept of indigeneity as referring to the colonial workforce and marking Indigenous peoples as 'uncivilized';
2. The development of the question of the rights of dependent populations after World War Two, and the introduction of an international dimension to the Indigenous problem as one of underdeveloped societies;
3. The construction of a 'modern' concept of 'Indigenous peoples' as culturally distinct nations/peoples with particular rights.

Each period, as defined here, can be seen as a response to certain historical events which generated specific ways of dealing with the relations between

---

precise definition for the purposes of using it as a way to determine the scope of application of relevant legal instruments, and a constructivist one, identified by a process of continuous evaluation of particular relevant claims, cases, and practices as a way to negotiate the kind of norms that would accommodate these claims, where a fundamental question behind defining attempts is justification of particular programmes emerging from the recognition of 'Indigenous peoples' as a legal category. Kingsbury, "'Indigenous Peoples' in International Law."

[26] Luis Rodríguez–Piñero, *Indigenous Peoples, Postcolonialism, and International Law: The ILO Regime, 1919–1989*, 46, 339.

[27] Thornberry, *Indigenous Peoples and Human Rights*, 38.

Indigenous groups and their states.[28] However, the fundamental legal criteria employed in each period to define populations as 'Indigenous peoples' emerged from the historical context of these populations in a state of dependency and reflect the inclination of states and of the international community to offer different types of humanitarian assistance to these populations, including assistance in addressing the protection of their rights.

In addition to international legal standards, the content of which, as Anthony Anghie suggests, should be looked upon as a response to a question of order among different social (cultural) entities, norms focused on treatment of indigenous groups were established within the jurisdictions of individual nation-states.[29] These norms differ from state to state and at times could contradict international-level principles for the treatment of groups understood as 'Indigenous peoples.' This can be illustrated by the Russian Federation. Under current law, Russia recognizes as 'Indigenous peoples' only those ethnic groups living on the territories of their ancestors, enjoying a 'traditional life-style', and whose populations remain under 50,000 individuals.[30] Hence, references to 'Indigenous peoples' in Russia are as "small," "numerically small peoples," or "small-numbered peoples"[31] (in Russian, *korennye maloch-*

---

[28] Similarly to the development of social theories of indigeneity, the laws focused on 'Indigenous peoples' originated in the contexts of nineteenth-century colonial expansion, Cold War confrontation, and the simultaneously emerging political power of the so-called Third World.

[29] Anthony Anghie, "Finding the Peripheries: Sovereignty and Colonialism in Nineteenth Century International Law," *Harvard International Law Journal* 40 (1999): 1–71.

[30] See The Federal Law of April 30, 1999 #82-FZ "Guarantees of the rights of Small Indigenous Peoples of the Russian Federation"; The Federal Law of July 20, 2000 #104-FZ "General Principles of Organization of Communities of Small Indigenous Peoples of the North, Siberia and the Far East of the Russian Federation"; The Federal Law of May 07, 2001 #49-FZ "About Territories of Traditional Land-use of Small Indigenous Peoples of the North, Siberia and the Far East of the Russian Federation." For further discussion of legal standards pertinent to Russian indigenous peoples, see Gail Osherenko, "Indigenous Rights in Russia: Is Title to Land Essential for Survival?" *Georgetown International Environmental Law Review* 13.3 (2000): 695–734, and Alexandra Xanthaki, "Indigenous Rights in the Russian Federation: The Case of Numerically Small Peoples of the Russian North, Siberia, and Far East," *Human Rights Quarterly* 26.1 (February 2004): 74–105.

[31] These groups consist of 45 distinct peoples (see the "Unified List of Indigenous Numerically Small Peoples of the Russian Federation" confirmed by Decree 255 of the

islennye narody). The treatment of Russian ethnic minorities as 'Indigenous peoples', as Russian legal scholars suggest, is an issue of 'positive discrimination,' which is a form of recognition in which the federal government must take specific protective measures for the benefit of ethnic groups whose existence as distinct peoples is threatened, as evidenced by the low demographic numbers.[32] What is peculiar is that the history of the 'positive discrimination' approach in the Russian context is rooted in the initial policies of the 1920s by the Soviet state aimed at assisting the 'backward' nations of the North.[33] This approach does not respond to the way in which Indigenous

---

Russian Government, 24 March 2000). Using data from the 2002 Russian Census, the number of Indigenous (i.e. "small-numbered") peoples can be estimated to be about 280 thousand: i.e. less than half a percent of the whole population of over 145 million people. At the same time, however, as Donahoe et al. suggest, of 200 nationalities living in Russia close to 130 could claim Indigenous status, which would amount to about 14% of the Russian population, or about 20 million people. See Anon., "All-Russian Census of Population" (2002): www.perepis2002.ru (accessed 29 September 2010); Brian Donahoe, Joachim Habeck, Agnieszka Halemba & István Sántha, "Size and Place in the Construction of Indigeneity in the Russian Federation," *Current Anthropology* 49.6 (2008): 993–1020; Irina Stoyanova, "Theorizing The Origins and Advancement of Indigenous Activism: The Case of the Russian North" (doctoral dissertation, George Mason University, 2009), http://digilib.gmu.edu:8080/bitstream/1920/5614/1/Stoyanova_Irina.pdf (accessed 29 September 2010); Aleksandr Shapovalov, "Straightening Out the Backward Legal Regulation of 'Backward' Peoples' Claims to Land in the Russian North: The Concept of Indigenous Neomodernism," *Georgetown International Environmental Law Review* 17.3 (2005): 435–69. For information on statistics, see Sergei Sokolovski, "The 2002 Census: Games According to Wittgenstein," *Anthropology and Archeology of Eurasia* 44.1 (2005): 25–33; Valery Stepanov, "Russian Experience in the North Indigenous Statistics for Statistical Questions," in *Proceedings of the Workshop on Data Collection and Disaggregation for Indigenous Peoples, New York, January 19–21* (United Nations Department of Economic and Social Affairs, Secretariat of the Permanent Forum on Indigenous Issues 2004, UN Doc. PFII/2004/WS.1/5).

[32] As suggested by Vladimir Kriazkov, "Prava Korennih Malochislennih Narodov Rossii: Metodologia Regulirovania," *Gosudarstvo i Pravo* 1 (1996): 18–24; and Igor Ponkin, "O poniatii 'korennoy narod'," *Mir* 4 (2008).

[33] Other important questions include the degree to which current protective measures actually help small peoples, since, as suggested by Nikolai Vakhtin, they are very modest in their effects. Nikolai Vakhtin, "Native Peoples of the Russian Far North," in *Polar Peoples: Self Determination and Development* (London: Minority Rights Group

politicians – who claim that states have historically denied basic rights to Indigenous groups by various means, including the imposition of an identity upon them – envisage the guiding principles for relations between states and their Indigenous peoples.[34]

The genealogical roots of the legal concept of 'Indigenous peoples' can be found in colonial policies of the last quarter of the nineteenth century, and, as such, as a number of theorists have suggested, are primarily a product of these policies. The first reference to the rights of 'natives' or 'aborigines' (the legal terms used at the time in reference to Indigenous populations) appeared within the framework of the doctrine of trusteeship (sometimes referred to as the 'tutelage doctrine'[35]). This doctrine authorized European nations to guard over 'uncivilized' peoples, as their responsibility as senior members of the 'family of nations.' Specifically, it provided a 'dual mandate' for European states to exercise political and economic control in the form of guardianship over the less developed peoples, subjected to the control of Europeans in the form of colonies.[36]

Report 5, 1992): 6–37. Another related aspect would be the reasons why individuals identify themselves as members of indigenous groups, and to what degree their self-identification (reflected in numbers) can be taken as an indication of a 'traditional' life-style as suggested by legal descriptions (where 'traditional' itself a very questionable concept).

[34] In general, Indigenous politicians working on the level of the United Nations as well as within separate state systems reject any attempt by their governments to create definitions for Indigenous peoples, as expressed, for example, in the following clause: "We, the Indigenous Peoples present at the Indigenous Peoples Preparatory Meeting on Saturday, 27 July 1996, at the World Council of Churches, have reached a consensus on the issue of defining Indigenous Peoples and have unanimously endorsed Sub-Commission resolution 1995/32. *We categorically reject any attempts that Governments define Indigenous Peoples*"; United Nations Permanent Forum on Indigenous Issues Secretariat, "The Concept of Indigenous Peoples (Background Paper)," note 5. (My emphasis.)

[35] "The right of underdeveloped races, like the right of undeveloped individuals, is a right not to recognition as what they are not, but to guardianship – that is to guidance – in becoming that of which they are capable, in realizing their special ideals." James Lorimer, as quoted in Rodríguez–Piñero, *Indigenous Peoples, Postcolonialism, and International Law: The ILO Regime, 1919–1989*, 19, note 14.

[36] *Indigenous Peoples, Postcolonialism, and International Law: The ILO Regime, 1919–1989*, 20.

The key events of this period were the international conferences of the last quarter of the nineteenth century in which policies regarding the rights of colonized populations were developed (Berlin Conference, Brussels and Saint Germain Conferences). The doctrine of trusteeship received authoritative status with Articles 22 and 23 of the 1919 Covenant of the League of Nations. Article 22, which specifically categorized 'dependent' territories according to the "stage of [their] development," assigned administration over them to be exercised by mandated "advanced nations" in accordance with their "civilized duties," and provided that governance of colonial territories was a "sacred trust of civilization."[37] The emergence of the term 'Indigenous' in international legal parlance occurred in the 1930s in the drafting of standards for disciplining colonial workforces under the auspices of the International Labour Organization. These standards included the definition of 'Indigenous workers' in Article 2 of the 1936 Recruitment of Indigenous Workers Convention. The definition referred to populations broadly understood within the ILO context as native labour, which included colonial subjects as well as populations of self-governing states existing in a position of dependency (such as Native Americans of the USA). As suggested by Luis Rodríguez–Piñero and Ian Brownlie, the preference for the expression 'Indigenous' (from the French *indigène*) as opposed to 'native' was due to the negative connotations associated with the latter term.[38] The definition was evidently based on general propositions of social evolution theory and served to regulate relations between colonized populations and their colonizers, and to legitimize colonization as being a humanitarian and philanthropic mission.

The second phase influencing the formation of the legal concept of 'Indigenous peoples' was the post-World War Two period, specifically events leading to the composition of the first Convention focused on the rights of In-

---

[37] Article 22 of the Covenant of the League of Nations (1919) reads: "To those colonies and territories which as a consequence of the late war have ceased to be under the sovereignty of the States which formerly governed them and which are inhabited by peoples not yet able to stand by themselves under the strenuous conditions of the modern world, there should be applied the principle that the well-being and development of such peoples form a sacred trust of civilization and that securities for the formance of this trust should be embodied in this Covenant" The League of Nations, "The Covenant of the League of Nations," *The Avalon Project* (2008), http://avalon.law.yale.edu/20th_century/leagcov.asp (accessed 29 September 2010).

[38] Rodríguez–Piñero, *Indigenous Peoples, Postcolonialism, and International Law: The ILO Regime, 1919–1989*, 46, note 148.

digenous peoples in 1957 under the auspices of the International Labour Organization. Rodríguez–Piñero characterizes this period as the time during which the 'Indigenous problem' gained an international dimension as a problem of development. He sees a specific connection between the Convention's purpose in bringing the development programmes to Ecuador, Peru, and Bolivia under the "Andean Indian Programme" (1952–62) and the interests of these states in assimilating their Indigenous populations.[39] He points out that the Convention was intended as a guide to help other states in assimilating their Indigenous populations.[40] In his analysis of the 1957 ILO Convention, Rodríguez–Piñero notes that during the postwar period Indigenous groups were considered to be territories politically and socially integrated within their respective states. Perceived as locations 'owned' by their states, Indigenous peoples became a 'technical category' at which international development

[39] One of the intriguing areas of possible study is the relationship between the emergence of the first instrument focused on the rights of Indigenous peoples (*ILO Convention 107*, 1957) and the aims of the US government in supporting development projects (such as the "Andean Indian Programme") in the climate of the Cold War.

[40] The writing of the Convention should also be analysed as falling within the context of the UN decolonization regime, officially beginning with the UN General Assembly Resolution 1514, Declaration on the Granting of Independence to Colonial Countries and Peoples (14 December 1960). At that time, emerging new nations faced with the problem of constructing their national identities and political systems came under the influence of Western political models and scientific theories. This influence is most evident in the way they started seeing their culturally different Indigenous groups as problematical (i.e. backward). Rogers Brubaker describes this period in the following manner: "[This was] a moment of high political confidence in Western models of political development and their transferability to the developing world, sustained by robust epistemological confidence in a generalizing style of social science capable of discovering universal patterns of social and political development and of validating policies aimed at promoting such development." The "nation-building" literature of the 1960s, according to Rogers Brubaker, had as a central idea that "the 'nation' is simply the citizenry, to the extent that it becomes a unit of identity and loyalty [... where]. [...] nationhood [...] was seen as strengthened [...] by [...] modernizing forces [...] [while ...] ethnicity could be understood as a potentially serious *impediment* to nation-building and national integration." Rogers Brubaker, *Nationalism Reframed: Nationhood and the National Question in the New Europe* (Cambridge: Cambridge UP, 1996): 80–82.

projects were directed.[41] A version of Indigenous peoples' rights, promoted by the Convention,[42] considered only their rights as citizens of the new nations. Their cultural practices and unique social and political organizations mattered only in relation to how they facilitated (or impeded) the aims and processes of assimilation. In the postcolonial context, indigeneity remained a characteristic that essentially denoted an inferior and temporary social state of peoples, now approached as territories possessed by their respective states.

The modern concept of indigeneity emphasizes the cultural differences of Indigenous groups from the dominant social and political structures of the nation-states within which they are located. The diverse forms of Indigenous social settings are no longer perceived as a disappearing abnormality but, rather, as a crucial part of the world's cultural diversity. Groups understood as 'Indigenous peoples' remain parts of their states, and by their location are subject to the political and economic influences of those states. Nonetheless, the internationally recognized standards of treatment for Indigenous peoples – the most important one being the UN Declaration on the Rights of Indigenous Peoples – stress the right of Indigenous individuals to control and maintain social and cultural differences of their groups by living according to their own historically developed ways of life.[43]

---

[41] As Rodríguez–Piñero explains, "The ILO standards were conceived as a set of technical guidelines that should guide state developmental policies toward these peoples, where indigenous cultures mattered only as factors in the success or failure of these policies, and where the international legal form was only meant to represent the international community's moral commitment to the solution of the 'indigenous problem'." Rodríguez–Piñero, *Indigenous Peoples, Postcolonialism, and International Law: The ILO Regime, 1919–1989*, 144.

[42] The preamble to the *ILO Convention 107* states that "adoption of general international standards on the subject will facilitate action to assure the protection of the populations concerned, their progressive *integration* into their respective national communities, and the improvement of their living and working conditions" (emphasis in the original). International Labour Organization, "Convention Concerning Indigenous and Tribal Populations," *Convention 107* (1957), http://www.ilo.org/images/empent/static/coop/pdf/Conv107.pdf (accessed 30 September 2010).

[43] A prominent Native American scholar, Duane Champagne, notes that the way in which the Declaration codifies these special rights of Indigenous peoples responds to the vision of Indigenous individuals as being citizens of their nation-states, as opposed to being members of political systems different from those exercised by nation-states, yet enjoying the same level of political authority (government-to-government relation-

The adoption of the Declaration by the UN General Assembly in 2007 was to a large extent a result of political activism among Indigenous groups and efforts by organizations working on their behalf. The political struggles of Indigenous peoples in different countries[44] influenced changes in the relations between these peoples and their states (i.e. of the questioning of states' sovereign powers), leading to the emergence of new norms and principles upon which peaceful relations among and within nation-states should rest. The milestones of this development included the 1971 Resolution of the UN Human Rights Commission to conduct a study focused on the problems of Indigenous populations. This study, known now as the Martinez–Cobo "Study of the Problem of Discrimination Against Indigenous Populations," took over a decade to complete.[45] A further step was taken at the 1977 NGO Conference on Discrimination against Indigenous Populations in Geneva, which led to the

---

ship). Hence, Champagne maintains, the key claims of Indigenous politicians, which for him are essentially from *the indigenous perspective* – those claims being political autonomy and self-government – are not recognized by the international community (and are not codified in the text of the Declaration). This is not to say that the emergence of the Declaration does not signify progress in the development of relations between indigenous groups and their states; rather, the content of the Declaration is, significantly, a result of what it was possible to achieve at this moment in history. (Champagne, personal communication). Champagne's critique helps to map out future developments of Indigenous politics on the international and state levels.

[44] Some of the best-known episodes in these struggles are the formation of the American Indian Movement, an organization established during the 1960s in the USA; the 1966 cattle workers' strike by the Gurindji peoples at Wave Hill, Australia; and the 1975 international conference in British Columbia of Indigenous representatives from North, Central, and South America, Australia, New Zealand, and Scandinavia, resulting in the formation of the World Council of Indigenous Peoples, one of the first UN-associated NGOs of international scope and influence. See, generally, Ken Coates, *A Global History of Indigenous Peoples: Struggle and Survival*; Douglas Sanders, "Background Information on the World Council of Indigenous Peoples," *World Council of Indigenous Peoples* (April 1980), http://www.halcyon.com/pub/FWDP/International/wcipinfo.txt (accessed 29 September 2010). For Australian history, refer to, for example, John Summers, "The Parliament of the Commonwealth of Australia and Indigenous Peoples 1901–1967," *Vision in Hindsight* (31 October 2000), http: //www.aph.gov.au/library/Pubs/rp/2000-01/01RP10.htm (accessed 29 September 2010).

[45] José Martínez–Cobo, "Study of the Problem of Discrimination against Indigenous Populations," United Nations Economic and Social Council, 1986, UN Doc E/CN.4/Sub.2/1986/7.

creation in 1982 of a UN Working Group on Indigenous Populations that functioned, until recently, as an open international forum for Indigenous leaders to bring their grievances and aspirations into the international arena. The most significant achievement of the Working Group was the composition in 1994 of a Draft Declaration on the Rights of Indigenous Peoples, upon which the text of the Declaration on the Rights of Indigenous Peoples rests. The Declaration was adopted by the UN General Assembly on 13 September 2007. Conceived and drafted with the active participation of Indigenous politicians, the Declaration is the primary internationally recognized instrument for the protection of Indigenous rights. The significance of the Declaration is limited to the ethical exercise of power, as it has no status as a legally binding treaty. The Declaration is not so much a set of standards by which indigenous peoples are to be treated by their respective states (such as preceding the ILO Convention 169 [1989]) as a "generalized indigenous 'position' on many issues," articulated from the point of indigenous peoples.[46]

The development of international legal standards for Indigenous peoples was evidently based on specific social-science ideas. At the same time, laws (as well as specific legal concepts which defined indigeneity) were established to address specific social problems arising from relations between states and populations with the status of 'Indigenous peoples.' Thus, these laws are a product of theorizing emanating from contexts very different from those of the social sciences. However, the two contexts are closely related and must be considered as such in any analysis of the realities of groups having the status of Indigenous peoples or self-identifying as such, and in the development of a comprehensive way of conceptualizing these realities.

The political perspective

Finally, the perspective I call 'political' refers to the way in which indigeneity is characterized by individuals self-identifying as Indigenous, most specifically those who represent the interests of Indigenous groups at United Nations political forums. This mode of thought[47] emerged somewhat recently from a

---

[46] Thornberry, *Indigenous Peoples and Human Rights,* 35.

[47] This mode of thinking should not be seen as limited to the group identified in this section and can be further studied by looking at the way in which indigeneity is defined by those who claim Indigenous identity outside of the UN human rights forums on state and local levels.

group previously isolated from political activities on the level of the UN.[48] This group is composed of Indigenous politicians, by whom I mean individuals who claim Indigenous identity and who work on behalf of Indigenous peoples' organizations or who are elected by members of a particular group to represent the group's interests at higher levels.[49] This is a heterogeneous

---

[48] Despite their isolation, these activities have been the subject of a number of studies. Ronald Niezen provides an account of what he terms Indigenous identity-formation within the contexts of late 'modernity', where identity emerges out of three interrelated factors: historiography, which provides a group with a unique vision of its self; moral imperatives, which allow perception of exclusiveness of a collective sense of identity; and use of identity for the political goals of a group. With regard to Indigenous experiences, Niezen refers to the commonality of a colonial past that the Indigenous peoples have experienced, and to the commonality of a vision of a different future. These commonalities have allowed representatives of diverse groups and cultures to come together in a shared vision of themselves as 'Indigenous peoples'. For Niezen, identity becomes a source of membership that allows perception of personal grievances to be elevated as a part of global history. Niezen, *Origins of Indigenism*.. Compare with Margaret Moore's conception of identity emerging from studies of nationalism, where it is, broadly speaking, a vision of self "concerned with the political community with which one identifies. [...] [and inpired by ...] political or institutional recognition of this community." Moore, cited in Siobhan Harty & Michael Murphy, *In Defence of Multinational Citizenship* (Vancouver & Toronto: U of British Columbia P, 2005): 14. Similar to this are conceptualizations of Indigenous peoples by the scholars of Indigenous political movements. For example, Seán Eudaily, while refusing to use the concept 'Indigenous peoples', resorts to the names that Indigenous groups use for self-identification (such as First Nations, Aboriginal Australians, Torres Strait Islanders). Nevertheless, he adds that in his use of 'Indigenous' he "alludes" to "peoples who are engaging in practices of resistance" which makes them Indigenous to "power relations" rather than to territories / lands they occupy. Seán Eudaily, *The Present Politics of the Past: Indigenous Legal Activism and Resistance to (Neo)liberal Governmentality*, 2. Ken Coates, likewise, notes that membership of an Indigenous organization "defines" the meaning of indigenous for that organization. Coates, *A Global History of Indigenous Peoples: Struggle and Survival*, 2–8.

[49] This study does not address the question of the nature and level of influence Indigenous organizations have exercised at the level of the UN. This question can be addressed using the framework suggested by Peter Willetts's analysis of the history of NGOs in the UN system. Willetts argues that an NGO can only influence UN programmes when it is both an organization and part of a wider social movement. Willetts, "From Consultative Arrangement to Partnership: The Changing Status of

group composed of individuals from different geo-political settings and socio-economic backgrounds, and whose states may or may not recognize their indigeneity claims.[50] Some of these individuals are supported by the governments of their states; others may employ international settings to voice particular injustices done to their peoples by the political authorities of states of which they are a part. What is of interest to this study, however, is the type of thought that emerges when these individuals form coalitions at UN political forums with the intention of influencing the policy-making process affecting indigenous groups. By nature, these coalitions are separate from Indigenous communities and groups, yet are organized to represent the political interests of Indigenous groups on the international level.[51] Indigenous participation on this level is coordinated by UN agencies and is often financially supported by funds donated by the states' governments. Settings within which the political perspective is formed are necessarily political, characterized, to use Karl Mannheim's words, by "conflict […] for social predominance."[52] Hence, the driving force behind definitions of indigeneity is the motivation to achieve practical consequences deemed to follow from them. These consequences are generally the creation of opportunities for Indigenous politicians to influence (1) the political settings in which they claim membership to oversee policy and law, and (2) the theory upon which the current legitimacy of policy-

---

NGOs in Diplomacy at The UN," *Global Governance* 6 (2000): 191–212. Thus, a key question in the analysis of the political power of an indigenous organization within the UN could be: *To what degree is this organization part of a larger social upheaval?*

[50] As suggested by Rosemary Coombe, "The Recognition of Indigenous Peoples' and Community Traditional Knowledge in International Law," *St. Thomas Law Review* 14 (2001): 275–286.

[51] The key arenas for indigenous political participation at the level of the UN since 1982 have been the Working Group on Indigenous Populations (established in 1982), and the UN Permanent Forum on Indigenous Issues (established in 2000). Indigenous participation in these forums is supported by the Voluntary Fund for Indigenous Populations, established in 1985 and funded by voluntarily contributions from governments and the private sector. Its administration is currently assisted by a Board of Trustees composed of indigenous politicians. Other relevant bodies include various advisory groups functioning as parts of particular UN organizations concerned with indigenous issues. Example are the CBD Working Group on Article 8(j), and the WIPO Intergovernmental Committee on Intellectual Property and Genetic Resources, Traditional Knowledge and Folklore.

[52] Mannheim, *Ideology and Utopia*, 36.

makers rests. The social positioning of Indigenous politicians as knowledge-makers is peculiar to their group, and very different from that of scientists and law-makers. Their mode of thinking is connected to their existence as thinking subjects. Not only are Indigenous politicians in a position to establish knowledge about their own reality, but simultaneously they are also the focus of investigation and of policy-/law-making activities conducted by others. This social positioning defines a form of consciousness expressed in various types of statements to international agencies/organizations and governments composed of Indigenous individuals as collective statements. The content of these statements emanates from relevant scientific investigations and legal provisions, used *strategically*. Some theories and legal provisions are used to support specific claims, while others serve as a basis for critique, where the aim pursued by Indigenous politicians is the development of a new way of looking at Indigenous reality that can potentially serve their interests.[53] In

---

[53] The way in which identity itself is defined within this perspective can be clearly seen in the discussion of identity by Duane Champagne. Specifically, he writes: "*Indian identity is a matter of social and cultural action as well as self-identification*" (19). Champagne discusses the relationship between imposed legal and formal identities and the way in which these outside forms of defining groups as Indigenous has facilitated possibilities for political actions and acceptance (or rejection of) these forms of identity by individuals living in American Indian communities (US context). On the international level, however, indigeneity becomes an analytical instrument to explain and further develop forms of political governance of groups that may or may not enjoy legal status as 'Indigenous peoples', yet have forms of political governance, the development of which has followed routes other than those which determined the growth of the current legal and political systems practised by nation-states. This is not to imply that indigeneity defines 'non-Western' political structures, which, as implied, should be essentially similar; rather, what makes Indigenous forms of political governance a category is their inferior state as forms of political power when compared with the level of political leverage of nation-states (in which these groups are, as a rule, encapsulated). Indigenous forms of political governance are multiple, responding to the way in which Indigenous communities have historically developed. Yet, as a rule, they do not enjoy the same degree of political influence within their community and on the state level as the political authority of the state in which these groups are located; nor do they have an independent voice in the international arena. At the same time, nation-states do not have analytical and legal tools to conceptualize and legally codify the way in which the political and social power of Indigenous communities functions both internally and as a part of state and international systems. Until, however, we do have such instruments as scholars of indigeneity and, in a broader sense, social scientists,

Foucauldian terms, this perspective involves the *strategic* use of the existing knowledge *about* indigeneity. The knowledge thus constructed is an instrument employed to further specific interests on behalf of Indigenous groups.[54]

The key knowledge components that compose the content of 'indigeneity' within this perspective are those that support the vision of the 'historical continuity' of Indigenous groups with the cultures/societies of their predecessors[55] and which results from one or more factors, such as: a) occupation of ancestral land; b) common ancestry with those who originally settled/occupied the land; c) cultural (social) characteristics which distinguish these groups from the social and political systems of which they are geographically a part, such as tribal/traditional/customary system of governance; d) language; and e) residence. These factors emerge from a set of formal documents focused on the protection of the rights of indigenous peoples, most significantly Martinez–Cobo's "Study of the Problem of Discrimination Against Indigenous Populations." His definition of 'Indigenous peoples' was adopted as a "working one" by the UN Working Group on Indigenous Populations.[56]

Champagne insists, our social theory cannot be seen as responding to current social realities which this theory is intended to explain and predict. — Duane Champagne, *Notes From the Center of Turtle Island.* Also personal communication.

[54] Mannheim brilliantly captures this reciprocal relationship between what he calls "theory and practice" in his discussion of the principles of political sociology. He argues that political settings (unlike academic ones) are characterized by a constant process of becoming, where thought is shaped by an active participation which influences the thought-process itself and determines the content of new knowledge thus created. The validity of this new knowledge is determined by the agreement of a group (as opposed to abstract criteria traditionally used by science), which makes a concept (i.e. an intellectual point of view) be primarily the product of group consensus. Because the authoritative status of a conceptualization results from the particular socio-historical situatedness of the group doing it, any concept is accordingly subject to change if the power-constellation changes. Mannheim, *Ideology and Utopia*, 22.

[55] This is captured in the following widely known definition by the World Council of Indigenous Peoples: "The term indigenous people refers to people living in countries which have a population composed of differing ethnic or racial groups who are descendants of the earliest populations living in the area and who do not as a group control the national government of the countries within which they live." Douglas Sanders, "Background Information on the World Council of Indigenous Peoples."

[56] See paras. 380–82 in Martínez–Cobo, "Study of the Problem of Discrimination Against Indigenous Populations."

These factors also received formal recognition in the ILO 1989 Convention (see Article 1). In order to achieve coherence in their statements, Indigenous politicians abstract from the concrete realities their groups face and compose their collective claims as resting upon common experiences, most often represented as a history of denial of basic human rights resulting from colonial conquest and/or other events leading to a current dependency of an Indigenous group on a metropolitan power which exploits Indigenous land and peoples.[57] In this context, what essentially emerges under the concept of 'indigeneity' and related terms is less *a property* which describes a particular trait of social reality, and/or which may require a particular legal action, than a theoretical tool to be used for specific actions.

III. Conclusion

The concept of 'Indigenous peoples', as considered in this essay, emerges from a Marxist perception of concept as a form of understanding and conceptualizing reality. Social scientists, in their attempts to construct a general concept, tend to produce generalized notions of indigeneity (such as 'native to a particular place' or 'essentially different from "modernity"'). These generalizations arrange particular experiences of very different groups under a vague idea, the use of which, as an instrument for scientific communication across disciplines, is questionable. Legal theorists resort to ideas from the social sciences, yet focus on problems of relations between Indigenous peoples (as specific subjects of law) and the states within which they are located. This approach makes legal notions of indigeneity subject to social-science theory, yet uses these theories in contexts very different from those out of which they originally emerged and for aims different from those originally intended. Finally, Indigenous politicians focus on specific political objectives, which define the scientific and legal sources and conceptions they employ.

One cause of difficulty in navigating through existing concepts of indigeneity and related terms is that these concepts are, as Marxist commodities, social constructions that exist apart from the context in which they originated. Legal scholars may borrow social-science ideas, while scientists may employ legal instruments in their studies. Legal and scientific knowledge can result in gains of political power when presented to support the interests of specific groups if used in a tactical manner. My modest conclusion is the proposition

---

[57] Thornberry, *Indigenous Peoples and Human Rights*, 48. Kingsbury, "'Indigenous Peoples' in International Law: A Constructivist Approach to the Asian Controversy."

that the concept of indigeneity and, emanating from it, the concept of 'Indigenous peoples' were formed and continue to exist as a product of three interrelated modes of thought, each illuminating a particular aspect of the reality of Indigenous groups, yet existing in a reciprocal relation to the others. Only if one considers these three perspectives as elements that constitute essentially one group of problems could one possibly approach the problem of understanding Indigenous realities in a deeper and more penetrating fashion.

WORKS CITED

Agrawal, Arun. "Dismantling the Divide Between Indigenous and Scientific Knowledge," *Development and Change* 26 (1995): 413–39.

Anaya, James. *Indigenous Peoples in International Law* (Oxford & New York: Oxford UP, 2004).

——, ed. *International Law and Indigenous Peoples* (Oxford & New York: Oxford UP, 2003).

Anghie, Anthony. "Finding the Peripheries: Sovereignty and Colonialism in Nineteenth Century International Law," *Harvard International Law Journal* 40 (1999): 1–71.

Anon. "All-Russian Census of Population" (2002), www.perepis2002.ru (accessed 29 September 2010).

Banuri, Tariq, & Frédérique Apffel–Marglin, ed. *Who will Save the Forests? Knowledge, Power and Environmental Destruction* (London: Zed, 1993).

Barnard, Alan. "Kalahari Revisionism, Vienna and the 'Indigenous Peoples' Debate," *Cultural Anthropology* 14.1 (2007): 1–16.

Brubaker, Rogers. *Nationalism Reframed: Nationhood and the National Question in the New Europe* (Cambridge: Cambridge UP, 1996).

Champagne, Duane. *Notes From the Center of Turtle Island* (New York, Toronto & Plymouth: AltaMira, 2010).

Clifford, James. "Varieties of Indigenous Experience: Diasporas, Homelands, Sovereignties," in *Indigenous Experience Today,* ed. Marisol de la Cadena & Orin Starn (Oxford & New York: Berg, 2007): 197–225.

Coates, Ken. *A Global History of Indigenous Peoples: Struggle and Survival* (New York: Palgrave Macmillan, 2004).

Coombe, Rosemary. "The Recognition of Indigenous Peoples' and Community Traditional Knowledge in International Law," *St. Thomas Law Review* 14 (2001): 275–86.

Dahl, Jens, Jack Hicks & Peter Jull, ed. *Nunavut: Inuit Regain Control of Their Lands and Their Lives* (Copenhagen: International Working Group on Indigenous Affairs, 2000).

Donahoe, Brian, Joachim Habeck, Agnieszka Halemba & István Sántha. "Size and Place in the Construction of Indigeneity in the Russian Federation," *Current Anthropology* 49.6 (2008): 993–1020.

Dove, Michael. "Indigenous People and Environmental Politics," *Annual Review of Anthropology* 35 (2006): 191–208.

Ellingson, Ter. *The Myth of the Noble Savage* (Berkeley & Los Angeles: U of California P, 2001).

Eudaily, Seán. *The Present Politics of the Past: Indigenous Legal Activism and Resistance to (Neo)liberal Governmentality* (New York & London: Routledge, 2004).

Fliert, Lydia, ed. *Indigenous Peoples and International Organizations* (Nottingham: Russell, 1984).

Godoy, Ricardo, Victoria Reyes–Garcia, Elizabeth Byron, William Leonard & Vincent Vadez. "The Effect of Market Economies on the Well-Being of Indigenous Peoples and on Their Use of Renewable Natural Resources," *Annual Review of Anthropology* 34 (2005): 121–38.

Grewe, Wilhelm. *The Epochs of International Law,* tr. Micahel Byers (*Epochen der Völkerrechtsgeschichte*, 1984; Berlin & New York: Walter de Gruyter, 2000).

Harty, Siobhan, & Michael Murphy. *In Defence of Multinational Citizenship* (Vancouver & Toronto: U of British Columbia P, 2005).

International Labour Organization. "Convention Concerning Indigenous and Tribal Populations," *Convention 107* (1957), http://www.ilo.org/images/empent/static/coop/pdf/Conv107.pdf (accessed 21 February 2011).

International Work Group for Indigenous Affairs. *The Indigenous World 2000/2001* (Copenhagen: International Work Group for Indigenous Affairs, 2001).

Ivison, Duncan, Paul Patton & Will Sanders, ed. *Political Theory and the Rights of Indigenous Peoples* (Cambridge UK & New York: Cambridge UP, 2000).

Kingsbury, Benedict. "'Indigenous Peoples' in International Law: A Constructivist Approach to the Asian Controversy," *American Journal of International Law* 92.3 (July 1998): 414–57.

Kothari, Brij. "Theoretical Streams in Marginalized Peoples' Knowledge(s): Systems, Asystems, and Subaltern Knowledge(s)," *Agriculture and Human Values* 19.3 (2002): 225–37.

Kriazkov, Vladimir. "Prava Korennih Malochislennih Narodov Rossii: Metodologia Regulirovania," *Gosudarstvo i Pravo* 1 (1996): 18–24.

Kuper, Adam. "The Return of the Native," *Current Anthropology* 44.3 (June 2003): 389–402.

The League of Nations. "The Covenant of the League of Nations," *The Avalon Project* (2008), http://avalon.law.yale.edu/20th_century/leagcov.asp (accessed 29 September 2010).

Li, Tania Murray. "Articulating Indigenous Identity in Indonesia: Resource Politics and the Tribal Slot," *Comparative Studies in Society and History* 42.1 (January 2000): 149–79.

Lorimer, James. *Studies National and International* (Edinburgh: William Green & Sons, 1890).

Mannheim, Karl. *Ideology and Utopia,* tr. Louis Wirth & Edward Shils (*Ideologie und Utopie*, 1929; London & New York: Harcourt, Brace, 1936).

Maybury–Lewis, David. "Indigenous Peoples," in *The Indigenous Experience: Global Perspectives,* ed. Roger Maaka & Chris Andersen (Toronto: Canadian Scholars, 2006): 17–29.

Martínez–Cobo, José. "Study of the Problem of Discrimination against Indigenous Populations," United Nations Economic and Social Council, 1986, UN Doc E/CN.4/Sub.2/1986/7.

Niezen, Ronald. *Origins of Indigenism* (Berkeley, Los Angeles & London: U of California P, 2003).

Osherenko, Gail. "Indigenous Rights in Russia: Is Title to Land Essential for Survival?" *Georgetown International Environmental Law Review* 13.3 (2000): 695–734.

Pelican, Michaela. "Complexity of Indigeneity and Autochthony: An African Example," *American Ethnologist* 36.1 (February 2009): 52–65.

Ponkin, Igor. "O poniatii 'korennoy narod'," *Mir* 4 (2008).

Quane, Helen. "The Rights of Indigenous Peoples and the Development Process," *Human Rights Quarterly* 27.2 (May 2005): 652–82.

Rodríguez–Piñero, Luis. *Indigenous Peoples, Postcolonialism, and International Law: The ILO Regime, 1919–1989* (New York: Oxford UP, 2005).

Russian Federation. *Ob Obschikh Printsipakh Organizatsii Obschin Korennykh Malochislennykh Narodov Severa, Sibiri i Dal'nego Vostoka Rossiyskoy Federatsii* [On General Principles for the Organization of *Obschinas* of the Indigenous Numerically Small Peoples of the North, Siberia, and the Far East of the Russian Federation]. Russian Federal Law 104-F3, 20 July 2000.

Said, Edward W. *Orientalism* (New York: Vintage, 1978).

Sanders, Douglas. "Background Information on the World Council of Indigenous Peoples," *World Council of Indigenous Peoples* (April 1980), http://www.halcyon.com/pub/FWDP/International/wcipinfo.txt (accessed 29 September 2010).

Shapovalov, Aleksandr. "Straightening Out the Backward Legal Regulation of "Backward" Peoples' Claims to Land in the Russian North: The Concept of Indigenous Neomodernism," *Georgetown International Environmental Law Review* 17.3 (2005): 435–69.

Sieder, Rachel. *Multiculturalism in Latin America: Indigenous Rights, Diversity, and Democracy* (New York: Palgrave Macmillan, 2002).

Smith, Eric, & Mark Wishnie. "Conservation and Subsistence in Small-Scale Societies," *Annual Review of Anthropology* 29 (2000): 493–524.

Sokolovski, Sergei. "The 2002 Census: Games According to Wittgenstein," *Anthropology and Archeology of Eurasia* 44.1 (2005): 25–33.

Spencer, Herbert. *The Study of Sociology* (London: Henry S. King, 1873).

Stamatopoulou, Elza. "Indigenous Peoples and The United Nations: Human Rights as a Developing Dynamic," *Human Rights Quarterly* 16.1 (February 1994): 58–81.

Stepanov, Valery. "Russian Experience in the North Indigenous Statistics for Statistical Questions," in *Proceedings of the Workshop on Data Collection and Disaggregation for Indigenous Peoples, New York, January 19–21*, United Nations Department of Economic and Social Affairs, Secretariat of the Permanent Forum on Indigenous Issues, 2004, UN Doc. PFII/2004/WS.1/5.

Stoyanova, Irina. "Theorizing the Origins and Advancement of Indigenous Activism: The Case of the Russian North" (doctoral dissertation, George Mason University, 2009), http://digilib.gmu.edu:8080/bitstream/1920/5614/1/Stoyanova_Irina.pdf (accessed 29 September 2010).

Summers, John. "The Parliament of the Commonwealth of Australia and Indigenous Peoples 1901–1967," *Vision in Hindsight* (31 October 2000), http://www.aph.gov.au/library/Pubs/rp/2000-01/01RP10.htm (accessed 29 September 2010).

Tennant, Chris. "Indigenous Peoples, International Institutions, and the International Legal Literature from 1945–1993," *Human Rights Quarterly* 16 (1994): 1–57.

Thornberry, Patrick. *Indigenous Peoples and Human Rights* (Manchester: Manchester UP, 2002).

United Nations. "State of The World's Indigenous Peoples" (2009), http://www.un.org/esa/socdev/unpfii/documents/SOWIP_web.pdf (accessed 29 September 2010).

United Nations Convention on Biological Diversity. "Knowledge, Innovations and Practices of Indigenous and Local Communities: Implementation of Article 8(j) (Note by Executive Secretary)," in *Proceedings of Conference of the Parties to the Convention on Biological Diversity, Third meeting, Buenos Aires, Argentina, 4–15 November, 1996,* United Nations Convention on Biological Diversity, 1996, UN Doc UNEP/CBD/COP/3/19.

United Nations Convention to Combat Desertification. "Traditional Knowledge. Addendum. Building Linkages Between Environmental Conventions and Initiatives (Note by the Secretariat)," in *Proceedings of Conference of the Parties. Committee on Science and Technology. Third Session, Recife 16–18 November 1999*, United Nations Convention to Combat Desertification, 1999, UN Doc ICCD/COP(3)/CST/3/Add.1.

United Nations Development Programme. "UNDP and Indigenous Peoples: A Policy of Engagement," (2001), http://www.undp.org/partners/civil_society/empowering_indigenous_peoples.shtml (accessed 29 September 2010).

United Nations Economic and Social Council, Commission on Human Rights. "Indigenous Issues: Report of the Working Group. Chairperson-Rapporteur: Mr. José Urrutia (Peru)," in *Proceedings of the Fifty Third Session, December 10, 1997*, United Nations Economic and Social Council, 1997, UN Doc E/CN.4/1997/102.

United Nations Permanent Forum on Indigenous Issues Secretariat. "The Concept of Indigenous Peoples (Background paper)," in *Proceedings of the Workshop on Data Collection and Disaggregation for Indigenous Peoples, New York, 19–21 January 2004*, United Nations Permanent Forum on Indigenous Issues 2004, UN Doc PFII/2004/WS.1/3.

Vakhtin, Nikolai. "Native Peoples of the Russian Far North," in *Polar Peoples: Self Determination and Development* (London: Minority Rights Group Report 5, 1992): 6–37.

Warren, Michael. *Using Indigenous Knowledge in Agricultural Development* (Washington DC: World Bank, 1991).

——, Jan Slikkerveer & David Brokensha, ed. *The Cultural Dimension of Development: Indigenous Knowledge(s) Systems* (London: Intermediate Technology, 1995).

Wickstrom, Stefanie. "The Politics of Development in Indigenous Panama," *Latin American Perspectives* 30.4 (2003): 43–68.

Willetts, Peter. "From Consultative Arrangement to Partnership: The Changing Status of NGOs in Diplomacy at the UN," *Global Governance* 6 (2000): 191–212.

World Bank. "Indigenous knowledge for development: Framework for action" (1998), http: //www.worldbank.org/afr/ik/ikrept.pdf (accessed 29 September 2010).

——. "Operational Directive 4.20: Indigenous Peoples," in *Indigenous Peoples and International Organizations*, ed. Lydia Fliert (Nottingham: Russell, 1991): 83–88.

Xanthaki, Alexandra. "Indigenous Rights in the Russian Federation: The Case of Numerically Small Peoples of the Russian North, Siberia, and Far East," *Human Rights Quarterly* 26.1 (February 2004): 74–105.

Young, Elspeth. *Third World in The First: Development and Indigenous Peoples* (New York: Routledge, 1995).

Zimmerman, Barbara, Carols Peres, Jay Malcolm & Terence Turner. "Conservation and Development Alliances with the Kayapo of South-Eastern Amazonia, a Tropical Forest Indigenous People," *Environmental Conservation* 28.1 (2001): 10–22.

⌘

Screening Indigenous Peoples' Genes
The End of Racism, or Postmodern Bio-Imperialism*

FRANK KRESSING

Introduction

IN THE LATE-TWENTIETH CENTURY, VARIOUS ATTEMPTS WERE MADE to establish a universal pedigree of human populations, combining presumably new findings concerning the long-range genetic relationship of languages (for example, Joseph Greenberg, Merritt Ruhlen) and the identification of major genetic clusters of mankind by means of DNA analysis (for example, Luigi Lucca Cavalli–Sforza). These attempts to establish a universal pedigree of human populations in the sense of a 'global phylogeny' have received widespread public attention and were presented as a 'new synthesis' of genetic, linguistic, and archaeological data, often being accepted as innovative and being grounded on sound science. The global phylogeny model evolved in the context of a 'general turn to genetics' in the life sciences and is closely linked to projects that focus on a 'genomization' of indigenous peoples: namely, the former Human Genome Diversity Project and the recent Genographic Project.

I will start with a few remarks regarding the recent history (1990–2010) of genetic screening pertaining to Indigenous peoples, asking why biomapping

* I would like to express my gratitude for the generous funding of the project Classification and Evolution in Biology, Linguistics, and the History of Science by the German Federal Ministry of Education and Research (BMBF), to which I am very much indebted for the help it offered me in the preparation of this essay. I would also like to thank Heiner Fangerau, Matthis Krischel, Stefanie Schütz, and Anja Weigel from the Institute of the History, Philosophy, and Ethics of Medicine (Ulm University) for their indispensable advice and generous assistance.

these particular populations seems to have been so attractive. After a brief outline of the Human Genome Diversity Project and the Genographic Project, I shall highlight their close link to the model of 'global phylogeny', considering the scientific legacy of modern genetic 'clustering' of humans and raising the question of the extent to which the 'global phylogeny' resembles ideas of a biological, linguistic, and cultural co-evolution of humans that can be traced back to the times of the Enlightenment. The early-twentieth-century decline of evolutionary theory in the sciences and humanities will then be contrasted with the re-emergence of co-evolutionary ideas in late-twentieth-century 'global phylogeny' and biomapping models, emphasizing the perpetuation and continuation of a primordialist base-line in Western perceptions of Indigenous peoples. Finally, practical implications of recent biomapping projects with regard to the contemporary political and social situation of Indigenous peoples will be considered.

In the context of this essay, the term 'Indigenous peoples' is broadly used in accordance with the United Nations definition,[1] characterizing these peoples by the features of (1) priority of settlement in their territory, (2) voluntary maintenance of a distinct culture, (3) self-identification, and (4) a marginal status within the dominant society into which they have been incorporated (often resulting from cultural or even physical genocide).

The criterion of priority of settlement raises certain difficulties, since Indigenous peoples have been subject to migration and replacement in the same way as larger, linguistically, socially, and politically dominant populations. I will argue that it is precisely the misleading primordial connotation of the term 'Indigenous peoples' that has given rise to serious misinterpretations, and that the assumption of these peoples' continuous, uninterrupted settlement in their present territories represents one of the shortcomings of the 'ideology of genomics' that will be presented below.

The Attractiveness of Mapping Indigenous Peoples' Genes

Already in 1984, geneticists from the Genetics Department of Yale University and Washington State University "started a pilot program to produce cell lines

---

[1] See José Martínez–Cobo, *Discrimination against Indigenous People*, United Nations Document No. E/CN.4/Sub.2 (1986): 87, and Erica–Irene A. Daes, *Working Paper by the Chairperson-Rapporteur, Mrs. Erica–Irene A. Daes, on the Concept of Indigenous People*, UN Document E/CN.4/Sub.2/AC.4(1996): 2.

from a number of Indigenous populations throughout the world."[2] After the inauguration of the Human Genome Project in October 1990, the renowned geneticist Luigi Luca Cavalli–Sforza and a number of colleagues called for a "worldwide survey of human genetic diversity."[3] They envisaged an extension of the Human Genome Project which was started in 1991 under the name Human Genome Diversity Project (HGDP) by the Morrison Institute of Stanford University – the academic institution where the two most prominent representative of the 'super-grouper approach' in comparative linguistics, the late Joseph Greenberg, and Luigi Luca Cavalli–Sforza, had been working in close interdisciplinary contact from the 1970s onwards. From its inception, the Human Genome Diversity Project was severely criticized, primarily by Indigenous organizations who perceived it as a "vampire project," "biopiracy," "legal theft of indigenous genes," or "a new wave of bio-colonialism," calling the proponents "gene hunters" and "predatory researchers."[4] In opposition to the project, the Indigenous Council on Bio-Colonialism was founded in 1993. Apart from moral and legal questions (such as access, consent, and benefit-sharing) that are raised by the sampling of the genetic diversity of Indigenous peoples, the Human Genome Diversity Project clearly perpetuates the historically developed perception of indigenous peoples as 'pristine populations', even labelling them "isolates of historic interest"[5] – or, in the words of the anthropologist Jonathan Marks,

> The major goal in this effort, unfortunately, is thus also guided by an archaic idea: the establishment of the ultimate genetic phylogeny of human groups. In pursuit of this objective, advocates are obliged to

---

[2] Matthew Rimmer, "The Genographic Project: Traditional Knowledge and Population Genetics," *Indigenous Law Review* 11.2 (2007): 34.

[3] Luigi Luca Cavalli–Sforza, Allan C. Wilson, Charles R. Cantor, Robert M. Cook–Deegan & Mary–Claire King, "Call for a Worldwide Survey of Human Genetic Diversity: A Vanishing Opportunity for the Human Genome Project," *Genomics* 11 (1991): 490–491.

[4] Constance MacIntosh, "Indigenous Self-Determination, and Research on Human Genetic Material: A Consideration of the Relevance of Debates on Patents and Informed Consent, and the Political Demands on Researchers," *Health Law Journal* 13 (2005): 224.

[5] "The Genographic Project: Frequently Asked Questions," https://genographic.nationalgeographic.com/genographic/lan/en/faqs_about.html (accessed 25 October 2010).

> maintain that non-European human populations are generally 'pure', and have been spared the vagaries of history, of contact, and of gene flow – assumptions that are certainly gratuitous […]. Some generations ago, it was generally assumed that only Europeans were socially cosmopolitan, and other peoples were generally isolated and pristine – that Europeans had history, and others did not. Anthropologists now appreciate the difficulty of that assumption: other cultures are not "frozen in time", and other peoples are not "completely isolated" from one another, except in very extreme cases.[6]

Due to scientific and moral critique as well as due to political resistance, the start of the project was halted, and it was unable to enlist the support of the United Nations. Therefore, research had to be carried out on a much smaller scale than originally envisaged. Even though laboratories worldwide contributed cell lines from populations they had studied to the Centre d'Étude Polymorphism Humain in Paris, the success of the project was limited.[7]

The Genographic Project – A Perpetuation of the Human Genome Diversity Project

As a successor to the Human Genome Diversity Project – which apparently could not meet all the expectations of its initiators – the privately funded *Genographic Project* was initiated by the National Geographic Society, the IBM Corporation, and the Waitt Family Foundation in April 2005. One month before, another human genetics project, the HapMap Project, had completed its first phase. Like the Genographic Project, the HapMap Project reflects aspects of the Human Genome Diversity Project, but takes into account new findings of epigenetics by enlarging the scope of research to "DNA variants interacting with environmental factors"[8] (haplotypes) and restricts its samples to four populations (Yorùbá from Ibadan, Nigeria, US Americans from Utah, Han Chinese from Beijing, and Japanese from Tokyo). Since the scope of its research does not explicitly pertain to Indigenous peoples, the HapMap Project will not be given further consideration in this essay.

---

[6] Jonathan Marks, *Human Biodiversity. Genes, Race and History* (New York: Aldine de Gruyter 1995)

[7] MacIntosh, "Indigenous Self-Determination, and Research on Human Genetic Material," 214.

[8] The International HapMap Consortium, "A Haplotype Map of the Human Genome," *Nature* 437 (October 2005): 1299.

Reasons for regarding the Genographic Project as the direct successor to the HGDP can be found on a personal as well as on a scientific level: again, the project's goal is a long-range study in mapping historical human migration patterns by gathering and analysing DNA samples from indigenous peoples, because

> DNA from indigenous populations contains key genetic markers that have remained unaltered for hundreds of generations. Reliable indicators of shared lineage, these markers can be used to trace the movement of humans across the globe.[9]

Another argument put forward by the Genographic Project is the endangered status of many Indigenous populations, which is regarded as a justification for immediate research rather than as a reason to take action in the defence of these peoples, linking the urgency of the project to cultural loss:

> Many indigenous populations around the world are facing strong challenges to their cultural identities. The Genographic Project will provide a "snapshot" of human genetic variation before we lose the cultural context necessary to make sense of the genetic data. (223)

Like the Human Genome Diversity Project before it, the Genographic Project met with strong protests from representatives of Indigenous Peoples and other non-governmental organizations. Representatives of the Indigenous Council on Biocolonialism feared that the project would treat Indigenous peoples as a mere "subject for scientific curiosity,"[10] and that it could endanger land rights and other benefits. In May 2006, the United Nations Permanent Forum on Indigenous Issues (UNPFII) recommended that the project be suspended.[11]

⌘

[9] MacIntosh, "Indigenous Self-Determination, and Research on Human Genetic Material," 213. Further page reference in the main text.

[10] Tina Butler, *Indigenous groups oppose National Geographic, IBM project* (9 May 2005), http://news.mongabay.com/2005/0509a-tina_butler.html (accessed 25 October, 2010): 3.

[11] Indigenous Peoples Council on Biocolonialism, *United Nations Recommends Halt to Genographic Project* (26 May 2006), http://ipcb.org/issues/human_genetics/htmls/unpfii_rec.html (accessed October 4 2010).

Biomapping and 'Global Phylogeny': The Reinvention of "Racial History"?[12]

Both the Human Genome Diversity Project and the Genographic Project reflect the attempt of geneticists to investigate the full range of human biodiversity. They place special emphasis on Indigenous populations, considering them to preserve genomes that have remained unaltered for many generations, or even seeing Indigenous peoples as 'genetic isolates'. Since the respective research programmes largely elaborate on the work of Luigi Luca Cavalli–Sforza and his collaborators, a closer look at the scientific premisses of their 'new synthesis' of genetic, linguistic, and archaeological data will be presented. It will be shown that these premisses perpetuate primordialist paradigms that evolved in the biology, linguistics, and anthropology of the nineteenth century.

Cavalli–Sforza and his collaborators gradually shifted from a 'racial' typology of humans based on anthropometric data to a classification based on human genotypes.[13] In a famous and since then widely distributed scheme of 1988, Cavalli–Sforza et al. presented a direct correlation between genetic clusters and linguistic macro-phyla of humankind.[14] This so-called 'new synthesis' was based on a long-range linguistic comparison and also involved archaeological data.[15] The resulting model presented a genetic differentiation

---

[12] Jonathan Marks, *What is the Background of the Human Genome Project?*, http://personal.uncc.edu/jmarks/hgdp/hgdp1.html (accessed 4 October 2010): 1.

[13] Luigi Luca Cavalli–Sforza & Walter F. Bodmer, *The Genetics of Human Populations* (San Francisco: W.H. Freeman, 1971).

[14] Luigi Luca Cavalli–Sforza, Alberto Piazza, Paolo Menozzi & Joanna Mountain, "Reconstruction of Human Evolution: Bringing Together Genetic, Archaeological, and Linguistic Data," *Proceedings of the National Academy of Sciences USA* 85 (August 1988): 6002–6006. Luigi Luca Cavalli–Sforza, Paolo Menozzi & Alberto Piazza, *The History and Geography of Human Genes* (Princeton NJ: Princeton UP, 1994).

[15] Joseph H. Greenberg, *Language in the Americas* (Stanford: Stanford UP, 1987). Joseph H. Greenberg & Merritt Ruhlen, *An Amerind Etymological Dictionary* (Department of Anthropological Sciences, Stanford CA: Stanford U, Version 12: 4 September 2007), http://www.merrittruhlen.com/files/AED5.pdf (accessed 29 September 2010). Merrit Ruhlen, *A Guide to the World's Languages* (Stanford: Stanford UP, 1987). Sergei A. Starostin, "Nostratic and Sino-Caucasian," in *Explorations in Language Macrofamilies*, ed. Vitaly Shevoroshkin (Bochum: Brockmeyer, 1989): 42–66. Aharon Dolgopolsky, *The Nostratic Macrofamily and Linguistic Palaeontology* (Oxford: McDonald Institute for Archaeological Research, 1988) Colin Renfrew,

of mankind along bifurcating trees, calculated time horizons for the respective bifurcations, and related these splits to prehistoric migrations.

Indigenous Peoples as Fossils of Human Co-Evolution?

In their 1988 paper, Cavalli–Sforza et al. gave the reason for focusing on presumably 'genetically pure' populations, and, for the correlation of genes and languages in Indigenous peoples, they took the following for granted:

> Languages evolve more rapidly than genes. They can also undergo rapid replacements, even if the new language is imposed by an invading minority, providing this minority has adequate political and military organization, as in the "elite dominance" model (Renfrew 1987). When this happens it may be difficult to find genetic traces of the invasion. Elites have developed only recently, however, rarely being older than 5000 years, and therefore episodes of rapid language replacement are relatively recent and often accounted for historically. In the more remote past, replacement was more rare, justifying the stability of the relation between linguistic phyla and genetic clusters.[16]

Thus, Cavalli–Sforza focused on a primordialist conception of ethnicity, claiming that communities of indigenous peoples have been genetically and linguistically stable for hundreds or even thousands of years, and generally tended to neglect cross-cultural interaction, gene flow, and linguistic borrowings between indigenous peoples. We can see that the 'new synthesis' perpetuates two old myths: (1) the idea of biological and linguistic co-evolution; and (2) the ethnographical myth of pristine 'natives' living in isolated communities, representing genetic 'purity'.[17]

Other assumptions in Cavalli–Sforza's work have also been criticized.[18] Apart from the fact that his definition of 'aboriginal peoples' remains rather vague ("no major migration prior to 1492"), at least five of the forty-two "aboriginal populations" studied in Cavalli–Sforza's early works (1988–94)

Archaeology and Language: The Puzzle of Indo-European Origins (Cambridge: Cambridge UP, 1987).

[16] Cavalli–Sforza et al., "Reconstruction of Human Evolution," 6005.

[17] Eric R. Wolf, *Europe and the People without History* (Berkeley: U of California P, 1982).

[18] April & Robert McMahon, *Language Classification by Numbers* (Oxford: Oxford UP, 2005): 119.

were first identified according to linguistic criteria (Nilotic, Dravidian, Uralic, northern Turkic, and southern Na-Dene groups) – thereafter, a direct correlation between genes and languages was claimed. Although Cavalli–Sforza is prepared to admit the shortcomings of his model and although even he himself points out contradictions and limitations, listing exceptions to a full correlation between the genetic tree and the classification of languages by families, this does not prevent him from claiming that a general correlation between genetic, linguistic, and archaeological data *does* exist.[19]

Historical and Theoretical Considerations: The Legacy of Primordialism

After having shown that late-twentieth-century and early-twenty-first-century biomapping projects focused on indigenous peoples' genes because it was assumed that (1) these peoples represent 'pristine societies of the past', and (2) that they possess genetic traits distinguishing them from dominant, majority populations, I shall now argue that these two viewpoints closely resemble notions of 'racial deviance' and of 'primitive cultures' that have a centuries-long tradition in the European perception of Indigenous peoples.

The tradition of identifying linguistic with 'racial' affinities of a human population is deeply rooted in a "neo-romantic concept"[20] defining *Volk* through a combination of language, 'race', and culture.[21] Such primordialist views of ethnicity were closely connected with evolutionist attitudes:

> In those theoretical frameworks strongly influenced by evolutionism, ethnicity is usually conceptualized as based in biology and determined by genetic and geographical factors.[22]

In the constant interdisciplinary contact between scholars of the humanities and their peers in the sciences, a close link was developed between linguistic

---

[19] Cavalli–Sforza et al., *The History and Geography of Human Genes*, 93–105.

[20] Sergey Sokolovskii & Valery Tishkov, "Ethnicity," in *Encyclopedia of Social and Cultural Anthropology,* ed. Alan Barnard & Jonathan Spencer (London & New York: Routledge, 1996): 191.

[21] Johann Gottfried Fichte, *Reden an die Deutsche Nation* (Munich: Bayerische Akademie der Wissenschaft, 1807/1808). Wilhelm. Friedrich Schlegel, *Ueber die Sprache und Weisheit der Indier. Ein Beitrag zur Begründung der Alterthumskunde* (Cologne & Heidelberg: Mohr & Zimmer, 1808). Johann Gottfried Herder, *Abhandlung über den Ursprung der Sprache* (Berlin: Voss, 1772).

[22] Sokolovskii & Tishkov, "Ethniciity," 190–91.

and 'racial' classifications,[23] and various systems of classifying the major 'races' of mankind were established from the eighteenth century onwards.[24] At that time, criteria of classification were confined to anthropometric data and human phenotypes.

Even before the classification of human 'races' became widespread, languages had been classified according to a 'genetic' model of descent from the seventeenth century onwards. Already in 1647, the Dutch scholar Marcus Zuerius van Boxhorn had identified the family of languages nowadays called Indo-European and formulated a theory of a "family of genetically related languages deriving from a common ancestral language and distinct from other linguistic families."[25] In the eighteenth century, language families such as Uralic (including Finno-Ugric) and Indo-European were established.[26] In the nineteenth century, these families of 'genetically' related languages were increasingly depicted in either a monophyletic or a polyphyletic model of the 'language tree'.[27] When the newly developed 'science' of comparative histori-

[23] Ruth Römer, *Sprachwissenschaft und Rassenideologie in Deutschland* (Munich: Wilhelm Fink, 1989).

[24] Alice M. Brues, *People and Races* (New York. Macmillan, 1977). Carolus Linnaeus, *Systemae Naturae* (Leiden: De Groot, 1735). Johann F. Blumenbach, *De generis humani varietate nativa liber* (Göttingen: Rosenbusch, 1775).

[25] George van Driem, "Sino-Austronesian vs. Sino-Caucasian, Sino-Bodic vs. Sino-Tibetan, and Tibeto-Burman as default theory," in *Contemporary Issues in Nepalese Linguistics*, ed. Yogendra Prasada, Govinda Bhattarai et al. (Kathmandu: Linguistic Society of Nepal, 2005): 285–338.

[26] Philip Johan von Strahlenberg, *Das Nord- und Östliche Theil von Europa und Asia, in so weit solches das gantze Russische Reich mit Sibirien und der grossen Tatarey in sich begriffet* (Stockholm: The author, 1730). Janos Sajnovics, *Demonstratio idioma Ungarorum et Lapponum idem esse* (Copenhagen, 1770). William Jones, "The Third Anniversary Discourse delivered 2 February, 1786, by the President, at the Asiatick Society of Bengal," *Asiatic Researches* (1788): 415–31.

[27] Sylvain Auroux, "Representation and the Place of Linguistic Change before Comparative Grammar," in *Leibniz, Humboldt, and the Origins of Comparativism*, ed. Tullio De Mauro & Lia Formigaria (Amsterdam: John Benjamins, 1990): 231–38. August Schleicher, "Die ersten Spaltungen des indogermanischen Urvolkes," *Allgemeine Monatsschrift für Wissenschaft und Literatur* (1853): 786–87/101–102. August Schleicher, *Compendium der vergleichenden Grammatik der indogermanischen Sprachen: Kurzer Abriss der indogermanischen Ursprache, des Altindischen,*

cal linguistics developed laws of regular sound correspondences and successfully reconstructed a number of proto-languages,[28] the perception of language as an indicator of race became an increasingly established viewpoint.[29] A primordialist definition of *Volk* served as a prerequisite for correlating and equating linguistic and 'racial' classifications of humans.[30]

All these eighteenth- and nineteenth-century 'racial' and linguistic classifications envisaged more than a linguistic taxonomy of languages or biological taxonomy of phenotypes. They also established a hierarchical scheme that intended to show the development from 'lower' to 'higher races', from 'inferior" to 'superior' languages. With the application of Darwin's theory of evolution to man,[31] the human species tended to be seen as part of a broader evolutionary process. Following August Schleicher's famous *Stammbaum* or 'tree of languages' (1853),[32] frequent attempts were made to match language trees with the development of human biological and cultural diversity. Darwin himself points repratedly to similarities between established theories of language change and his theory of descent with modification:

> The natural system is genealogical in its arrangement, like a pedigree. It may be worthwhile to illustrate this view by taking the case of languages. If we possessed a perfect pedigree of mankind, a genealogical arrangement of the races of man would afford the best classification of the various languages now spoken throughout the world.[33]

Altiranischen, Altgriechischen, Altitalischen, Altkeltischen, Altslawischen, Litauischen und Altdeutschen (Weimar: Böhlau, 1861/62).

[28] Franz Bopp, *Über das Conjugationssystem der Sanskritsprache in Vergleichung mit jenem der griechischen, lateinischen, persischen und germanischen Sprache* (Frankfurt am Main: Andreäsche Buchhandlung, 1816). Max F. Müller, *Languages of the Seat of War in the East, with a Survey of the Three Families of Language, Semitic, Arian, and Turanian* (London: Williams & Norgate, 1855).

[29] Joseph Arthur de Gobineau, *L'essai sur l'inégalité des races humaines* (Paris: Firmin–Didot, 1853/55). Gustave Le Bon, *Lois psychologiques de l'évolution des peuples* (Paris: Félix Alcan, 1894). Georges Vacher de Lapouge, *L'aryen et son rôle social* (Paris: Payot, 1899).

[30] Schlegel, *Ueber die Sprache und Weisheit der Indier*, 60–70.

[31] Charles Darwin, *The Descent of Man, and Selection in Relation to Sex* (London: John Murray, 1871).

[32] Schleicher, "Die ersten Spaltungen des indogermanischen Urvolkes," 786/102.

[33] Charles Darwin, *On the Origin of Species by Means of Natural Selection* (London: John Murray, 5th ed. 1869): 342.

Thus, we can assume that equating pedigrees of human languages with pedigrees of human races seems to have been firmly established by the mid-nineteenth century. This claim is exemplified by the strong personal and professional ties between the biologist Ernst Haeckel (1834–1919) – who was Darwin's foremost popularizer in the German-speaking areas – and the linguists William Bleek (1827–75) and August Schleicher (1821–68). In his famous *Sendschreiben an Herrn Dr. Ernst Häckel* (1863), Schleicher advanced the view that the history of language development constitutes one of the main aspects of the history of human development and that languages can be compared to natural organisms which develop without human interference. In Schleicher's view, languages could be directly equated with biological species, dialects with subspecies, and families of languages with families of plants or animals.[34] According to Haeckel, Schleicher, and Bleek, both the evolution of human languages and that of human 'races' led from lower to higher stages on an evolutionary ladder, with inferior or superior positions being assigned to certain populations.[35] Thus, by the second half of the nineteenth century, two major strands of thought had developed, both closely bound up with a colonialist ideology:

- —The idea that races and languages were intrinsically bound to each other and human populations constituted fixed entities which could by characterized and identified by criteria of race and language alike. Thus, ethnicity was determined by language, 'race', and culture, or, in more modern terms: both culturally and genetically bound.
- —The idea that a hierarchy of races, peoples, and languages existed which could either be derived from the plan set out by divine creation or which – in a popularized version of Darwinian thought – had developed due to the mechanism of natural selection and the survival of the fittest – principles which were not only applied to plants and animals but to human populations alike. In accordance with colonialist thinking and the idea of white supremacy, racial superiority or inferiority was linked to the assumed position of a particular population in the evolutionary tree of primates and this population's distance from other, non-hominid primates.

---

[34] Schleicher, *Die Darwinsche Theorie und die Sprachwissenschaft – Offenes Sendschreiben an Herrn Dr. Ernst Haeckel* (Weimar: H. Böhlau, 3rd ed. 1873): 13.

[35] Wilhelm H.I. Bleek, *Über den Ursprung der Sprache*, intro. Ernst Haeckel (Weimar: H. Böhlau, 1868): iii–viii.

Thus, the classification of human populations in physical anthropology as well as in linguistics had come to imply a hierarchical, evolutionary order – a basic idea that was also advocated by the emerging academic discipline of cultural and social anthropology.[36]

The Decline and Re-Emergence of Evolutionist Thinking

I shall now focus on the idea of linguistic and 'racial' co-evolution is rooted in a primordialist perception of ethnicity and that this was well-developed in European-based thought of the nineteenth century, attention shall now be devoted to the backlash undergone by evolutionist thinking in both the sciences and the humanities at the beginning of the twentieth century, when evolutionism suffered a severe decline, not only in biology[37] but also in sociology, linguistics, and anthropology.[38] In cultural anthropology, the idea of primordialism was fiercely criticized by the emerging cultural relativists and particularists fostered by Franz Boas and his disciples. Thee latter emphatically rejected speculative ideas of cultural evolution and claimed that culture developed independently of biological characteristics of human populations, with culture, 'race', and language constituting mutually independent and unrelated determinants of human existence.[39]

---

[36] Edward B. Tylor, *Primitive Culture: Researches into the Development of Mythology, Philosophy, Religion, Language, Art, and Custom* (London: John Murray, 1871). Lewis H. Morgan, *Ancient Society, or: Researches in the Lines of Human Progress from Savagery through Barbarism to Civilization* (New York: Henry Holt, 1877). Friedrich Engels, *Der Ursprung der Familie, des Privateigentums und des Staats: Im Anschluß an Lewis H. Morgans Forschungen* (Zürich: Schweizerische Volksbuchhandlung, 1884).

[37] Peter J. Bowler, *Evolution: The History of an Idea* (Berkeley & Los Angeles: U of California P, 1983).

[38] Émile Durkheim, *Les formes élémentaires de la vie religieuse* (Paris: Félix Alcan, 1912); Marcel Mauss, "L'ethnographie en France et à l'étranger," *Revue de Paris* 20.5 (1913): 537–60, 815–37; Ferdinand de Saussure, *Cours de linguistique générale* (Paris & Lausanne: Payot, 1916); Bronislaw Malinowski, *The Trobriand Islands* (London: Routledge & Kegan Paul, 1915); Alfred Radcliffe–Brown, *The Andaman Islands* (London: Routledge & Kegan Paul, 1922).

[39] Franz Boas, *Kultur und Rasse* (Berlin: de Gruyter, 1913). Franz Boas, *Race, Language, Culture* (New York: Macmillan 1940); Bernhard Streck, "Kulturanthropologie," in *Wörterbuch der Völkerkunde*, ed. Bernhard Streck (Wuppertal: Hammer, Edition Trickster, 2000): 142.

In the 1930s and 1940s, however, evolutionism enjoyed renewed interest in the sciences and the humanities. In biology, the modern evolutionary synthesis that combined the results of Mendelian population genetics with Darwinian evolutionary theory gained general acceptance.[40] When Leslie A. White (1900–75) and Julian H. Steward (1902–72) revived interest in evolutionism within American anthropology, they restricted their approach to cultural and social phenomena, not linking models of the development of cultures and societies in successive stages to the evolution of human linguistic and biological diversity.[41] Only in the last quarter of the twentieth century did ideas of co-evolution in human linguistic, cultural, and biological (by now termed 'genetic') diversity re-emerge in the shape of the new synthesis of genetic, linguistic, and archaeological data labelled 'global phylogeny'.

Evaluation of Global Phylogeny from an Historical Perspective

As indicated above, the premisses of the 'global phylogeny' model deny the meanwhile well-established fact that ethnicity is merely considered to represent a contemporary state of individual and collective awareness as opposed to a lifelong affiliation that is eventually passed on to the next generation. Constructivist anthropology has explicitly refuted the assumption that ethnic identity is based on the common biological origin of its representatives.[42] But according to the 'global phylogeny' model, ethnic groups tend to be seen as steady, inalterable groups of humans instead of as dynamic clusters constantly changing their affiliation and identity. Ethnicity is once again defined biologically – despite the fact that, dating back to the times of Franz Boas, language, 'race', and culture have been perceived as independently transmitted

---

[40] Theodosius Dobzhansky, *Genetics and the Origin of Species* (New York: Columbia UP, 1937); Julian Huxley, *Evolution: The Modern Synthesis* (London: Allen & Unwin, 1942).

[41] Henri J.M. Claessen, "Evolution and Evolutionism," in *Encyclopedia of Social and Cultural Anthropology*, ed. Alan Barnard & Jonathan Spencer (London & New York: Routledge, 1996): 213–18; Leslie A. White, *The Science of Culture: A Study of Man and Civilization* (New York: Farrar, Straus & Cudahy, 1949); Julian H. Steward, *Theory of Culture Change. The Methodology of Multilinear Evolution* (Urbana: U of Illinois P, 1955); Marshall D. Sahlins & Elman R. Service, *Evolution and Culture* (Ann Arbor: U of Michigan P, 1960).

[42] Frederik Barth, *Ethnic Groups and Boundaries: The Social Organization of Culture Difference* (Oslo: Universitetsforlaget, 1969).

replicators of human existence. Nevertheless, geneticists still tend to insist on the 'purity' of ethnic groups – for example, in the case of the Diné (Navajo), who have a history of extensive migration and inter-ethnic admixture with neighbouring Hopi, Ute, Walapai, Apache, and Pueblo groups.[43]

It can be argued that, after the 'racial' classification of humans was discredited in the aftermath of the Second World War[44] and a general 'turn to genetics' in the life sciences had been accomplished,[45] notions of 'race' have increasingly become replaced by notions of genes, DNA, or 'genetic clusters' of humankind. The new emphasis on genetics also promoted the construction of an intimate link between the physical and the linguistic traits of a given population by substituting traditional anthropometric data with genetic data. The 'genomization of ethnic identity' has effected a revival of an essentialist and primordialist base-line and perpetuated the old idea of the 'racial purity' of aboriginal groups, with Indigenous peoples being perceived as 'cultural and genetic fossils' living in a pristine state of isolation.

Contemporary Hazards for Indigenous Peoples Resulting from Biomapping

Apart from the fact that recent biomapping projects focusing on Indigenous peoples can be seen as a revival of nineteenth-century essentialist and primordialist base-lines that raise serious doubts about their academic credibility, the 'genomic era' may also set up serious obstacles to the struggle of Indigenous peoples for land rights as well as for their political and social participation in the framework of modern nation-states. The pressure to have their genes screened might seriously compromise ethnic groups with mixed ancestry – for example, the Seminoles of Florida, the Misquito of Nicaragua, and the Garífuna or Black Caribs of Belize, Guatemala, and Honduras, who all share an Afro-Amerind heritage. Testifying to their genetic purity is also especially difficult for groups like the Haudenosaunee (Iroquois) who have obviously been subject to intense genetic admixture with Euro-Americans since early

---

[43] Antonio Torroni et al., "Native American Mitochondrial DNA Analysis indicates that the Amerind and the Nadene Populations were founded by two Independent Migrations," *Genetics* 130 (1992): 139–52; Marks, *Human Biodiversity,* 177.

[44] United Nations Educational, Scientific and Cultural Organization. *The Race Concept: Results of an Inquiry* (Westport CT: UNESCO, 1952).

[45] Richard Lewontin, *Biology as Ideology: The Doctrine of DNA* (New York: HarperCollins, 1991).

colonial times and have extensively adopted members of other foreign ethnic groups – a practice that was common throughout North America.[46] In the case of the Lubicon Cree of northern Alberta, one argument put forward against their land claim was that their settlement in their presently occupied territory did not antecede precolonial times.[47] According to this viewpoint, ethnic groups whose members are considered either to be genetically mixed or to be recent migrants to their present area of settlement are not deemed eligible for Indigenous status.

Another hazard that Indigenous peoples might increasingly have to face is that the claim to ethnically defined rights, for example Scheduled Tribe Status in India, might increasingly depend on the ability to provide the 'right genome frequency'. The 'genomization of ethnicity' might also be misused in the course of ethnic and nationalist mobilization or in favour of national homogenization strategies. In both cases, the right of self-determination in identifying with an ethnic group or Native nation could be severely undermined and replaced by an imposed, biologically based definition of ethnicity.

Conclusion

When the 'genomic era' began increasingly to affect Indigenous peoples from the 1980s onwards, two old paradigms dating back to the nineteenth century were revived: (1) the idea of the co-evolution of languages, 'races' (either defined by phenotype or by genotype), cultures, and social organization;[48] and (2) the idea that Indigenous peoples preserved the 'aboriginal' features of mankind – in terms of genomes or indigenous knowledge – that had been lost in the course of history by the 'more advanced', demographically dominant, and socially stratified cultures and societies and can be used for scientific and medical purposes (both the Human Genome Diversity Project and the HapMap Project have strong medical implications). The emphasis on genomics can easily lead to the idea that ethnic identity is predominantly genetically deter-

---

[46] Francis Jennings, *The Ambiguous Iroquois Empire: The Covenant Chain Confederation of Indian Tribes with English Colonies from Its Beginnings to the Lancaster Treaty of 1744* (New York: W.W. Norton, 1984): 95.

[47] John Goddard, *The Last Stand of the Lubicon Cree* (Vancouver: Douglas & McIntyre, 1991); James G.E. Smith, "Chipewyan, Cree and Inuit Relations West of Hudson Bay, 1714–1955," *Ethnohistory* 28.2 (Spring 1981): 133–56.

[48] For a recent example, see Quentin D. Atkinson, "The prospects for tracing deep language ancestry," *Journal of Anthropological Sciences* 88 (2010): 231–33.

mined and exert pressure on individuals as well as collective groups to justify their cultural and social status genetically. When the 'ideology of the gene' is applied to non-Western, Indigenous peoples, primordialist thinking of the nineteenth century that was centred on 'race' (both physically and culturally defined) is perpetuated. Concomitantly, modern attempts in biomapping neglect not only the modern constructivist perception of ethnicity[49] but also the struggle for cultural and often even physical survival in which Indigenous peoples are still engaged in many parts of the world. Instead, the world's aboriginal populations are reduced to objects of scientific interest – a state of affairs that gives sufficient reason for applying the term 'genomic bio-colonialism' to recent research programmes attempting to screen Indigenous peoples' genes.

WORKS CITED

Anon. "The Genographic Project. Frequently Asked Questions," https://genographic.national-geographic.com/genographic/lan/en/faqs_about.html (accessed 25 October 2010).

Atkinson, Quentin D. "The prospects for tracing deep language ancestry," *Journal of Anthropological Sciences* 88 (2010): 231–33.

Auroux, Sylvain. "Representation and the Place of Linguistic Change before Comparative Grammar," in *Leibniz, Humboldt, and the Origins of Comparativism*, ed. Tullio de Mauro & Lia Formigaria (Amsterdam: John Benjamins, 1990): 231–38.

Barth, Frederik. *Ethnic Groups and Boundaries: The Social Organization of Culture Difference* (Oslo: Universitetsforlaget, 1969).

Bleek, Wilhelm Heinrich Immanuel. *Über den Ursprung der Sprache*, intro. Ernst Haeckel (Weimar: H. Böhlau, 1868).

Blumenbach, Johann Friedrich. *De generis humani varietate nativa liber* (Göttingen: Rosenbusch, 1775).

Boas, Franz. *Kultur und Rasse* (Berlin: De Gruyter, 1913).

——. *Race, Language, Culture* (New York: Macmillan, 1940)

Bopp, Franz. *Über das Conjugationssystem der Sanskritsprache in Vergleichung mit jenem der griechischen, lateinischen, persischen und germanischen Sprache* (Frankfurt am Main: Andreäsche Buchhandlung, 1816).

Bowler, Peter J. *Evolution: The History of an Idea* (Berkeley & Los Angeles: U of California P, 1983).

Brues, Alice Mossie. *People and Races* (New York: Macmillan, 1977).

---

[49] Barth, *Ethnic Groups and Boundaries* (1969).

Butler, Tina. *Indigenous groups oppose National Geographic, IBM project* (9 May 2005), http://news.mongabay.com/2005/0509a-tina_butler.html (accessed 25 October, 2010): 3.

Cavalli–Sforza, Luigi Luca, & Walter F. Bodmer. *The Genetics of Human Populations* (San Francisco: W.H. Freeman, 1971).

——, Alberto Piazza, Paolo Menozzi & Joanna Mountain. "Reconstruction of Human Evolution: Bringing Together Genetic, Archaeological, and Linguistic Data," *Proceedings of the National Academy of Sciences USA* 85 (August 1988): 6002–6003.

——, Allan C. Wilson, Charles R. Cantor, Robert M. Cook–Deegan & Mary–Claire King. "Call for a Worldwide Survey of Human Genetic Diversity: A Vanishing Opportunity for the Human Genome Project," *Genomics* 11 (1991): 490–91.

——, Paolo Menozzi & Alberto Piazza. *The History and Geography of Human Genes* (Princeton NJ: Princeton UP, 1994).

Claessen, Henri J.M. "Evolution and Evolutionism," in *Encyclopedia of Social and Cultural Anthropology*, ed. Alan Barnard & Jonathan Spencer (London & New York: Routledge, 1996): 213–18.

Cobo, José–Martínes. *Discrimination against Indigenous People* (UN Nations Document No. E/CN.4/Sub.2, 1986).

Daes, Erica–Irene A. *Working Paper by the Chairperson–Rapporteur, Mrs. Erica–Irene A. Daes, on the Concept of Indigenous People* (UN Document E/CN.4/Sub.2/AC.4, 1996): 2.

Darwin, Charles. *The Descent of Man, and Selection in Relation to Sex* (London: John Murray, 1871).

——. *On the Origin of Species by Means of Natural Selection, or the Preservation of Favoured Races in the Struggle for Life* (London: John Murray, 5th ed. 1869).

Dobzhansky, Theodosius. *Genetics and the Origin of Species* (New York: Columbia UP, 1937).

Dolgopolsky, Aharon. *The Nostratic Macrofamily and Linguistic Palaeontology* (Oxford: McDonald Institute for Archaeological Research, 1988).

Driem, George van. "Sino-Austronesian vs. Sino-Caucasian, Sino-Bodic vs. Sino-Tibetan, and Tibeto-Burman as Default Theory," in *Contemporary Issues in Nepalese Linguistics*, ed. Yogendra Prasada, Govinda Bhattarai, Ram Raj Lohani, Balaram Prasain & Krishna Parajuli (Kathmandu: Linguistic Society of Nepal, 2005): 285–338.

Durkheim, Émile. *Les formes élémentaires de la vie religieuse* (Paris: Félix Alcan, 1912).

Engels, Friedrich. *Der Ursprung der Familie, des Privateigentums und des Staats: Im Anschluß an Lewis H. Morgans Forschungen* (Zurich: Schweizerische Volksbuchhandlung, 1884).

Fichte, Johann Gottfried. *Reden an die Deutsche Nation* (Munich: Bayerische Akademie der Wissenschaft, 1807/1808).

Frazer, James G. *The Golden Bough: A Study in Magic and Religion* (London: Macmillan, 1890).

Gobineau, Joseph Arthur de. *L'essai sur l'inégalité des races humaines* (Paris: Firmin–Didot, 1853/55).

Goddard, John. *The Last Stand of the Lubicon Cree* (Vancouver: Douglas & McIntyre, 1991).

Greenberg, Joseph Harold. *Language in the Americas* (Stanford CA: Stanford UP, 1987).

——, & Merritt Ruhlen. *An Amerind Etymological Dictionary* (Stanford CA: Department of Anthropological Sciences, Stanford University, Version 12: 4 September, 2007), http://www.merrittruhlen.com/files/AED5.pdf (accessed 29 September 2010).

Herder, Johann Gottfried. *Abhandlung über den Ursprung der Sprache* (Berlin: Voss, 1772).

Huxley, Julian. *Evolution: The Modern Synthesis* (London: Allen & Unwin, 1942).

Indigenous Peoples Council on Biocolonialism. *United Nations Recommends Halt to Genographic Project* (26 May 2006), http://ipcb.org/issues/human_genetics/htmls/unpfii_rec.html (accessed 4 October 2010).

The International HapMap Consortium. "A Haplotype Map of the Human Genome," *Nature* 437 (October 2005): 1299–1320.

Jennings, Francis. *The Ambiguous Iroquois Empire: The Covenant Chain Confederation of Indian Tribes with English Colonies from Its Beginnings to the Lancaster Treaty of 1744* (New York: W.W. Norton, 1984).

Jones, William. "The Third Anniversary Discourse delivered 2 February, 1786, by the President, at the Asiatick Society of Bengal," *Asiatic Researches* (1788): 415–31.

Lapouge, Georges Vacher de. *L'aryen et son rôle social* (Paris: Payot, 1899).

Le Bon, Gustave. *Lois psychologiques de l'évolution des peuples* (Paris: Félix Alcan, 1894).

Lewontin, Richard. *Biology as Ideology: The Doctrine of DNA* (New York: HarperCollins, 1991).

Linnaeus, Carolus. *Systemae Naturae* (Leiden: De Groot, 1735).

MacIntosh, Constance. "Indigenous Self-Determination, and Research on Human Genetic Material: A Consideration of the Relevance of Debates on Patents and Informed Consent, and the Political Demands on Researchers," *Health Law Journal* 13 (2005): 213–51.

McMahon, April, & Robert McMahon. *Language Classification by Numbers* (Oxford: Oxford UP, 2005).

Malinowski, Bronislaw. *The Trobriand Islands* (London: Routledge & Kegan Paul, 1915).

Marks, Jonathan. *Human Biodiversity. Genes, Race, and History* (New York: Aldine de Gruyter, 1995).

——. *What is the Background of the Human Genome Diversity Project?*, http: //personal.uncc.edu/jmarks/hgdp/hgdp1.html (accessed 4 October 2010).

Mauss, Marcel. "L'ethnographie en France et à l'étranger," *Revue de Paris* 20.5 (1913): 537–60, 815–37.

Mayr, Ernst. *Systematics and the Origin of Species* (New York: Columbia UP, 1942).

Morgan, Lewis Henry. *Ancient Society, or: Researches in the Lines of Human Progress from Savagery through Barbarism to Civilization* (New York: Henry Holt, 1877).

Müller, Max Friedrich. *Languages of the Seat of War in the East, with a Survey of the Three Families of Language, Semitic, Arian, and Turanian* (London: Williams & Norgate, 1855).

Radcliffe–Brown, Alfred. *The Andaman Islands* (London: Routledge & Kegan Paul, 1922).

Renfrew, Colin. *Archaeology and Language: The Puzzle of Indo-European Origins* (Cambridge: Cambridge UP, 1987).

Rimmer, Matthew. "The Genographic Project: Traditional Knowledge and Population Genetics," *Indigenous Law Review* 11.2 (2007): 33–54.

Römer, Ruth. *Sprachwissenschaft und Rassenideologie in Deutschland* (Munich: Wilhelm Fink, 1989).

Ruhlen, Merrit. *A Guide to the World's Languages* (Stanford CA: Stanford UP, 1987).

Sahlins, Marshall D., & Elman R. Service. *Evolution and Culture* (Ann Arbor: U of Michigan P, 1960).

Sajnovics, Janos. *Demonstratio idioma Ungarorum et Lapponum idem esse* (Copenhagen, 1770).

Saussure, Ferdinand de. *Cours de linguistique générale* (Paris & Lausanne: Payot, 1916).

Schlegel, Karl Wilhelm Friedrich von. *Ueber die Sprache und Weisheit der Indier. Ein Beitrag zur Begründung der Alterthumskunde* (Cologne & Heidelberg: Mohr & Zimmer, 1808).

Schleicher, August. "Die ersten Spaltungen des indogermanischen Urvolkes,“ *Allgemeine Monatsschrift für Wissenschaft und Literatur* (1853): 786–87/101–102.

——. *Compendium der vergleichenden Grammatik der indogermanischen Sprachen. Kurzer Abriss der indogermanischen Ursprache, des Altindischen, Altiranischen, Altgriechischen, Altitalischen, Altkeltischen, Altslawischen, Litauischen und Altdeutschen* (Weimar: H. Böhlau, 1861/62).

——. *Die Darwinsche Theorie und die Sprachwissenschaft – offenes Sendschreiben an Herrn Dr. Ernst Haeckel* (Weimar: H. Böhlau, 3rd ed. 1873).

Smith, James G.E. "Chipewyan, Cree and Inuit Relations West of Hudson Bay, 1714–1955," *Ethnohistory* 28.2 (Spring 1981): 133–56.

Sokolovskii, Sergey, & Valery Tishkov. "Ethnicity," in *Encyclopedia of Social and Cultural Anthropology*, ed. Alan Barnard & Jonathan Spencer (London & New York: Routledge, 1996): 190–91.

Starostin, Sergei Anatolevich. "Nostratic and Sino-Caucasian," in *Explorations in Language Macrofamilies*, ed. Vitaly Shevoroshkin (Bochum: Brockmeyer, 1989): 42–66.

Steward, Julian Haynes. *Theory of Culture Change: The Methodology of Multilinear Evolution* (Urbana: U of Illinois P, 1955).

Strahlenberg, Philip Johan von. *Das Nord- und Östliche Theil von Europa und Asia, in so weit solches das gantze Russische Reich mit Sibirien und der grossen Tatarey in sich begriffet* (Stockholm: The author, 1730).

Streck, Bernhard. "Kulturanthropologie," in *Wörterbuch der Völkerkunde*, ed. Bernhard Streck (Wuppertal: Hammer, Edition Trickster, 2000): 141–44.

Torroni, Antonio, Theodore G. Schurr, Chi-Chuan Yang, Emoke J.E. Szathmary, Robert C. Williams, Moses S. Schanfield, Gary A. Troup & William C. Knowler. "Native American Mitochondrial DNA Analysis Indicates that the Amerind and the Nadene Populations were Founded by two Independent Migrations," *Genetics* 130 (1992): 139–52.

Tylor, Edward B. *Primitive Culture: Researches into the Development of Mythology, Philosophy, Religion, Language, Art, and Custom* (London: John Murray, 1871).

United Nations Educational, Scientific and Cultural Organization, *The Race Concept: Results of an Inquiry* (Westport CT: UNESCO, 1952).

White, Leslie Alvin. *The Science of Culture: A Study of Man and Civilisation* (New York: Farrar, Straus & Cudahy, 1949).

Wolf, Eric R. *Europe and the People without History* (Berkeley: U of California P, 1982).

⌘

No Matter How White or Black the Skin, How Pure the Blood

Cherokee Identity and the 2007 Vote

SÉVERINE GAUTHIER–LABOUROT

NATIVE AMERICAN IDENTITY HAS ALWAYS BEEN A VERY DIVISIVE ISSUE and has come to be inextricably linked with the notions of race and blood. The questions raised imply that a high native blood quantum correlates with cultural authenticity or ethnic identity. Similarly, race-mixing has been inevitably associated with cultural loss. Today there is much talk among Indian tribes about how science and DNA can determine tribal enrolment and prove Indian ancestry, as DNA testing companies offer their services to Indian tribes.[1] If some tribal leaders remain sceptical and express their concern at seeing the complex nature of Indian identity reduced to a DNA test and a strictly biological connection, others express interest in DNA testing at a time when Indian tribes struggle to control identities and tribal resources. Throughout the Cherokee tribe's history, the issue has proven particularly divisive. The Cherokees have long been reluctant to consider blood a valid criterion for tribal identification, choosing instead, in their 1975 constitution, to rely on genealogy.

This increased the number of people identified as Cherokees from fewer than 10,000 in the late 1950s to more than 300,000 today. This was seen by many as diluting what was left of 'real' Cherokee heritage. To the Cherokees, it meant the exact opposite. It meant preserving the Cherokee people and the ways of the traditionally inclusive Cherokee society. Over centuries, attacks

---

[1] 'Indian' is used in the sense commonly employed in the USA to refer to First Nations peoples.

on their indigenousness have led them to redefine their tribal identity several times in order to survive as a people. These attacks eventually got the better of the Cherokees' cultural definition of their tribal identity and led them to adopt in 2007 more radical requirements for tribal membership. The heated debate in 2007 over the inclusive nature of Cherokee identity reveals just how complex and intricate Indian identity is.

Blood has not always been so meaningful to the Cherokees, and they first indigenized the white concepts of blood and race rather than fully embrace them. In the late-eighteenth and early-nineteenth centuries, hundreds of white men lived among the southeastern Indian tribes. Caleb Swan estimated that among a Creek population of "25,000 or 26,000 souls, [...] it may be conjectured with safety, that [...] the whites of every description throughout the country [...] amount to nearly 300 persons." Swan reported the presence of traders, saying that "every town and village has one established white trader in it," and adding that "each trader commonly employs one or two white packhorse men." He also talked of many white families having found refuge in "this asylum of liberty." Swan considered the presence of these "nearly 300 persons" sufficient to culturally "contaminate all the natives."[2] According to the Dutch naturalist Bernard Romans, the presence of white traders among Indian tribes considerably increased the number of intermarriages and thus the number of "half breeds."[3] In 1809, the Indian agent Return J. Meigs counted 341 whites married to Indian women in a Cherokee population of about 12,395 individuals.[4] At the beginning of the nineteenth century, the Mohawk John Norton reported the presence among the Cherokees of many Europeans and Indians from other tribes, and he stated in his journal that "there appears less regularity of features in [the Cherokees] than in the other Nations." He then explained that "this [...] may have proceeded from their warlike character and their universal custom of adopting in their own Nation

[2] Caleb Swan, "Position and State of Manners and Arts in the Creek, or Muskogee Nation in 1791," in Henry Rowe Schoolcraft, *Historical and Statistical Information Respecting the History, Condition, and Prospects of the Indian Tribes of the United States* (Philadelphia PA: Lippincott, Gambo, 1851–57), vol. 5: 279–81.

[3] Bernard Romans, *A Concise Natural History of East and West Florida*, ed. Kathryn E. Holland Braund (Tuscaloosa: U of Alabama P, 1999): 137.

[4] William G. McLoughlin & Walter H. Conser, Jr., "The Cherokee Censuses of 1809, 1825 and 1835," in McLoughlin, *The Cherokee Ghost Dance: Essays on the Southeastern Indians, 1789–1861* (Macon GA: Mercer UP, 1984): 218, 240.

the captive females and youths they had taken from hostile tribes."[5] Adoption of individuals or sometimes whole groups into the tribe seemed a well-established practice among the Cherokees. The naturalist William Bartram, who travelled the Cherokee country in the 1770s, wrote that "some of their young women are nearly as fair and blooming as European women."[6] Like Bartram, many white observers assumed that the Cherokees were a tribe of half-bloods, most individuals having white ancestors, and according to him, many traders gave them "the byname of The Breeds, supposing them to be mixed with the White People."[7] Many interpreted the presence of these foreigners, missionaries, government agents, settlers, traders, and adopted captives among the Indians as a factor of cultural loss and a disruption of Cherokee ways. To a white eye, the Cherokee people lacked homogeneity, and the presence of all these non-Cherokees blurred any clear conception of tribal identity. To a white mind, Cherokee social and political structure suggested a lack of cohesion. The Cherokees spoke different dialects and lived in five distinct settlements, each settlement representing a coalition of distinct towns, each town working as an independent political entity.[8] And yet they had a perception of their identity as a distinct people.

Cherokee identity was based on kinship. Being Cherokee meant belonging to one of the seven Cherokee matrilineal clans. Cherokees were members of a clan – and thus were Cherokee – only if they had a Cherokee mother. Kinship overrode racial considerations and only birth or adoption into a clan conferred tribal citizenship. An individual who did not belong to a clan had no ties in the community and was an enemy. Only adoption into a clan could turn an enemy into a member of the community. The adopted foreigner literally became Cherokee, regardless of his blood or race, two notions which were wholly alien to the Cherokees. Through the ceremony of adoption, the Cherokees imposed their ways on foreigners coming from the chaos outside the tribe. Adoption ensured the preservation of tribal harmony.

---

5 John Norton, *The Journal of Major John Norton, 1816,* ed. Carl Frederick Klinck & James John Talman (Toronto: Champlain Society, 1970): 134.

6 *William Bartram on the Southeastern Indians*, ed. Gregory A. Waselkov & Kathryn E. Holland Braund (Lincoln: U of Nebraska P, 1995): 111.

7 *William Bartram on the Southeastern Indians*, ed. Waselkov & Braund, 150.

8 James Mooney, *History, Myths and Sacred Formulas of the Cherokees* (Asheville NC: Bright Mountain, 1992): 15, 182.

If the adopted foreigners became Cherokees, the intermarried white traders did not, as only birth or adoption into a clan could confer Cherokee identity. In the eighteenth century, the Indian tribes enjoyed unquestionable political and cultural sovereignty on their land, and the chiefs made sure that the presence of the traders did not disrupt Cherokee ways. The traders were subject to the authority of the chiefs, who strictly controlled their activities and made sure they did not take advantage of the Indians. The Cherokees did not "suffer [the traders] to cultivate much land upon the supposition that if the traders raise produce themselves, they will not purchase the little they have to sell."[9] Respecting Cherokee ways allowed an honest trader to receive tribal protection. Dishonest traders had much to lose indeed. Bartram reported that some dishonest traders "have been ruined, their property seized and themselves driven out of the country or slain by the injured, provoked natives."[10] Traders and other foreigners, unable to cultivate the soil or raise cattle, depended on the Indians for food, and food was the tribal prerogative of Cherokee women. The only way a foreigner could get Indian land to cultivate was to marry a woman of the tribe. At the beginning of the eighteenth century, John Lawson noted that "the English traders are seldom without an Indian Female for his bed-fellow."[11] These traders and the other Europeans who intermarried into the tribe rapidly perceived the advantages of these unions. Their marriages created ties of friendship with the Cherokee clans of their Indian wives and they were protected and their businesses kept safe. Bartram wrote:

> there are but few instances of [Indian wives] neglecting or betraying the interests and views of their temporary husbands; they labor and watch constantly to promote their private interests, and detect and prevent any plots or evil designs which may threaten their persons or operate against their trade.[12]

But, even if the white husbands benefitted from these unions, they were subject to Cherokee ways. The Indian agent Benjamin Hawkins affirmed that "a white man marrying an Indian woman […] so far from bettering his condition

---

[9] Caleb Swan, "Position and State of Manners and Arts in the Creek, or Muskogee Nation in 1791," 282.

[10] *William Bartram on the Southeastern Indians*, ed. Waselkov & Braund, 79.

[11] John Lawson, *A New Voyage to Carolina,* ed. Hugh Talmage Lefler (Chapel Hill: U of North Carolina P, 1967): 35–36.

[12] *William Bartram on the Southeastern Indians*, ed. Waselkov & Braund, 55.

becomes a slave to her family."[13] The interest of their tribe and family came first. With these marriages, the Cherokees placed the foreigners under the control and social order of the tribe. The tribe also gained new spiritual powers when Indian women gave birth to the children of white men. The Indians considered the child Cherokee, and not white or even half-blood. In Cherokee culture it was the women who gave flesh and blood to the child, the father only the bones. The tie between the mother and the child was the only blood tie that existed for the Cherokees. In the matrilineal Cherokee society, intermarriages did not threaten the balance or the continuity of the tribe, because the children born of these unions were unquestionably Cherokees. On the contrary, with these marriages, the Cherokees asserted their superiority by turning the children of white men into Cherokees. Europeans, regarding their own culture as superior to Indian culture and considering the role of women in that culture as the sign of its superiority, could not understand a matrilineal society in which women were the basis of clan organization and guaranteed the survival of the tribe. To them, Indians were inferior to whites, and women inferior to men. They interpreted Cherokee society in terms of their own culture and came to regard it as a "wanton female government."[14]

But observers and historians, traders and government agents chose to focus on the children born of the unions of whites and Indians, on the clothes they wore, the houses they built, their conversion to Christianity, and the political and economic innovations they brought to the tribes. They tried to show how these children lived like whites and completely ignored the way they really were Indians. Gradually, for these observers, the presence of half-bloods in a tribe became an indication of its degree of acculturation or of civilization. The more half-bloods in a tribe, the more civilized the tribe. Assuming that the presence of intermarried whites and their mixed-blood offspring made a tribe more civilized and superior to any other tribe is problematical. This idea equates culture and blood, and links miscegenation and cultural loss, associating miscegenation with civilization, implying that European culture is superior to Native American culture, and thus that half-bloods behave, by virtue of their white blood, more like Europeans than Indians. Whites used this association to foster tension in Indian communities, offering privileges to some and

---

[13] Benjamin Hawkins, *Letters, Journals, and Writings of Benjamin Hawkins*, volume 1, ed. C.L. Grant (Savannah GA: Beehive, 1980): 239.

[14] James Adair, *Adair's History of the American Indians*, ed. Samuel Cole Williams (New York: Promontory, 1930): 133–35.

discrediting others. Native American societies were thus racialized in a way that is quite alien to their traditions.

The number of half-bloods in Southeastern tribes rapidly increased in the eighteenth and early-nineteenth centuries, and some of the half-blood children became important land-owners or even chiefs. Their success was naturally ascribed to their European blood, to the superiority of that blood over Indian blood, and to the influence of their European fathers. Intermarriages were at first seen as a way to civilize the Indians. William Byrd wrote that "a sprightly lover is the most prevailing missionary that can be sent amongst these."[15] Like him, the Indian agent Return J. Meigs calculated that "by this measure civilization is farther advanced than in any other way."[16] Science concurred, and it was commonly admitted that "the offspring of the European and the North American Indian [...] exceeds in sagacity, his white and [Indian] parent." In the 1820s, scientists thought that half-blood children were more intelligent than their Indian parent, but less so than their white parent. They also thought that these children were naturally attracted to European culture because they recognized its natural superiority. For Dr Caldwell, the only hope of civilizing the Indians was to "cross the breed."[17] At the time of the debate over the transfer of the tribes west of the Mississippi, new categories appeared in the hierarchy of races, and the half-bloods were deemed superior to the full-bloods but inferior to the whites. To the Cherokees, the children of white men and Cherokee women were unquestionably Cherokee. To them, there was no difference between a half-blood and a full-blood Cherokee. The missionaries established among them confirmed that although "a considerable part of the tribe are of mixed blood; yet all, who are partly Indian, are spoken of as Cherokees." When the Cherokees did use the word 'half-blood', they imparted to it a cultural meaning, not a biological one. They thus considered a member of the community a full-blood regardless of his racial identity or Indian blood quantum. To them, the full-bloods were the individuals who

---

[15] William Byrd, *The Westover Manuscripts: Containing the History of the Dividing Line Betwixt Virginia and North Carolina and a Journey to the Land of Eden, 1733* (Whitefish MT: Kessinger, Legacy Reprint, 2007): 3.

[16] Return J. Meigs, quoted in Theda Perdue, *Mixed Blood Indians: Racial Construction in the Early South* (Athens: U of Georgia P, 2003): 76.

[17] Charles Caldwell, quoted in Reginald Horsman, *Race and Manifest Destiny: The Origins of American Racial Anglo-Saxonism* (Cambridge MA: Harvard UP, 1981): 118.

were dedicated to the preservation of Cherokee ways: i.e. those who were culturally Cherokee. Similarly, a mixed-blood was a member of the tribe who had given up Cherokee ways to adopt various aspects of white culture. In the early-nineteenth century, the Cherokees indigenized the concepts of blood and race, and it was the attitude of an individual, not his race, that determined his identity.

Hawkins and Meigs assumed that the Cherokee women married to white men would soon adopt the superior white way of life of their husbands and would thus set an example to the rest of the tribe. But the culturally inferior Indians did not automatically give up their ways, customs, and beliefs to adopt the superior white culture. At the end of the eighteenth century, Bernard Romans observed that the Indians "would think themselves degraded in the lowest degree, were they to imitate us in any respect whatever," and that they "look down on us and all our manners with the highest contempt." The poor results of the civilization programme to culturally transform the Indians into American citizens troubled American politicians. It seemed that, if the superior white culture should have won out over the inferior Natives' culture, in reality the native people often had more influence on the whites. As Henry Schoolcraft explained in his report to the Secretary of the interior in 1855,

> time has, indeed, passed to the tribes who have kept themselves in the forest, as if it had no value. Three centuries have produced, apparently, no more effect than three years might be expected to do; and were Colombus [...] to return tomorrow he would be astonished to find the forest tribes so essentially like their forefathers at their eras.[18]

The Indians adopted only those aspects of civilization that seemed useful to them but rejected those which threatened to disrupt their traditional values and beliefs. After the American Revolution, the rapid growth of the American population increased the demand for more land. Converting the tribes to civilization and turning the Indian hunters into subsistence farmers promised the sale of thousands of acres of tribal hunting grounds. In 1803, Thomas Jefferson asked agent Hawkins "to promote among the Indians a sense of the superior value of little land, well cultivated, over a great deal, unimproved." But Jefferson refused to take the Indian land by force, convinced that it would be

---

[18] Henry Rowe Schoolcraft, *Historical and Statistical Information Respecting the History, Condition, and Prospects of the Indian Tribes of the United States* (Philadelphia PA: Lippincott, Gambo, 1851–57), vol. 5: 33.

"useless and even disadvantageous," still considering "the whole human race as brothers."[19]

But soon, faced with growing Indian resistance to American expansion, it became necessary for the government to discredit the tribes and to question their right to the land they claimed as their own. In 1775, Bernard Romans maintained that the Indians were a completely different race of men and that they were "incapable of civilization."[20] Many politicians and Americans generally shared Romans' opinion, in particular those living west of the Appalachians who succumbed to the rampant anti-Indian racism. Lewis Cass, Governor of the Michigan territory and Andrew Jackson's Secretary of War, wrote:

> there is a principle of repulsion in ceaseless activity operating through all their institutions, which prevents them from appreciating or adopting any other modes of life, or any other habits of thought or action, but those which have descended to them from their ancestors.[21]

The opinions of politicians were supported by science and phrenology, which ostensibly demonstrated that "the existing races of native American Indians show skulls inferior in the moral and intellectual development to those of the Anglo-Saxon race," the scientific conclusion being that, "morally and intellectually, these Indians are inferior to their Anglo-Saxon invaders, and have receded before them."[22] Science thus justified the political intentions of those who operated to transfer the eastern tribes west of the Mississippi. Phrenologists like George Combe or Dr Charles Caldwell were positive that "when the wolf, the buffalo and the panther shall have been completely domesticated, [...] then, and not before, may we expect to see the full-blooded Indian civilized."[23] Soon, the scientists convinced even the strongest supporters of the civilization of the Indians that their assimilation into white society was impossible. Once scientifically proven that the Indians were indeed "incapable of civilization," the idea that the whites living among the tribes could

---

[19] *The Writings of Thomas Jefferson*, ed. Paul Leicester Ford (New York: G.P. Putnam's Sons, 1892–99), vol. 8: 394.

[20] Romans, *A Concise Natural History of East and West Florida*, 110–12.

[21] Lewis Cass, "Removal of the Indians," *North American Review* 30 (1830): 67.

[22] George Combe, quoted in Reginald Horsman, *Race and Manifest Destiny: The Origins of American Racial Anglo-Saxonism* (Cambridge MA: Harvard UP, 1981): 58–59.

[23] Charles Caldwell, quoted in Reginald Horsman, *Race and Manifest Destiny*, 118.

influence the Indians towards civilization was called into question, as was the capacity of the half-bloods to be more civilized than their Indian parent.

The evolution of race theories revealed a paradox, since, despite what science said, some Indian tribes had become civilized, some Indians had become farmers and skilled traders, had adopted republican governments and the Christian religion, dressed like the whites, and spoke English. This did not fit well with the new conviction that they were naturally inferior and incapable of civilization, and the advances of the southeastern tribes were attributed to the half-bloods living among them. Even if it was true that the half-bloods played an important role in the economy and political life of tribal governments in the early-nineteenth century, they now attracted all the attention and their success was seen as the direct consequence of race-mixing. Half-bloods were now considered more progressive or more civilized than the 'real' Indians. The attention of politicians focused particularly on the Cherokee tribe because they established a constitutional government in 1827 and published a bilingual newspaper as early as 1828. Governor Wilson Lumpkin of Georgia admitted that "in the Cherokee country many families [enjoy] all the common comforts of civil and domestic life, and [possess] the necessary means to secure these enjoyments." Yet Lumpkin maintained that civilization among the Cherokees was "confined to the blood of the white man, either in whole or in part," and that "very few of the real Indians participate largely in these blessings." He claimed that "a large portion of the full-blooded Cherokees still remain a poor degraded race of human beings." Those "real Indians" could not live among the whites. They "struggle with want and misery, without hope of bettering their condition by any change but that of joining their brethren west of the Mississippi."[24] Lewis Cass agreed, stating that "the great body of the Cherokee people are in a state of helpless and hopeless poverty." He doubted that "there is, upon the face of the globe, a more wretched race than the Cherokees, as well as the other southern tribes."[25] The removal of these tribes west of the Mississippi seemed to be the only way to save them.

President Andrew Jackson, eager to transfer the southeastern tribes west of the Mississippi, was infuriated by the Indians' refusal to sell their land. He

[24] "Speech of the Hon. Wilson Lumpkin of Ga," in "Committee of the Whole House, on the State of the Union, on the Bill Providing for the Removal of the Indians," in Wilson Lumpkin, *The Removal of the Cherokee Indians from Georgia* (Wormsloe GA: The author & New York: Dodd, Mead, 1907): 77.

[25] Cass, "Removal of the Indians," 71.

was particularly incensed by the "base and designing white men [...] and half-breeds, who by intrigue and corruption have got into the council of the nation and have turned out the old chiefs."[26] According to him, those white men and half-bloods manipulated the "real Indians," who would gladly sell their land east of the Mississippi to move west, as they "appear very solicitous for an exchange." Jackson believed that they were "overawed by the council of some whitemen and half-breeds, who have been and are fattening upon the annuities, the labours, and folly of the native Indian." Jackson concluded that "there cannot be [...] as corrupt and Despotic government as now exists in this nation."[27] In the 1820s, racializing Indian societies in this way became a useful tool with which politicians could discredit chiefs who opposed the removal of their tribes. Not only were the half-bloods no longer more civilized or more progressive than the "real Indians," they were also more immoral, dishonest, venal, and self-interested. The politicians focused on the corruption of tribal governments and claimed that the designs of the half-bloods threatened not only the Cherokee people but also the republican institutions of the USA. Governor Lumpkin considered Cherokee rulers "the most violent and dangerous enemies of our civil institutions."[28] The only way, then, to protect the "real Indians" was to extend the laws of the USA to Cherokee country.

The notion of identity based on race was gradually instilled in the tribe through their contact with the slaveholding white population surrounding them in the Southeastern USA, and through the whites' treatment of their black slaves. There were also slaves in the Cherokee Nation. Slavery existed among the Cherokees prior to contact with the Europeans, but it had nothing in common with the white institution of slavery as we know it. The Cherokees obtained slaves in warfare, and frequently adopted prisoners of war into their clans to replace kinsmen who had died or had been killed. The Cherokees encountered Africans as early as they did Europeans, as there were black slaves

---

[26] U.S. Commissioners to Secretary Graham, 8 July 1817, in *Correspondence of Andrew Jackson*, volume 2, ed. John S. Bassett (Washington DC: Carnegie Institution, 1926–35): 300.

[27] Jackson to Butler, 21 June 1817, in *Correspondence of Andrew Jackson*, volume 2, ed. John S. Bassett (Washington DC: Carnegie Institution, 1926–35): 299.

[28] "Speech of the Hon. Wilson Lumpkin of Ga," in "Committee of the Whole House, on the State of the Union, on the Bill Providing for the Removal of the Indians," in Lumpkin, *The Removal of the Cherokee Indians from Georgia*, 77.

in the Spanish expedition of Hernando de Soto, for instance. As they nearly always encountered Africans with Europeans, it is possible that the Cherokees at first did not distinguish between the two races. The enslavement of Indians brought them into extensive contact with the African slaves, and the Cherokees no doubt realized that the whites regarded their black slaves as inferior, and that as slaves of the whites they would receive the same treatment.[29]

The English settlers brought the first African slaves from Africa at approximately the same time as they enslaved the first Indians.[30] The American politicians put an end to the Indian slave trade after the Revolution, the black slaves proving to be a better investment. Far from their homeland, in unknown country, they were less likely to escape than Indian slaves, who received the help of members of their tribes in their escape. The enslavement of Indians made negotiating with the tribes all the more difficult and the settlers also feared a revolt of the neighbouring Indian tribes. The Cherokees also captured black slaves during raids against enemy tribes. The Cherokees soon engaged in the black slave trade, capturing or stealing black slaves and selling them, seldom keeping the slaves for themselves. Some runaway slaves also found refuge in the mountainous Cherokee country, and the treaty of 1730 between the Cherokees and the English stated:

> if any Negroe Slave shall run away into the Woods from their English Masters, the Cherokee Indians shall endeavour to apprehend them and either bring them back to the Plantation from whence they run away, or to the Governor, and for every Negroe so apprehended and brought back, the Indian who brings him shall receive a Gun and a Match-coat.[31]

---

[29] Richard R. Wright, "Negro Companions of the Spanish Explorers," *American Anthropologist* 4 (1902): 217–28; *Narratives of the Career of Hernando de Soto*, 2 vols., ed. Edward Gaylord Bourne (New York: A.S. Barnes, 1943): 163; Lowery Woodbury, *The Spanish Settlements Within the Present Limits of the United States, 1513–1561* (New York: Putnam, 1911): 165–67.

[30] Almon W. Lauber, *Indian Slavery in Colonial Times Within the Present Limits of the United States* (New York: Columbia University, 1913): 118; Kenneth M. Stampp, *The Peculiar Institution: Slavery in the Ante-Bellum South* (New York: Alfred A. Knopf, 1956): 18.

[31] "Articles of Friendship and Commerce, prepared by the Lord Commissioners for Trade and Plantations to the Deputies of the Cherokee Nation in South Carolina by His Majesty's Order, on Monday the 7th Day of September 1730," in Vicky Rozema,

Afraid that the runaway slaves might establish their own black communities in Cherokee mountains, slave owners offered rewards to the Cherokees who brought the slaves back. Afraid, too, that the black slaves and the Cherokees might join forces against them, the whites fostered hate and hostility between the two peoples, using African slaves in their attacks against Indian tribes.[32] By the late-eighteenth century, in response to these manoeuvres, the Cherokee had internalized the European view of racial difference and racial prejudice. They felt that they had to assert their difference if they did not want to end up like the blacks. This led them to voluntarily distance themselves from the blacks and regard them as being more suitable as slaves than as adopted kinsmen. They felt the need to codify their distinct identity in terms of race if they wanted the Euro-Americans to respect their claim to tribal sovereignty.

The Cherokees also modelled their national ideology and structure on those of the USA. The traditional matrilineal clan system eroded as the US government implemented its programme to 'civilize' the Indians, and, by 1808, the Cherokees began to adopt the European patrilineal system. The traditional Cherokee sexual division of labour was also disrupted as the Cherokees were encouraged to adopt Euro-American farming techniques. The Cherokees began to replicate the racial ideologies and practices of the whites, and to model their institutions and social structure on those of the USA, making Cherokee identity increasingly complex. It took many possible forms, as the traditional ways continued to exist, side by side with the new. But the appearance and growth of plantation slavery brought new inequalities into Cherokee society and a growing class antagonism between slaveholding and non-slaveholding Indians. Cherokee slaveholders composed the upper class of the Nation. In 1835, only 17% of all the people living in the Cherokee Nation had any white ancestors, whereas 78% of the members of slaveholding families did. So the antagonism between slaveholding and non-slaveholding Cherokees also tended to fall along racial lines and the terms 'mixed-blood' and 'full-blood' were soon used to refer to the members of the different groups. At the basis of this growing antagonism in the Cherokee Nation was the definition of Cherokee identity. The wealthy slaveholding mixed bloods embraced the white values of individualism and self-reliance, whereas the so-called 'full bloods'

Cherokee Voices, Early Accounts of Cherokee Life in the East (Winston–Salem NC: John F. Blair, 2002): 7–11.

[32] Verner Crane, *The Southern Frontier, 1670–1732* (Durham NC: Duke UP, 1928): 113–14.

believed they were abandoning sacred traditions and pushing the Nation towards the whites. They wanted to preserve traditional culture and values.

This antagonism threatened to irredeemably divide the Cherokees into two different nations during the crisis of their removal to the Indian Territory. In the late-1830s, a group of wealthy mixed-blood Cherokee slave-owners, convinced that the efforts to resist removal were useless, thought that their only hope was to sign a removal treaty. In December 1835, they signed the Treaty of New Echota with the USA on behalf of the whole Cherokee tribe. The signing of the treaty and the forced removal of the Cherokees to the Indian Territory in 1838 led to seven years of internal war. Some leaders among the mixed-bloods, the treacherous signers of the removal treaty, were assassinated. The crisis worsened, and the mixed-bloods claimed that there was no possible way to reconcile the opposed factions and their two different visions of Cherokee identity. For them, the only solution was to provide a territorial division of the country into two separate Nations. They tried to destabilize the Cherokee government and created terrorism in the Nation. The situation was such that outsiders believed a genuine civil war was in progress in the Cherokee Nation. Cherokee unity was finally preserved thanks to Chief John Ross's dedication to sustaining the coherence and sovereignty of the tribe at all costs. Peace returned to the Nation, but factionalism was far from dead.

The issues of slavery and secession further divided the Cherokees. In the 1850s, a concern for southern rights arose among the wealthy mixed-blood slaveholding Cherokees. At the dawn of the American Civil War, the Cherokee Nation was in a difficult situation, located between the Deep South and Kansas. The two opposed groups of Cherokees worked at cross-purposes, as the mixed-bloods believed that the best course of action was an alliance with the southerners in the defence of the institution of slavery. Regarding the full-bloods as backward and ignorant, they believed that they, the mixed-bloods, deserved to lead the Nation. Chief John Ross tried to favour a neutralist approach to the growing cleavage between whites of the north and south. Two groups embodied these conflicting definitions of Cherokee identity. The Knights of the Golden Circle, mostly the mixed-blood elite of the Cherokee Nation, promoted slavery and advocated the removal of abolitionists from the Nation. Their leader, Stand Watie, was a Cherokee full-blood and a colonel in the Confederate army. By contrast, the Keetoowah Society tried to defend traditional Cherokee religion and beliefs to preserve the unity of the Cherokee people and the old ways. They promoted patriotism and nationalism. The rituals and activities associated with the Keetoowah Society were designed to unite the conservative Cherokees for political action. They represented the full-blood majority, and believed they should direct the destiny of their people. Their leader, Principal Chief John Ross, was only one-sixteenth Cherokee. The question they asked was: Could a true Cherokee support the institution of slavery and personal material self-aggrandizement and still be true to the spirit of his ancestors? Ross and Watie respectively embodied the full-blood and the mixed-blood factions, but these terms had only a loose correspondence with racial ancestry. There were slave-holding and non-slave-holding Cherokees, as well as Cherokees of mixed racial ancestry on both sides. The term mixed-blood simply designated the elite class of slave-holding Cherokees who used their wealth and influence to force the tribal government to support slavery, even at the expense of national sovereignty.

Ross and the Cherokees sided with the North, and civil war became a reality for the Cherokee Nation. In February 1863, the Cherokees loyal to the Union passed an Act emancipating all slaves in the Cherokee Nation. However, this Act freed few slaves, as most of them belonged to the Cherokee mixed-bloods allied with the Confederacy. Interestingly, the freed slaves were not adopted into the tribe according to Cherokee tradition. Racial hierarchy was now so deeply rooted in Cherokee minds that even the Keetoowah Cherokees disowned the inclusive nature of Cherokee society in terms of absorbing other individuals from other tribes or other races. The loyal Cherokees now shared the view of most northern whites: they were willing to end the institution of slavery but not to have former slaves living in their vicinity. They were, however, willing to exploit the former slaves as cheap labour. At the end of the Civil War, Watie and many 'mixed-blood' leaders were convinced that the only solution was a division of the Cherokees into two separate nations, and their delegates in Washington negotiated their own treaty with the USA to try to obtain this division. They almost succeeded, but the Keetoowahs finally signed a treaty with the American government on 19 July 1866 that guaranteed the survival of the Cherokees as one people.

Article 9 of the treaty stated:

> all freedmen who have been liberated by voluntary act of their former owners or by law, as well as all free coloured persons who were in the country at the commencement of the rebellion, and are now residents

> therein, or who may return within six months, and their descendants, shall have all the rights of native Cherokees.[33]

The treaty made the freedmen Cherokees, but the question of their citizenship and identity still proves problematical today. At the time of the signing, the Cherokee Nation numbered 17,000 members and slightly fewer than 2,000 blacks.[34] More than half the blacks living in the Cherokee Nation fled or died during the war. Under the treaty, the freedmen had the same status as the native Cherokees, adopted Indians, and white citizens of the tribe. The traditionally inclusive ideal of Cherokee identity seemed to prevail. But there remained the problem of the freedmen who came back to the Nation after the six-month period had expired. At the start of the war, some Cherokee slaves or free blacks established in the Nation left for Kansas, Arkansas, Texas or even Mexico, and many found it hard to make it back on time, if they heard about the six months at all. This was a major problem for the Cherokees, many of these blacks having been forced to follow their former masters in their flight out of the Nation. Besides, some runaway slaves from the neighbouring states had sought refuge in the Cherokee Nation during the war. These 'intruders', who did not live in the Nation at the outbreak of the war, were willing to stay. They tried to take advantage of the terms of the treaty and of the confusion that reigned in the Nation to lodge false claims to citizenship, pretending they had resided in the Nation for many years. Knowing that their enemies would exploit whatever internal conflict against them, the Cherokees united to survive the attacks from settlers, railroad companies, and speculators. They elected Lewis Downing as their Principal Chief in 1867. Downing was the first full-blood to be elected since the Constitution of 1827. He advocated unity and inclusion and a return to the traditional Cherokee way of identifying its members, recommending that the Cherokees be "one" and that "every line of distinction be blotted out."[35] The Indian agent William Davis observed in his 1868 annual report that "the Cherokees may be

---

[33] *Indian Affairs: Laws and Treaties*, ed. Charles J. Kappler (Washington DC: Government Printing Office, 1904), vol. 2: 946.

[34] Daniel F. Littlefield, *The Cherokee Freedmen: From Emancipation to American Citizenship* (Westport CT: Greenwood, 1978): 28; Russel Thornton, *The Cherokees: A Population History* (Lincoln: U of Nebraska P, 1990): 102.

[35] Lewis Downing, quoted in John Bartlett Meserve, "Chief Lewis Downing and Chief Charles Thompson," *Chronicles of Oklahoma* 16 (September 1938): 320.

regarded as one people, all working harmoniously for the advancement and prosperity of their tribe."[36]

In spite of the terms of the 1866 treaty, the freedmen have never been fully accepted into the tribe and their status remains uncertain today. To the Indian commissioner Hiram Price, the freedmen were undoubtedly Cherokee culturally. He stated in 1882:

> many of the colored people speak the Cherokee language, and having been brought up among the Cherokees and accustomed to their ways, it would be a hardship to remove them from that country, and remaining in the nation, they should be accorded all their rights.[37]

The freedmen themselves felt they belonged to the Cherokee tribe, men such as Joseph Rogers:

> Born and raised among these people, I don't want to know any other. The green hills and blooming prairies of this Nation look like home to me. [...] I look around and I see Cherokees who in the early days of my life were my playmates in youth and early manhood, my companions, and now as the decrepitude of age steals upon me, will you not let me lie down and die your fellow citizen?[38]

This cultural identity would have been enough for them to be considered Cherokee full-bloods in the past.

In 1887, the Dawes Act organized the allotment of tribal territory and promised to free thousands of acres of 'vacant' land for American settlers and speculators. The Cherokee Nation and other 'civilized' tribes were first unaffected by the Dawes Act, but in March 1893, the Indian Appropriation Act arranged for the allotment of Cherokee land and detribalization. That year, the Dawes Commission was given the responsibility of taking a census of the Cherokee population and of negotiating the terms of the allotment of tribal land. The members of the Dawes Commission drew up three lists of Cherokee citizens, one of freedmen, one of intermarried whites, and one of Cherokees

[36] William Davis, quoted in William G. McLoughlin, *After the Trail of Tears, The Cherokees' Struggle for Sovereignty, 1839–1880* (Chapel Hill: U of North Carolina P, 1993): 245.

[37] *Annual Report of the Commissioner of Indian Affairs to the Secretary of the Interior for the year 1882* (Washington DC: Government Printing Office, 1882): 57.

[38] *The Cherokee Advocate* (9 September 1876).

by blood. This third list bore mention of the individual's blood quantum. The Dawes Rolls listed 41,798 Cherokee citizens, including 4,924 freedmen.[39]

According to the 1975 Cherokee Constitution, "all members of the Cherokee Nation must be citizens as proven by reference to the Dawes Commission Rolls."[40] So the freedmen listed on the Dawes Rolls are Cherokee citizens, but in reality the Cherokees only grant citizenship to the individuals listed on the "Cherokee by blood" list and their descendants. Indeed, an individual applying for Cherokee citizenship must obtain a Certificate of Degree of Indian Blood (CDIB), issued by the American government. If he can prove that he is the descendant of a Cherokee listed on the citizens-by-blood list, then his blood quantum is determined and he is automatically granted citizenship. Contrary to most Native tribes in the USA, the Cherokees do not limit their citizenship to individuals with a minimum blood quantum, and the blood quantum of Cherokees varies from full-blood to 1/2048. In February 1996, out of 175,326 members, only 37,420: i.e. twenty-one percent, had one-quarter Cherokee blood or more. This strictly racial definition of Cherokee identity dates back to the 1880s, to the lists of the Dawes Commission drawn according to racial criteria. Today, via the Bureau of Indian Affairs, the federal government continues to use racial considerations to deal with Indian tribes. The CDIB is the document used by the Bureau to determine Indian identity, now only a biological identity.

The inclusive nature of Cherokee society makes it the second-largest tribe in the USA, with more than 1,500 applications a year. Miscegenation in the tribe has triggered much controversy. Most Cherokees today confer on blood an ideological value, and blood is linked to the question of knowing who is or is not a 'real' Cherokee. Most Cherokees speak of blood as a powerful thing that connects them all. They feel Cherokee because of blood. To them today, someone who has no Cherokee blood does not belong to the community. That is why many refuse to regard the freedmen as Cherokees. In 1984, Vice-Chief Wilma Mankiller declared that the freedmen "should not be given membership in the Cherokee Nation. That is for people with Cherokee blood."[41] Many Cherokees share her opinion, such as Jimmy Philips, who declared in the *Baltimore Sun*: "Whether they are white, black or red, if they've got the

---

[39] Littlefield, *The Cherokee Freedmen: From Emancipation to American Citizenship*, 238.

[40] 1975 Cherokee Nation Constitution, Article III, section 1, Membership.

[41] *Baltimore Sun* (29 July 1984).

blood then they are tribal members. Without it... no."[42] In 2007, the Cherokees acted on this and voted to adopt more radical requirements for tribal membership. Being listed on the Dawes Rolls was no longer enough; now a Cherokee must be a citizen by blood, and this excluded from the tribe the freedmen who could not prove their Indian ancestry. Ironically, a lot of the freedmen listed on the Dawes freedmen list do have Cherokee ancestors. The Dawes Rolls list 4,028 adult freedmen, about 300 having Cherokee ancestry as shown by their census cards. This means that about seven percent of the Cherokee freedmen on the Dawes Roll have Cherokee blood but are listed on the list of freedmen and not on the list of Cherokees by blood, because they are black. If a citizen is listed on the freedmen list, then he is an adopted citizen, not a citizen by blood. And yet, looking closely at the Dawes lists and census cards, we see that some adopted citizens are on the Cherokee-by-blood list. These individuals are all white. The members of Richard C. Byrd's family, descendants of the missionary Evan Jones, are listed on the Cherokee-by-blood list, but their census card states that they are "adopted whites" (census card # 10,925). His children, aged three and one, are also listed on the Cherokee-by-blood minors list. The Byrd family is not the only example of whites listed on the by-blood list.[43]

In the tri-racial American society of the early-twenty-first century, the status of mixed-bloods is not clear. Individuals are placed in categories that do not correspond to their personal experience or family ties. Individuals with any degree of 'black blood' are automatically listed as blacks on the freedmen list. The racial prejudices that result from the adoption of slavery by the Cherokees seem to be the best explanation for the way the freedmen are considered and treated, but also for the way black mixed-bloods, freedmen, and their descendants see their complex identity reduced to just 'black'. Some Cherokee leaders maintain that the descendants of freedmen are not Indian, because their ancestors have been listed on the Dawes lists as freedmen. To them, if these freedmen had Cherokee blood, and Cherokee ancestors, they would have been listed on the Cherokee-by-blood list. Some even say that if their claims were legitimate, they would apply for a CDIB. They ignore the story of the Dawes census and the confusion and prejudices that prevailed at the time.

---

[42] *Baltimore Sun* (29 July 1984).

[43] Department of the Interior, Commission to the Five Civilized Tribes, Claremore, I.T., 12 November 1900.

The case of Cynthia Lynch's family is quite telling with regard to the obliteration of the Cherokee identity of black-Cherokee mixed-bloods. Cynthia Lynch is listed on the Cherokee-by-blood list. According to her census card (# 31,968), her blood quantum is one-quarter Cherokee. Her four children are also listed on that card (#31,969 to # 31,972) as Cherokees by blood, with a 1/8 blood quantum. The father of Cynthia's children, Allen Lynch, is also listed on Cynthia's card. He is a freedman. The census cards of Cynthia's children as Cherokees by blood have consequently been annulled and new cards made, freedmen cards, where no mention of the children's Cherokee blood quantum can be found. On the second page of Andrew Lynch's census card, the name of his mother, Cynthia Lynch, can be read, but nowhere is it said that she is Cherokee by blood. Only the black identity of her children remains, and the name of the owner of their father, a John Lynch.[44]

On 3 March 2007, when more than seventy-six percent of Cherokee voters chose to limit Cherokee citizenship to Cherokees by blood and their descendants, the immediate consequence was the disenrolment of some 2,800 Cherokee freedmen descendants. The Cherokee vote can seem paradoxical in many ways, because after having indigenized the white concepts of blood and race to preserve their traditional ways, culture, and sovereignty, after having adapted and adopted the white criteria for identification, particularly in their relations to blacks, and after having given up parts of what originally was Cherokee identity in order to survive as a people, the Cherokee today claim that the only criterion for Cherokee identification must be Indian blood. It is no longer the metaphorical blood that unites the members of a same clan and creates kinship ties among all Cherokees, native or adopted, but the biological blood that runs in their veins and that proves they belong to the Cherokee race. This blood does not run in the veins of some of the descendants of the freedmen. When they radicalize the criteria for tribal identification like this, the Cherokees answer the attacks on their native identity and sovereignty. The paradox is evident when the Cherokees themselves define their identity, as Jason Terrell does in the *Cherokee Observer* in 1996. Isn't what the freedmen and their descendants feel today just what Joseph Rogers felt in 1876?

---

[44] *The Final Rolls of Citizens and Freedmen of the Five Civilized Tribes in Indian Territory, prepared by the Commission and Commissioner to the Five Civilized Tribes and approved by the Secretary of the Interior on or prior to March 4, 1907, compiled and printed under Authority conferred by the Act of Congress approved June 21, 1906* (Baltimore MD: Genealogical Publishing, 1907): 472–502.

> How do you explain to someone that there's no half-way point being Cherokee? You either are or you aren't. It's not a question of how many Europeans vs. how many Cherokees one has in the ole' family tree. Most all of us can play that game. It's not even a question of where you live. It IS a question of loyalty. You either have a loyalty to our people, or you don't. It IS a question of commitment. That means getting involved and not letting self-interested individuals take the people for a ride while you sit by. It means that no matter where you go, you come home to family and friends and you want to make a difference. It's the way you live and the way your family has lived. It's knowing who your relations are and where you fit into our society. You can't suddenly "become" Cherokee. It's not a club with a membership card and dues. It's something you're born with and if you really are Cherokee, it's something you can't ignore.[45]

Works Cited

Adair, James. *Adair's History of the American Indians*, ed. Samuel Cole Williams (New York: Promontory, 1930).

Annual Report of the Commissioner of Indian Affairs to the Secretary of the Interior for the year 1882 (Washington DC: Government Printing Office, 1882).

Bassett, John S., ed. *Correspondence of Andrew Jackson* 2 (Washington DC: Carnegie Institution, 1926–35).

Byrd, William. *The Westover Manuscripts: Containing the History of the Dividing Line Betwixt Virginia and North Carolina and a Journey to the Land of Eden, 1733* (Whitefish MT: Kessinger, Legacy Reprint, 2007).

Cass, Lewis. "Removal of the Indians," *North American Review* 30 (1830): 62–121.

Crane, Verner. *The Southern Frontier, 1670–1732* (Durham NC: Duke UP, 1928).

Ford, Paul Leicester, ed. *The Writings of Thomas Jefferson* (New York: G.P. Putman's Sons, 1892–99), vol. 8.

Gaylord Bourne, Edward, ed. *Narratives of the Career of Hernando de Soto* (New York: A.S. Barnes, 1943).

Hawkins, Benjamin. *Letters, Journals, and Writings of Benjamin Hawkins*, ed. C.L. Grant (Savannah GA: Beehive, 1980), vol. 1.

Horsman, Reginald. *Race and Manifest Destiny: The Origins of American Racial Anglo-Saxonism* (Cambridge MA: Harvard UP, 1981).

Kappler, Charles J., ed. *Indian Affairs: Laws and Treaties* (Washington DC: Government Printing Office, 1904), vol. 2.

---

[45] Jason Terrell, "Part Cherokee?" *Cherokee Observer* 4.4 (April 1996).

Lauber, Almon W. *Indian Slavery in Colonial Times Within the Present Limits of the United States* (New York: Columbia University, 1913).

Lawson, John. *A New Voyage to Carolina,* ed. Hugh Talmage Lefler (Chapel Hill: U of North Carolina P, 1967).

Littlefield, Daniel F. *The Cherokee Freedmen: From Emancipation to American Citizenship* (Westport CT: Greenwood, 1978).

Lumpkin, Wilson. *The Removal of the Cherokee Indians from Georgia*, 2 vols. (Wormsloe GA: The author & New York: Dodd, Mead, 1907).

McLoughlin, William G. *After the Trail of Tears, The Cherokees' Struggle for Sovereignty, 1839–1880* (Chapel Hill: U of North Carolina P, 1993).

——. *The Cherokee Ghost Dance: Essays on the Southeastern Indians, 1789–1861* (Macon GA: Mercer UP, 1984).

Meserve, John Bartlett. "Chief Lewis Downing and Chief Charles Thompson," *Chronicles of Oklahoma* 16 (September 1938): 315–25.

Mooney, James. *History, Myths and Sacred Formulas of the Cherokees* (Asheville NC: Bright Mountain, 1992).

Norton, John. *The Journal of Major John Norton, 1816,* ed. Carl Frederick Klinck & James John Talman (Toronto: Champlain Society, 1970).

Perdue, Theda. *Mixed Blood Indians: Racial Construction in the Early South* (Athens: U of Georgia P, 2003).

Romans, Bernard. *A Concise Natural History of East and West Florida*, ed. Kathryn E. Holland Braund (Tuscaloosa: U of Alabama P, 1999).

Rozema, Vicky. *Cherokee Voices: Early Accounts of Cherokee Life in the East* (Winston–Salem NC: John F. Blair, 2002).

Schoolcraft, Henry Rowe. *Historical and Statistical Information Respecting the History, Condition, and Prospects of the Indian Tribes of the United States*, 5 vols. (Philadelphia PA: Lippincott, Gambo, 1851–57).

Stampp, Kenneth M. *The Peculiar Institution: Slavery in the Ante-Bellum South* (New York: Alfred A. Knopf, 1956).

Terrell, Jason. "Part Cherokee?" *Cherokee Observer* 4.4 (April 1996).

Thornton, Russel. *The Cherokees: A Population History* (Lincoln: U of Nebraska P, 1990).

Waselkov, Gregory A., & Kathryn E. Holland Braund, ed. *William Bartram on the Southeastern Indians* (Lincoln: U of Nebraska P, 1995).

Woodbury, Lowery. *The Spanish Settlements Within the Present Limits of the United States, 1513–1561* (New York: Putnam, 1911).

Wright, Richard R. "Negro Companions of the Spanish Explorers," *American Anthropologist* 4.2 (April–June 1902): 217–28.

⌘

Tribal Communities and Genetic Research

Concerns and Expectations

MARIE–CLAUDE STRIGLER

LET US IMAGINE THIS: YOU HAVE DONATED A SAMPLE OF YOUR BLOOD to study the genetics of diabetes. The disease is common among your relatives and in your community, and you want to help the research. You feel it is your duty to find out the reasons and to develop a cure that everyone, including yourself, could benefit from. Then, later on, you learn that your DNA has been used for other studies – schizophrenia, human migration, inbreeding! Unbeknownst to you, scientists have been drawing conclusions about your community and your ancestors.

This in fact did happen to the Havasupaï, a small tribe that live at the bottom of the Grand Canyon. They are difficult to reach; in fact, they are accessible mainly by mule or horseback or by helicopter. Because they have been isolated for so long, the Havasupaï are supposed to have a 'pure' bloodline that has been undiluted by marriage – an ideal object of study for geneticists.

In 1989, two hundred tribal members gave blood to researchers to help with a diabetes study. No money changed hands, but fifteen Havasupaï were to take college-level summer-school classes for free. Then they learned that the study had been used for research about schizophrenia and inbreeding, and migration (ironically, the original study seeking a genetic basis for diabetes produced no significant results). This was particularly distressing to the tribe, as the research concerning migration contradicted one of their core spiritual beliefs: namely, that they originated in, and have always lived in, the Grand Canyon. They believe that they did not migrate from Asia via the Bering Strait as genetic research has in fact shown.

In 2004, they sued Arizona State University – the institution responsible for collecting their DNA – for not providing ethical oversight on the use of the samples. The case is still working its way through the courts. The tribe's lawyers asserted that the defendants had "violated their clients' privacy, as well as their cultural, religious, and legal rights." Many of the research results were stigmatizing and contradictory to tribal history and ancestry. The study, indeed, had drifted far afield from its original research on diabetes. Yet the plaintiffs' case was dismissed because they allegedly did not meet the requirements for filing a lawsuit. In 2008, the Arizona court of appeals cleared the way for the Havasupaï tribe to sue the state university system for improper use of the blood samples taken in 1989. The fifty-million-dollar lawsuit that the tribe is pressing for claims that the blood was not only taken under false pretences but was also used to undermine tribal members' beliefs about the world.

In *Moore vs Regents of the University of California*,[1] a patient sued his physician and a biotechnology company for using his biopsied tissue, his tangible personal property, without his consent and transforming it into a commercial cell line. The court sided with the interests of the defendants. Its reasoning was that giving the patient a property right to his tissue would impede progress and "destroy the economic incentive to conduct important medical research."

Western scientific research has rightly caused great unease and wariness among many Indigenous peoples. In the name of scientific knowledge, sacred stories and sacred sites have been made public, biological material has been used to contradict and stereotype peoples, and Indian[2] property has been stolen and displayed in museums all over the world; all of this perpetrated under the cover of scientific knowledge. Moreover, a history of broken promises and other misdeeds lies just below the surface – all of which accounts for a deep-seated mistrust of the federal government and other state institutions.

Joyce Oberly (Comanche, Osage, Chippewa/Cree) and Jocelyn Macedo (Yurok and Hupa) point out that, although it may seem to be a contradiction,

---

[1] *Moore vs Regents of the University of California*, 51 Cal. 3d 120, was a landmark California Supreme court decision which dealt with the issue of property rights on one's own body parts.

[2] 'Indian' is used in the sense commonly employed in the USA to refer to First Nations peoples.

research 'on' Indians historically excludes Indians, misappropriating, misinterpreting, and, ultimately, misrepresenting Native culture and knowledge.[3]

In Vine Deloria's words, past research in Indian country derives from biased colonizers using eurocentric views to 'analyse' and 'record' American Indian culture.[4] According to him, there is a legacy of racism and paternalism in studies of Native peoples from the late-nineteenth and early-twentieth centuries that persists today.[5]

Current strategies of exclusion and dismissal are more subtle, and operational in the design and implementation of the research. Felicia Schanche Hodge, Steven Weinmann, and Yvette Roubideaux use the phrase "helicopter research" for what occurs when researchers swoop in, collect data, and leave.[6]

There are four key problem areas in this type of research: inappropriate use of culturally sensitive information, especially spiritual information; commercial or other exploitive use of information; unauthorized infringement of individual, family or group ownership (for songs, stories or other information); and potential conflict or harm resulting from the research, including the harm resulting from inappropriate interpretation of information, inappropriate intrusion into community life, and breaches of confidentiality and friendship.

Due to small population size and high incidence of some diseases (such as diabetes or breast cancer), tribes are often seen as attractive research subjects.

The Human Genome Project

Completed in 2003, the Human Genome Project was a thirteen-year, multi-billion-dollar project coordinated by the US Department of Energy and the National Institutes of Health. It was primarily concerned, not with genetic difference, but with genetic sameness, a shared genetic heritage that the HGP is attempting to represent by mapping the human genome. When scientists first decoded the Human Genome in 2000, they were quick to depict it as proof of

---

[3] Joyce Oberly & Jocelyn Macedo, "The R Word in Indian Country: Culturally Appropriate Commercial Tobacco-Use Research Strategies," *Health Promotion Practice* 5.4 (October 2004): 354–62.

[4] Vine Deloria, *Custer Died for your Sins* (Norman: U of Oklahoma P, 1963).

[5] Vine Deloria, *Red Earth, White Lies: Native America and the Myth of the Scientific Fact* (Golden CO: Fulcrum, 1995).

[6] Felicia Schanche Hodge, Steven Weinmann, & Yvette Roubideaux, "Recruitment of American Indians and Alaska Natives into Clinical Trials," *Annals of Epidemiology* 10 (2000): S41–S48.

mankind's remarkable similarity. The DNA of two people is ninety-nine percent identical. From the outset, it had been planned to transfer the related technologies to the private sector and to address the ethical, legal, and social issues (ELSI) that might arise from the project. But the Harvard professor and geneticist Richard Lewontin warned that "the human DNA sequence will be a mosaic of some hypothetical average person corresponding to no one."[7] We are all, in fact, deviations from this abstracted norm.

Genetic research was first presented to tribal peoples as a means to try to find cures for some diseases, which was the primary goal of the Human Genome Project. The existing genetic tests enabled patients (and others) to learn their genetic risk of disease. Moreover, the tests led to an increasing ability to connect DNA variations with non-medical conditions, such as intelligence and personality traits, thus challenging society legally and ethically. My intention in this essay is to examine both sides of the problem, pointing out the reasons why most indigenous peoples oppose genetic research and what solutions, if any, can be found.

The Human Genome Diversity Project

HGDP started with Stanford University's Morrison Institute in 1990 when the Stanford population geneticist Luigi Luca Cavalli–Sforza proposed the creation of a database of genetic information about the world's distinct populations.

Contrary to popular belief, this is not related to the HGP or to the Genographic Project. The goal was to reconstruct the history of human evolution and the historical and geographical distribution of populations with the help of scientific research. Through this kind of research, the entire spectrum of genetic diversity to be found in the human species was to be explored, in the hope of achieving a better understanding of the history of mankind. An important part of the process consisted in taking blood and tissue samples from Indigenous populations, which are supposed to be 'purer'.

The small, isolated groups often studied by this science are now mobilizing themselves as political subjects, pressing sovereignty claims, and demanding control over the direction and interpretations of research. Negotiations between the geneticists and the people asked to donate their DNA have resulted not only in explicit bio-ethic protocols but also in diffuse anxiety over the

---

[7] Richard Lewontin, "The Dream of the Human Genome," *New York Review of Books* 39.10 (28 May 1992): 35.

incommensurability between expert and non-expert views about genetic evidence for identity claims.

The wide range of social meanings placed upon the body in defining community and reinforcing acceptable behaviour has converged in disputes surrounding the Human Genome Diversity Project, which was to be an international effort to map the genetic variability of some 4,000 to 8,000 human groups. It not only offered an opportunity to understand the genetic factors in health and illness, but it also promised to reveal important genetic information about the human condition in general.

It quickly became bogged down in controversies regarding how the sampling could be done in ways that were fair to the Indigenous peoples under study. In addition, HGDP has met with international controversy over the meaning of genetic information, the nature of community identity, and the role of racism in genetics and genetic research. According to the official website, it "seeks to understand the diversity and unity of the entire human species."

Since its inception, HGDP has met with criticism about ethical issues in its proposed research, particularly from advocacy groups representing Indigenous populations concerned about exploitation and abuse – so much so, that, in 1994, the representatives of HGDP sought the support of UNESCO. UNESCO's International Bioethics Committee rejected the request, as it was unclear how the project would benefit the groups sampled or how the researchers would address the question of intellectual property rights.

Advocates for Indigenous peoples argued on an even more basic level that the science of genetics violated the holistic world-view of many cultures in which human origins are a sacred part of ancient mythology.

In 1996, to pursue funding, the NSF (The National Science Foundation) commissioned a review of the project by the National Academy of Sciences. A variety of scientists and lay people testified, including several groups of Indigenous peoples of the Americas. They expressed their opposition to the project, based on a different ground including the experience of Western colonization: "The agenda of the non-indigenous forces has been to appropriate and manipulate the natural order for the purposes of profit, power, and control."[8]

---

[8] Declaration of Indigenous Peoples of the Western Hemisphere Regarding the HGDP, 1996.

In some cases, the objection to the project's collection of blood and tissue for DNA analysis reflects ritual beliefs in the communal importance of body tissue. Scientists wanted to 'immortalize' the cell lines of groups that were going to become extinct. But members of Indigenous groups feared that preservation of their DNA could eliminate the incentives to improve social conditions that would ensure their survival. In 1993, the World Council of Indigenous Peoples unanimously voted to "categorically reject and condemn the Human Genome Diversity Project as it applies to our rights, lives, and dignity."[9]

Today, thousands of cell lines are maintained in laboratories around the world. Each one of us is a potential source of these 'biologicals' (a term coined in the early 1980s).[10] Practices that create cell lines "make it increasingly difficult to say where the body is bounded in time, space, and form."[11] Some cell lines are patented, others are not. Those that are patented make it difficult to distinguish them from their utilitarian value in terms of the disparate goals of scientific inquiry, progress, and profit. One easily loses sight of the 'gift' of donors without whom no cell lines would exist. In order to procure a patent on a biological, it must be shown that, through the "process of their production," the "natural" object has been transformed into an "invention."[12] It is estimated that the profits for companies involved in mapping genetic diversity will be billions of dollars. Geneticists believe that valuable genes, those having the greatest utility, are most likely to be found among peoples that are geographically isolated, and consequently the 'Gene Giants' have been accused by political activists of promoting a particularly pernicious form of neocolonialism. This is because of a promise of the development of new 'designer' drugs from biological prospecting. Should such drugs ever materialize, their use would almost certainly be limited to those individuals who participate in well-funded health-care systems.

---

[9] World Council of Indigenous Peoples' Resolution on the HGDP.

[10] Hannah Landecker, "Between Beneficience and Chattel: the Human Biological in Law and Science," *Science in Context* 12.1 (1999): 203–25.

[11] Hannah Landecker, *Culturing Life: How Cells Became Technologies* (Cambridge MA: Harvard UP, 2007): 221.

[12] Alberto Cambrosio & Peter Keating, *Exquisite Specificity: The Monoclonal Antibody Revolution* (Oxford: Oxford UP, 1995).

How Indigenous Peoples Were Selected

A total of 722 groups of people were selected, without consultation, as 'genetic isolates' on the assumption that they were genetically pure, even though there is no agreement as to how many separate migrations took place across the Bering Strait in prehistoric times. Geneticists refer to those human groups as 'isolates of historic interest' because they represent groups that should be sampled before they disappear as integral units so that their role in human history can be preserved (which is reminiscent of the 'Vanishing Indian'). It implies that they are already soon-to-be-extinct cultural entities and therefore justifies the 'commodification' of their respective Indigenous knowledge, which now includes archiving their DNA for posterity. Their blood would be 'immortalized' and stored in facilities, mostly in America. Anyone who so desired, for a small fee, could gain access for experimental purpose. The Indigenous peoples quickly understood that even though their blood was going to be "immortalized," they themselves were to be allowed to continue on the road to extinction. Yet their DNA is valuable enough for some geneticists to want to patent it in their own names.

The Yuchi people of Oklahoma were among those targeted in the USA for DNA testing in the genomics scheme. However, their spokespersons said they would not cooperate unless they were granted federal recognition they had long been denied. One spokesperson, Corky Allen, said:

> Our DNA is regarded as a vital, irreplaceable part of the global heritage of humankind, yet we are denied federal acknowledgment which would give us a political standing, more clout in the fight to keep our language, our culture.[13]

As they are threatened with extinction, there is urgent need for the research to make biomedical discoveries using these peoples – discoveries they will not be around to benefit from. The Yuchi turned down the scientists.[14]

Instead of answers, mainly questions arise: What criteria are being used by the geneticists to target these groups and not others? What benefit do targeted indigenous peoples gain from genetic research? What bioethical implications in terms of basic human rights are emerging from these genomic agendas?

---

[13] Arturo J. Aldama, *Violence and the Body: Race, Gender, and the State* (Bloomington: Indiana UP, 2003): 179–80.

[14] Richard A. Grounds, "The Yuchi Community and the HGDP," *Cultural Survival Quarterly* 20.2 (Summer 1996): 64–68.

Whose genes are these in the first place? Who can be identified as an 'Indian'? Studying genes as a criterion has its roots in the 'blood quantum' criteria established by the federal government which are still being used by many tribes.

The National Congress of the American Indians (NCAI) continues to support a call for a moratorium as part of Indigenous Peoples of the Western Hemisphere, and stands in direct opposition to the few tribal entities that are negotiating for 'shareholding' interests which would allow them to share in whatever profit is realized from these genomic enterprises.

The Genographic Project

In 2005, a newcomer entered the jungle of genetic research: the Genographic Project, National Geographic Society's multimillion-dollar five-year research project to collect DNA from Indigenous groups around the world in the hope of reconstructing humanity's ancient migrations. Billed as the "moonshot of anthropology," the Genographic Project aimed at collecting 100,000 Indigenous DNA samples. It is supposed to answer the fundamental questions about human migration. Where did we come from? And how did we get here? And why do we all look so different? On the Project's website, we can read:

> You can contribute your DNA anonymously by purchasing a public participation kit. By participating in the project, you will learn interesting information about your family's deep ancestry over the ages. By purchasing the kit, you become an "associate explorer" on our team.

That kit may even be "a unique gift"![15]

The argument is that comparing the DNA of large numbers of American Indians might reveal whether their ancestors were from a single founding population and when they reached the Americas. Knowing the routes and timing of migrations within the Americas would provide a foundation for studying how people came to be so different so quickly.

But almost every federally recognized tribe in North America has declined or ignored the invitation to take part. Maurice Foxx, chairman of the Massachusetts Commission on Indian Affairs and a member of the Mashpee Wampanoag, said: "What the scientists are trying to prove is that we're the same as the Pilgrims except we came over several thousand years before."

---

[15] Amy Harmon, "DNA Gatherers Hit Snag: Tribes don't Trust Them," *New York Times* (10 December 2006).

Reasons for Opposing Genetic Research[16]

In the past, minority groups were penalized because they were thought to have inferior genes. Today, sadly, people who have suffered discrimination based on their purported inferiority may be excluded from compensation because their genes indicate that the discrimination they suffered was unjustified. A woman who looks like an Indian, who was raised on a reservation, might in the future be denied a scholarship because her genetic profile does not match the one that researchers claim identifies Native Americans.[17]

Moreover, genetic data may be used to take land back from Native American groups on the grounds they are not Native American enough. We are faced with a paradox: genetic data can be used to deny people minority-group membership in ways that cause a loss of social identity and social benefits.

Some findings can harm individuals and communities: concerns about research-related harm to identifiable communities have been spurred on by advances in genetic technologies. Studies of genetic human variation can present risks to all members of a social group, not only to those individuals who choose to participate in research. Findings that associate an ethnic group with a genetic predisposition to disease, for example, could lead to group discrimination or stigmatization. Individuals may have to pay higher insurance premiums, or face another more subtle form of discrimination, on the basis of the apparent association between those genetic variants and an increased risk of a disease.

Even experienced researchers and Internal Review Board members can fail to anticipate significant research-related risks before a study begins. Risks that are viewed as minor by researchers may be viewed by study participants as substantial. For example, studies of population histories and patterns of population migrations can affect the legal standing of claims made by sovereign Native American tribes for the repatriation of human remains or the return of tribal artifacts held in federal museums.[18] Similarly, studies involving genetic

---

[16] See Bettina M. Beech, *Race and Research: Perspectives on Minority Participation in Health Studies* (Washington DC: American Public Health Association, 2004).

[17] Lori B. Andrews, *Future Perfect: Confronting Decisions about Genetics* (New York: Columbia UP, 2001).

[18] Richard R. Sharp & Morris W. Foster, "Community Involvement in the Ethical Review of Genetic Research: Lessons from American Indians and Alaska Native Populations," *Environmental Health Perspectives* 110, Supplement 2 (April 2002): 147.

markers found more commonly in American Indian and Alaska Native populations can disrupt the social equilibrium that exists within a community by revealing that participants and their families are more 'European' in ancestry than they believed. Such findings have social consequences for many Indigenous communities, as the ability to occupy a political office often depends on establishing one's ancestry as sufficiently 'native'.

Identifying potential group harm is especially difficult when researchers never have direct contact with members of the study population, because many types of environmental health research can be done using information or biological samples that were collected for unrelated purposes. We can even wonder, in the case of isolated populations, whether sending outsiders to obtain DNA samples will not put participants at risk for disease, which may be the case for some Amazonian tribes.

So far, we have evoked tangible collective harm: discrimination or stigmatization, as well as loss of social opportunities. Dignitary harm to communities, by contrast, involves violation of collective rights or disrespectful treatment of the community. For example, by using stored biological materials in a manner that the group would find objectionable would constitute dignitary harm not only to the individuals who contributed the materials but also to the community as a whole. How to determine such harm to communities – and to ascertain whether it is significant or not – remains largely unexplored in the literature on community consultation in research.

Concerns of individuals from whom the cells are taken are to a large extent based on the continued indifference on the part of the dominant world to their condition. The political activist Aroha Te Pareake Mead, deputy convenor of the Māori Congress, says that all human genetic research must be viewed in the context of colonial imperialistic history.[19] She adds that human genes are being treated by science in the same way as indigenous 'artifacts' were gathered by museums; collected, stored, immortalized, reproduced, engineered, all for the sake of humanity and public education; or so we were made to believe.

She claims, moreover, that talk of ethics is simply deception. 'Informed' consent among peoples such as the largely non-literate Hagahai was probably obtained through sign language in the first place. The burden of proof should be on HGDP and other researchers to demonstrate how their project will benefit communities.

---

[19] Aroha Te Pareake Mead, "Genealogy, Sacredness and the Commodities Market," *Cultural Survival Quarterly* 20.2 (Summer 1996): 46–51.

Preserving the Integrity of the Individual[20]

In some cases, the objection to the collection of blood and tissue for DNA analysis reflects ritual beliefs in the communal importance of body tissue. Scientists want to 'immortalize' the cell lines of groups that are going to become extinct. But members of Indigenous groups fear that preservation of their DNA could eliminate the incentives to improve social conditions that would ensure their survival.

We have seen that Native Americans worry about the collection and use of human body tissue that invalidate the integrity of the body. Human tissue has always provided clues to health status. But if, today, DNA analysis of 'waste' tissue such as hair, blood or saliva can give information that may open beneficial therapeutic or remedial options, it can also open the possibility of employment or insurance discrimination,[21] and, according to recent scientific claims, human tissue can reveal information about behavioural traits, race, or sexual preference.[22]

Whereas scientists seek greater access to bodily materials, others defend their cultural values and individual rights. Apart from the actual use of human tissue, the reductionist language often used is objectionable, such as the commercial metaphors in the following phrases: body parts are "extracted" like minerals, "harvested" like a crop, "mined" like a resource. Tissue can be "procured," like a commodity provided for a client. Cells or tissues can be frozen, banked, marketed, patented, bought or sold. Pathologists called a collection of 50,000 blood samples at the Centers for Disease Control and Prevention a "treasure trove."[23]

Social Meaning of the Body

A person's control over his or her body is important to the individual's psychological development. But body tissue has social importance beyond the

---

[20] Testimonies collected by Robert Bensen in *Children of the Dragonfly: Native American Voices on Child Custody and Education*, ed. Benson (Tucson: U of Arizona P, 2001).

[21] Dorothy Nelkin & Laurence Tancredi, *Dangerous Diagnostics* (Chicago: U of Chicago P, 1995).

[22] Dean Hamer & Peter Copeland, *The Science of Desire* (New York: Simon & Schuster, 1995).

[23] Lori Andrews & Dorothy Nelkind, "Whose Body is it Anyway? Disputes over Body Tissue in a Biology Age," *Lancet* 351 (3 January 1998): 53–56.

individual. Blood, hair, placenta are important in social rituals and defining community identification. Sensitive questions arise when genetic analysis is used to reveal community identity. Some patients do not want their tissue used (even without their names attached) for research on race and intelligence, race and crime, or sex and mathematical ability, because the findings of this research could stigmatize their group.

The plaintiffs in body-tissue disputes talk of "violations." John Moore, the patient whose cell line was patented without his knowledge or consent, said he felt as though he had been "raped." Control over the use of body tissue is also critical for establishing religious identity. The Navajo believe that the placenta should be buried, and not regarded as 'waste' available for research. Moreover, given that body integrity is sacred in many cultures, how will participants feel about their cells travelling around the world and outliving their bodies?

Native Peoples Claim their Right to Genetic Materials as Cultural Property

Today, Native peoples claim they have the right to control access to genetic material which they deem to be cultural property. There undoubtedly have been abuses of genetic material in relation to Native peoples.

Indeed, article 27 of the International Covenant on Civil and Political Rights of 1966 recognizes the right of ethnic, religious, and linguistic "minorities" to enjoy their own cultures.[24] One of the difficulties is that "minority" is not defined; and Native peoples do not want to be treated as "minorities" but as distinct, sovereign peoples. The passage of laws such as the Indian Self-Determination and Education Act (PL101–601), and NAGPRA (PL 94–437) has strengthened tribal sovereignty. Sovereignty specifically assures tribes self-governance and the right to collect and report chosen data. In exercising their sovereignty, some tribes have instituted their own institutional reviews or privacy boards to review research in their communities, which is the case with the Navajo. With the creation of Data Sharing Agreements (DSA), data are turned over to the tribe once the study has been completed. Ownership of data means power, allowing the tribe to use them for health planning and to better serve their communities. Tribes and tribal members should receive payment for their participation. Tribes should know exactly where the data are

---

[24] G.A. Res. 2000A (XXI), 16 December 1966.

going and how outcomes will benefit them. Publication should avoid negative tribal representation, such as the numerous articles about Alaska alcoholics, including the many cases of foetal alcoholic syndrome (FAS).

One of the difficulties in finding a common ground is that Western thinking and language are most often expressed in abstract terms and are based on rational concepts, whereas Native thinking is affective and pertains to feelings.

Resistance from the 'Objects' of Investigation

Native American websites have shown for years letters of protest in response to the proposed Human Genome Diversity Project. For example, on 21 December 1993, Chief Leon Shenandoah, firekeeper of the Six Nations Confederacy, and the Onondaga Council of chiefs sent an e-mail to the National Scientific Foundation in Arlington, Virginia. They asked why the Project had progressed to its fifth meeting "without discussion or consent of the indigenous nations and peoples it affects." This letter followed an account given one month earlier of the proposed twenty-three million dollars in which up to 15,000 "specimens" would be collected, many from "isolates of historic interest," the terms used on the HGDP website. The objects of investigations were, it seems, viewed as "specimens," as items from our uncivilized past.

The ultimate goal is, quite simply, to find out "who we are as a species, and how we came to be," as expressed at the first organizational meeting in 1992. According to Margaret Lock, professor of anthropology at McGill University, "the scale of this project, its range through time and space, exhibits remarkable hubris."[25]

It is worth pointing out that the question of who exactly comes under the rubric of "indigenous" was never made clear; it was assumed that this was self-evident, a "factual category."

In 1993, the Rural Advancement Foundation International (RAFI), the Ottawa-based organization that had first alerted the World Council of Indigenous Peoples about the proposed HGDP, demanded that Indigenous peoples be involved at every stage of the project, to grant them veto powers, and to place the project under United Nations control – a request that was ignored for years.

---

[25] Margaret Lock, "The Alienation of Body Tissue and the Biopolitics of Immortalized Cells," *Body and Society* 7.2–3 (September 2011): 78.

Political and Legal Issues

Developments in genetic research have raised political and legal issues even within the US Senate. They include privacy concerns, inappropriate access and use of private genetic information, discrimination by employers or insurers based on genetic information, questions of who should have access to it, and debates over private versus public ownership of genetic information.[26]

The growing number of genetic databases has created many new problems: the scope of informed consent, the entry of information into databases using 'presumed consent', potential harm to individuals, communities, and groups, potential opportunities to stigmatize countries and groups, biological warfare.[27]

Many of the identified abuse, misuse, or unanswered question issues apply to Native peoples – taking genetic materials without informed consent or under false pretences. One example among many others is the use of blood samples drawn from the Navajo for paternity tests in speculative genetic tracking of ancient Navajo migration times and routes, when the Navajo blood donor hadn't consented to or even knew of such uses. Also, the failure to compensate donors, and the refusal to share discoveries with the providers of materials – particularly when it comes to medicines or treatment method – are issues that, if remedied, would help the Native peoples.

Some issues deserve special attention: the relationship issue has been a troublesome one for native peoples since 1492. Native peoples participated in the work of the World Commission on Environment and Development. The 1987 report defined them as "vulnerable groups."[28] Its conclusion is essential to the issue of the taking of native property:

> It is a terrible irony that, as formal development reaches more deeply into rain forests, deserts, and other isolated environments, it tends to destroy the only cultures that have proved able to thrive in these environments.

---

[26] James M. Jeffords & Tom Daschle (US Senate), "Political Issues in the Genome Era," *Science* 291/5507 (February 2011): 1249–51.

[27] James W. Zion, "The Rights of Native Peoples to Genetic Material as Cultural Property," paper presented at the Indigenous Institute for Indigenous Resource Management, January 2003.

[28] The World Commission on Environment and Development, *Our Common Future* (1987): 114–16.

One of the issues that deserve special attention is attack on native identity; journalists may misinterpret information. "Researchers find genetic markers unique to Africans"; "Asians biologically less susceptible to alcoholism"; "All Native Americans descended from a small number of founders." With such headlines, people are convinced that group differences are significant, which can bolster pre-existing prejudices and racism. As a matter of fact, the word 'race' is extremely common in the literature about genetic research.

There was a large controversy over 'Kennewick Man', found in the shallows of the Columbia River: did those remains that did not 'look Indian' prove that Indians were not the first inhabitants of the continent?[29] Another controversy arose when a member of the Vermont legislature introduced a bill to establish standards and procedures to do genetic testing to determine the identity of people as Indians.[30] For some, Native identity is a matter of cultural and political concern. It is doubtful that there is a 'native gene' that can be used as an identifier.[31] There is a danger that population-specific research will reinforce stereotypes about race and ethnic differences.[32]

Besides, the doctrine of 'prior informed consent' derives from medical ethics, where it concerns the patient's right to agree to or refuse certain medical treatments after being informed by the practitioner about the risks and benefits. The concept extends increasingly to other fields, notably to medical research using human tissue. The 2005 UNESCO Declaration on Bioethics and Human Rights provides that both scientific research and medical interventions "should only be carried out with the prior, free, express and informed consent of the person concerned." This approach would appear to require a patient's express consent in the event that samples taken in the course of the medical intervention are used for research purposes.

But a further issue then arises. What if genetic materials, taken from the human body and used as inputs for research, subsequently lead to biotechno-

---

[29] Kimberly TallBear, "Genetics, Culture and Identity in Indian Country," paper presented at the 7th International Conference of Ethnobiology in Athens, Georgia, 23–27 October 2000.

[30] TallBear, "Genetics, Culture and Identity in Indian Country".

[31] Brett Lee Shelton & Jonathan Marks, "Genetics 'Markers': Not a Valid Test of Native Identities," Indigenous Peoples Council on Biocolonialism Briefing Paper, May 2001.

[32] Dorothy Nelkin, "A Brief History of the Political Work of Genetics," *Jurimetrics* 42.2 (2002): 121–32.

logical inventions which are then patented? Should consent over use of research inputs also extend to the patent of research outputs? Should a separate consent then be obtained for each stage?

Free, prior, and informed consent is a cross-cutting theme, affecting biotechnological innovation in bio-prospecting. While the UNESCO Declaration sets prior informed consent in the context of human dignity and autonomy, the Convention on Biodiversity (CBD) links it to the sovereignty of nations over their resources and the interests of Indigenous and local communities.

The appropriate linkage between consent arrangements and the patent system is the subject of international debate and of several international legal processes.[33]

One of the issues for Native groups is jurisdiction over outsiders to control the taking of genetic materials. For example, there is a private blood-collecting business in Gallup, New Mexico (which is surrounded by the Navajo nation) where many Navajo give blood for money. To what extent do they know what will be done with their blood? To what extent are Navajo officials aware that research that will go around its Institutional Review Board will be done using those samples?

Aside from the government's being unable to control samples given at a blood bank or in paternity tests, they cannot control the sharing of materials outside of native areas.

Among the Navajo, elders stress that genetic research is not the same as other types of research, because it deals with the individual's body parts (blood, hair, and saliva) – very sacred to the Navajo. Moreover, most illness is attributed to the result of mishandling of body specimens separated from the body, something that the Navajo surgeon Lori Arviso Alvord confirms.

In most Native cultures, there is reverence for the dead, and it is sacrilegious to conduct research on the deceased.[34] Collecting hair, blood or nail-clippings may be seen as the intention to practise witchcraft.[35]

---

[33] Anja van der Ropp & Tony Taubman, "Bioethics and Patent Law, the Cases of Moore and the Hagahai," *WIPO Magazine* (World Intellectual Property Organization) (September 2006): 16–17.

[34] Dr Ben Muneta, Navajo, National Institute of General Medical Sciences and National Human Genome Research Institute, American Indian and Alaska Native Genetic Research Policy Formulation Meeting: Summary meeting report, 7–9 February 2001.

The HGDP Model Ethical Protocol for Collecting DNA samples states in section NA.2 that group consent is required and that, if a given group does not wish to participate, that will bind the researcher. Unfortunately, researchers may go to large collections of genetic material to get the refusing group's DNA.

As for property and cultural property, there is a great deal of discussion in contemporary international law regarding whether or not the law should recognize the wishes and desires of Native peoples and their consensus on the nature and kinds of protection they need, and whether or not it should be based on the recognition of the World Commission on Environment and Development, which says that Native peoples are "vulnerable" but hold keys to the future of mankind. There are some fourteen international or regional treaties and international agreements that cover the subject of "cultural protection."[36]

Cultural property is the most distinguishing form of a culture's expression. It is a culture's archaeological remains, ethnological materials, art, and architecture, its historically and politically important memorabilia, literature, traditional dance, customs, and ceremonies, and whatever else, tangible or intangible, is believed crucial for defining a people, community, or country. In its broadest sense, cultural property is whatever is thought to make a culture what it is – the forms of expression that consciously determine and identify it.[37]

The United Nations Declaration on the Rights of Indigenous Peoples addresses issues related to the return of cultural heritage and states "the right to the restitution of cultural, intellectual, religious and spiritual property taken without a people's free and informed consent or in violation of their laws, traditions and customs." The Declaration also includes the right to the use and control of ceremonial objects and the right to the repatriation of human remains.[38]

---

[35] Maureen Trudelle Schwarz, *Molded in the Image of Changing Woman: Navajo Views on the Human Body and Personhood* (Tucson: Arizona UP, 1997).

[36] Robert C. Ellickson, *Order without Law: How Neighbors Settle Disputes* (Cambridge MA: Harvard UP, 1991).

[37] Arti Kaur Rai, "Regulating Scientific Research: Intellectual Property Rights and the Norms of Science," *Northwestern University Law Review* 94.1 (1999): 77–88.

[38] Daniel Shapiro, "Repatriation: A Modest Proposal," *International Law and Politics* 31 (1998): 95–96.

The question of 'who owns genetic materials' is still unsettled. In most cases involving Indigenous people, the question is 'informed consent' before the taking of plant, animal, or human cells. In the situation where researchers obtain samples of material where the consent was for something else (such as paternity tests), the problem would be proving damages.

In the 1990s, the proliferation of disputes suggests that social conceptions of the body serve important purposes for individuals and society. Therefore, in more recent decisions, the federal government, the courts, and institutional review boards have begun to apply values other than mere scientific progress. The personal feelings of individuals about maintaining body integrity are increasingly recognized by the courts.

Cultural norms, too, are beginning to influence the treatment of body tissue. The North American Advisory group to the HGDP has emphasized the importance of sensitivity to cultural values.

Legal decisions about historical remains show how the social meaning of the body can take precedence over its potential scientific use. This is how NAGPRA redefined Native American remains as the personal or tribal property of descendants, requiring their repatriation.

In an effort to respond to its numerous critics, the North American Regional Committee of the HGDP published a proposed model ethical protocol in 1997 with a number of provisions including the following: any financial reward accruing from the specific analyses instigated by the HGDP should entitle individuals or populations who donate blood to receive monetary compensation. Permission is to be obtained from individuals and/or groups; a respected international body such as UNESCO should act as trustee or overseer of negotiations.

None of these suggestions, of course, could be made to apply to multinationals and privately sponsored gene-prospecting.

HGDP supporters claim that assistance may be given as training for local staff; it should be an opportunity to promote their "development," but serious concern has been expressed by several communities that such initiatives may just divert finances away from the implementation of urgently needed public-health projects.[39] They were considered as another form of colonialism.

HGDP planners have asserted that this project will also provide information on the genetic pattern of disease responsibility. However, the project

[39] Migues Baumann et al., *The Life Industry: Biodiversity, People and Profits* (London: Intermediate Technology Publications, 1996).

makes no provision for this, because there are no plans to collect information about the local environment, phenotypic data, individual life stories, and nutritional practices and disease histories to match with DNA samples – all essential information before disease susceptibility can be seriously researched. There is a significant danger that participants may be misled into believing cures for disease are imminent whereas research into therapeutics has never been among the objectives.

What Role Can Tribes Play to Protect Their Rights?

Increasingly, many racial and ethnic groups have demanded an active role in planning and carrying out research conducted on their members. For groups, increased involvement in genetic studies is often viewed as a means of ensuring that research adequately benefits the group, either by directly addressing its health care needs or by providing other benefits, such as employment, education, or infrastructure development. Fear of being financially exploited, based in part on past examples of 'sample and run' or 'helicopter' research, where researchers or corporations profited from patents or technologies obtained through research into racial or ethnic groups without compensating the group involved. As a result, some of those groups have chosen to limit their research involvement to studies strictly investigating health issues and have voiced opposition to large population genetics projects, such as the HGDP.

Concerns have been expressed that DNA may be patented by scientists or corporations. In some cases, groups found such patents particularly offensive because the groups attributed sacred status to the materials involved. Concerns over patenting may also reflect the fear that, in acquiescing to genetic research, groups risk losing control of their biological heritage and with it the ability to make claims about their identity and culture. They also reflect the view held by many Indigenous groups that their genetic material is a collective possession, the ownership of which is not relinquished through participation in research. In recent years, a number of groups participating in genetic research have sought benefit-sharing arrangements with researchers whereby the groups retain some portion of the intellectual property rights over information or products resulting from research.[40]

Fears exist across Native American tribes that genetic material collected for research purposes will be used in ways detrimental to the tribes, either to

[40] Donna Dickenson describes some of those agreements in *Who Owns Your Body?* (Oxford: Oneworld, 2004): 109–24.

profit researchers without compensating tribes or to undermine the sovereignty of tribes by countering claims that they are Indigenous to the regions they inhabit. They see relevant precedents in the patenting of indigenous plant varieties and in the assertions by geneticists and anthropologists that Kennewick Man is not genetically related to current Native American tribes, and thus is not protected by NAGPRA.

In response, tribes have been particularly stringent in reviewing proposals for genetic research, demanding that studies focus solely on identifying genetic aspects of particular diseases affecting the tribe and not on analysing the origins or ancestral composition of the tribe. Even so, the policies of the various tribes are totally different: the Navajo voted their Institutional Review Board (IRB) in 2001,[41] the Pima were willing to cooperate for diabetes research, whereas the Havasupaï sued ASU for misusing their blood samples. Like the Navajo, some Indian nations have enacted legislation to regulate and limit research. The Indigenous Peoples Council on Biocolonialism developed a model Indigenous Research Protection Act, and the American Indian Law Center, Inc. has a Model Tribal Research Code. These kinds of legislation are important because they give some control. There are enforcement difficulties, however. Given the existence of human gene banks outside native areas where massive amounts of genetic material have already been gathered, and given the means of casual access to native areas for bioprospecting, such codes can be difficult to enforce. There is no precise information on what repositories of DNA materials are in existence, because of an obvious secreting-away of DNA material, usually in the hope of profiting from its transformation into pharmaceutical materials. In 1995, there were 148 commercial or academic repositories in the USA and seventy-five in Canada. Enforcement mechanisms are also expensive in administrative terms. Another problem is that unless the researcher of a commercial firm actually enters a native area to conduct activities there, a native court would have no jurisdiction. There are also limitations on civil jurisdiction over non-members.

How can regulation be monitored and enforced, and at whose expense, particularly when so much research is initiated by the private sector?

The need for a clear set of rules is becoming urgent, given the proliferation of private corporations that promise consumers insight into their ethnic origins.

---

[41] The Navajo Nation Human Research Review Board's goal is "to support research that promotes the interests and the vision of the Navajo People."

Genetic ancestry tests, which can cost just a few hundred dollars and require only a simple cheek swab, are gaining popularity among genealogy hobbyists. For the moment, it is an unregulated no-man's land with little oversight and few industry guidelines to ensure the quality and interpretation of information sold.

While the federal Office of Human Research Protection requires researchers to obtain consent from donors of DNA, the rules are not clear about how scientists can then use these samples. In the Havasupaï case, for example, samples were not tagged to individuals' names, so the scientists assumed that they were free to use them for later studies. The problem is that, because science can now identify ancestry behind the DNA, such samples can be used to draw conclusions about small, possibly vulnerable, groups of people.

Developing a set of rules is challenging because of the diverse interests of the various groups involved in genetic testing, such as for-profit companies, academic scientists, casual consumers, Native American tribes, and specific ethnic or *racial* subsets of the population. Communication is obscured by unclear terminology and disagreements about the nature of concepts such as 'origin'. To geneticists, the word conjures up visions of genetic markers. But to Native Americans it is a location or landscape important to the tribe's cultural identity. It seems there is a need for stronger federal oversight.[42]

Conclusion

Who 'owns' genetic material? Individuals? Communities? Corporate organizations? Mankind? For some, DNA cannot belong to anyone; alternatively, it belongs to all of us. For others, ownership through patenting is essential if scientific research is to remain competitive.

Above all, it is questions of stigmatization, discrimination, and eugenics associated with investigations into genetic diversity that are the greatest source of anxiety. In North America, it seems unlikely that targeted groups will voluntarily cooperate with such research unless individual and group identity is rigorously protected. Furthermore, it must be made certain that ongoing legal negotiations with governments, most of them in connection with land claims, will in no way be jeopardized. Response by Native Americans to descriptions of Kennewick Man makes it clear just how political the issue is.

---

[42] Kimberly TallBear, "Genetics, Culture and Identity in Indian Country."

Local tabloids and radio talk shows referred to Kennewick man as "a white man" and suggested that this discovery "changes everything with respect to the rights of Native Americans."[43]

Given that situation, only those projects designed to investigate disease causation that promise therapeutic innovations are likely to be acceptable to most people.

The HGDP was designed solely for scientific knowledge, and profit was not the stake. *But*, in contrast to the HGDP, the activities of the 'Gene Giants' are rarely exposed to public scrutiny. Patenting actually promotes secrecy. Thus far, neither the United Nations Human Rights commission, the World Health Organization, nor UNESCO's bioethical committee has taken official positions on human DNA collection.

Yet today there is sufficient international interest in cultural rights that soft norms are now materializing rapidly, and there are possibilities of litigation to protect individual and group rights. Native peoples have a receptive audience in various governmental and non-governmental organizations, including scientific organizations that set standards for genetic research. The issues, however, remain serious.

WORKS CITED

Aldama, Arturo J. *Violence and the Body: Race, Gender, and the State* (Bloomington: Indiana UP, 2003).

Andrews, Lori B. *Future Perfect: Confronting Decisions about Genetics* (New York: Columbia UP, 2001).

Andrews, Lori B., & Dorothy Nelkind. "Whose Body is it Anyway? Disputes over Body Tissue in a Biology Age," *Lancet* 351 (3 January 1998): 53–56.

Baumann, Migues, Janet Bell, Florianne Koechlin & Michel Pimbert. *The Life Industry: Biodiversity, People and Profits* (London: Intermediate Technology Publications, 1996).

Beech, Bettina M. *Race and Research: Perspectives on Minority Participation in Health Studies* (Washington DC: American Public Health Association, 2004).

Bensen, Robert, ed. *Children of the Dragonfly: Native American Voices on Child Custody and Education* (Tucson: U of Arizona P, 2001).

Cambrosio, Alberto, & Peter Keating. *Exquisite Specificity: The Monoclonal Antibody Revolution* (Oxford: Oxford UP, 1995).

[43] Douglas Preston, "The Lost Man," *New Yorker* (16 June 1997): 70–81.

Declaration of Indigenous Peoples of the Western Hemisphere Regarding the HGDP (Phoenix AZ, 1996).

Deloria, Vine. *Custer Died for your Sins* (Norman: U of Oklahoma P, 1963).

——. *Red Earth, White Lies: Native America and the Myth of the Scientific Fact* (Golden CO: Fulcrum, 1995).

Dickenson, Donna. *Who Owns Your Body?* (Oxford: Oneworld, 2004).

Ellickson, Robert C. *Order without Law: How Neighbors Settle Disputes* (Cambridge MA: Harvard UP, 1991).

Grounds, Richard A. "The Yuchi Community and the HGDP," *Cultural Survival Quarterly* (Summer 1996): 64–68.

Hamer, Dean, & Peter Copeland. *The Science of Desire* (New York: Simon & Schuster, 1995).

Harmon, Amy. "DNA Gatherers Hit Snag: Tribes don't Trust Them," *New York Times* (10 December 2006).

Hodge, Felicia Schanche, Steven Weinmann, & Yvette Roubideaux. "Recruitment of American Indians and Alaska Natives into Clinical Trials," *Annals of Epidemiology* 10 (2000): S41–S48.

Jeffords, James M., & Tom Daschle (US Senate). "Political Issues in the Genome Era," *Science* 291/5507 (February 2001): 1249–51.

Landecker, Hannah. "Between Beneficience and Chattel: the Human Biological in Law and Science," *Science in Context* 12.1 (1999): 203–25.

——. *Culturing Life: How Cells became Technologies* (Cambridge MA: Harvard UP, 2007).

Lewontin, Richard. "The Dream of the Human Genome," *New York Review of Books* 39.10 (28 May 1992): 31–40.

Lock, Margaret. "The Alienation of Body Tissue and the Biopolitics of Immortalized Cells," *Body and Society* 7.2–3 (September 2001): 63–91.

Muneta, Ben (Dr.). American Indian and Alaska Native Genetic Research Policy Formulation Meeting, National Institute of General Medical Sciences and National Human Genome Research Institute (February 2001).

Nelkin, Dorothy. "A Brief History of the Political Work of Genetics," *Jurimetrics* 42 (2002): 121–32.

——, & Laurence Tancredi. *Dangerous Diagnostics* (Chicago: U of Chicago P, 1995).

Oberly, Joyce, & Jocelyn Macedo. "The R word in Indian Country: Culturally Appropriate Commercial Tobacco-Use Research Strategies," *Health Promotion Practice* (October 2004): 354–62.

Preston, Douglas. "The Lost Man," *New Yorker* (16 June 1997): 70–81.

Ray, Arti Kaur. "Regulating Scientific Research: Intellectual Property Rights and the Norms of Science," *Northwestern University Law Review* 94.1 (Fall 1999): 77–152.

Shapiro, Daniel. "Repatriation: A Modest Proposal," *International Law and Politics* 31 (1998): 95–96.

Sharp, Richard R., & Morris W. Foster. "Community Involvement in the Ethical Review of Genetic Research: Lessons from American Indians and Alaska Native Populations," *Environmental Health Perspectives* 110, Supplement 2 (April 2002): 145–48.

Shelton, Brett Lee, & Jonathan Marks. "Genetic 'Markers': Not a Valid Test of Native Identities," Indigenous Peoples Council on Biocolonialism Briefing Paper (May 2001).

TallBear, Kimberly. "Genetics, Culture and Identity in Indian Country," The World Commission on Environment and Development, *Our Common Future* also known as the *Bruntland Report* (1987).

Te Pareake Mead, Aroah. "Genealogy, Sacredness and the Commodities Market," *Cultural Survival Quarterly* 20.2 (Summer 1996): 46–51.

Trudelle Schwarz, Maureen. *Molded in the Image of Changing Woman: Navajo Views on the Human Body and Personhood* (Tucson: Arizona UP, 1997).

Van der Ropp, Anjo & Tony Taubman. "Bioethics and Patent Law, the Cases of Moore and the Hagahai," *WIPO Magazine* (World Intellectual Property Organization) (September 2006).

Zion, James W. "The Rights of Native Peoples to Genetic Material as Cultural Property," paper presented at the International Institute for Indigenous Resource Management (January 2003).

⌘

The Geneticization of Ethnicity and Ethnicization of Biomedicine

On the "Taiwan Bio-Bank"*

YU–YUEH TSAI

1 Introduction

TAIWAN IS AN IMMIGRANT, ISLAND SOCIETY where inter-ethnic marriages have been common. Its 'four great ethnic groups' (*sida zuqun*) – the Hoklo, Hakka, Mainlanders, and aboriginal peoples – exist only as a social construction that arose in the 1990s in a specific political-cultural context.[1] In 2005, a major government-sponsored research pro-

---

* This article is a revised version of my earlier article under the title "Geneticizing Ethnicity: A Study of the Taiwan Bio-Bank" as published in *East Asian Science, Technology and Society: An International Journal (EASTS)* 4.3 (2010): 433–55. An earlier version of this essay was presented at "The Society for Social Studies of Science (4S) and European Association for the Study of Science and Technology (EASST) Joint Annual Conference: Acting with Science, Technology and Medicine" (Rotterdam, August 20–23, 2008) and "Workshop on Medicine, Technology and Taiwanese Society" (The Institute of Sociology, Academia Sinica, Taipei, Taiwan, 20 November 2009). The author sincerely thanks Professors Kuo–ming Lin 林國明, Fu–Chang Wang 王甫昌, Sean Hsiang–Lin Lei 雷祥麟, and two anonymous reviewers for their useful comments and suggestions. My thanks also go to my postdoc advisor, Professor Steven Epstein at the Science Studies Program, University of California, San Diego, for his inspiring guidance and encouragement. This research is part of the project funded by the Taiwan Merit Scholarship from Taiwan's National Science Council (NSC-095–SAF–1–564-637–TMS).

[1] During the postwar period, the ethnic categorization of Taiwan's population was based on the 'original domicile' system. Derived from this criterion, the distinction

ject, the Taiwan Biological Sample Bank – or Taiwan Bio-Bank (hereafter TBB) – was organized by a group of scientists and physicians. The purpose of the project is to collect genetic data from the 'four great ethnic groups' of Taiwan in order to build a national database.[2] An ethnic categorization that emerged in the past fifteen years or so has been taken for granted in a cutting-edge study.

Focusing on the TBB, I hope to analyse the relationship between the rise of biomedicine and the geneticization of ethnicity in Taiwan by tracing the increasingly intimate ties between genetic discourse and the conception of the four great ethnic groups. My analysis is based mainly on primary data, including such archival records as the "Preliminary Feasibility Report on the Establishment of the Taiwan Biological Sample Bank," the "Final Report of the National Science Council's Research Grant Proposal: The Completion of the Taiwan Biological Sample Bank," "A Research Proposal for a Population Genetic Database for Taiwan: The Minutes of the 24th Academicians' Meeting," and the like. I also rely on such secondary sources as newspapers, magazines, and websites, emphasizing materials prepared by scientists themselves.

It is still early days for the TBB, which is (at the time of writing) in the midst of its pilot study and has not yet reached the laboratory stage. But it is important not to postpone an examination of how the project has been shaped politically and culturally and to point out its potential unintended social consequences. I argue that the TBB, meant to identify the definitive genetic markers of human similarity and difference, opens up new possibilities for the essentialization of ethnicity – even though biomedical scientists presumably do not mean to provide grist for the mill of ideological or political debates. This project, I argue, would make a significant contribution to the construction of Taiwanese identity-politics by encouraging the view that Taiwanese

---

between 'local Taiwanese' and 'mainlanders' has shaped the subsequent political and cultural dynamics. The original domicile system was abolished after major institutional changes in the early 1990s, and the classification of the 'four great ethnic groups' has largely replaced it. Still, the old distinction between local Taiwanese and Mainlanders lingers.

[2] Chen–Yang Shen 沈志陽, Yuan–Tsong Chen 陳垣崇, Der–Tsai Lee 李德財, & Chien–Te Fan. 范建得. "Final Report of the National Science Council's Research Grant Proposal: The Completion of the Taiwan Biological Sample Bank," *National Science Council*, NSC 94-3112-B-001-017 (2007). Anon., *The Preliminary Project of the Establishment of Taiwan Biological Sample Bank* (1996–2011), http://www.twbiobank.org.tw (accessed 23 April 2010).

people as a whole are genetically unique. My analysis of the TBB case shows that at a specific social and historical moment science and politics may be mutually constitutive. A laboratory project like the TBB may be built atop politically and culturally loaded conceptions that invalidate all claims to scienticity. When those involved make theoretical and methodological assumptions about the 'four great ethnic groups', they neglect the problematical history of Taiwanese society, particularly the long series of clashes over ethnic politics. Unless biomedical scientists address these complicated issues by drawing on findings from the social sciences, their project may come to serve as the unreliable foundation for dubious cultural and political causes.

2 Genes, Race, Ethnicity, and Identity Politics

Biomedicine and biotechnology have developed rapidly since the late-twentieth century. The Human Genome Project initiated in 1990 by the US Department of Energy and the National Institutes of Health was a significant moment. In the 2000s, scientists announced the first working draft sequence of the entire genome, and by April 2003 the project was complete. Since then, many countries have set about building national human genetic databases. Such projects, which involve vast numbers of genetic samples categorized by racial and ethnic labels, have brought about changes in contemporary identity-politics.

First of all, *it has become increasingly difficult to distinguish the concept of race from that of ethnicity in biomedical and medical studies*. Some context is necessary here. In the humanities and social sciences, the notion of race was replaced by the concept of ethnicity in the 1960s. Ethnicity accounts for human variations in culture, traditions, language, social patterns, and ancestry, while the discredited idea of race pictured a humanity divided into fixed, genetically determined, biological types.[3] Since the 1980s, there has been concern about a possible revival of genetic and biomedical research based on racial or ethnic categories, which would take for granted the discredited yet resilient idea that such groups have essential, biological, and immutable characteristics. In some cases, the use of ethnic categories simply served as a cover for familiar racial criteria (such as skin colour), racial categories (such

[3] Bill Ashcroft, Gareth Griffiths & Helen Tiffin, *Key Concepts in Post-Colonial Studies* (London: Routledge, 1998).

as continental groups), and racial labels (such as Caucasian).[4] When biomedicine began to develop rapidly, these two concepts were increasingly lumped together. Their conceptual borderline was blurred.

Can genes account for racial or ethnic differences? It depends on who you ask: the conceptualizations of race and ethnicity in the social sciences differ markedly from discussions in the medical field. In 1998, the following statement was adopted by the American Anthropological Association:

> With the vast expansion of scientific knowledge in this century, [...] it has become clear that human populations are not unambiguous, clearly demarcated, biologically distinct groups. Evidence from the analysis of genetics (e.g., DNA) indicates that most physical variation, about ninety-four percent, lies within so-called racial groups. Conventional geographic 'racial' groupings differ from one another only in about six percent of their genes.[5]

The statement is representative of the prevailing view in the contemporary social sciences. Many social scientists have questioned the assumption that race is a scientific or objective reality, contending that it is forged from the discourses of politics, society, and history.[6] However, as Alan H. Goodman points out in his article "Why Genes Don't Count (For Racial Differences in Health)," genetic variation continues to be used to explain alleged racial differences.[7] Goodman shows that such explanations require the acceptance of two disproved assumptions: that genetic variation explains disease, which is a

---

[4] Andrew Smart, Richard Tutton, Richard Ashcroft, Paul Martin, Andrew Balmer, Richard Elliot & George T.H. Ellison, "Social Inclusivity versus Analytical Acuity? A Qualitative Study of UK Researchers Regarding the Inclusion of Minority Racial/ Ethnic Groups in Biobanks," *Medical Law International* 9.2 (2008): 169–90.

[5] American Anthropological Association, "American Anthropological Association Statement on 'Race'" (17 May 1998), *American Anthropologist* 100.3 (1999): 712–13.

[6] Alan H. Goodman, "Why Genes Don't Count (For Racial Differences in Health)," *American Journal of Public Health* 11 (2000): 1699–702; Troy Duster, "Selective Arrests, an Ever-Expanding DNA Forensic Database, and the Specter of an Early Twenty-First-Century Equivalent of Phrenology," in *DNA and the Criminal Justice System: The Technology of Justice*, ed. David Lazer (Cambridge, MA: MIT Press, 2004): 315–34; Troy Duster, "Policy Forum: Race and Reification in Science," *Science* 307 (2005): 1050–51; and Robert S. Schwartz, "Racial Profiling in Medical Research," *New England Journal of Medicine* 344 (2001): 1392–393.

[7] Goodman, "Why Genes Don't Count (For Racial Differences in Health)."

form of geneticization; and that genetic variation explains group pathologies, which is a kind of racialization and scientific racism. A methodology that exaggerates the salience of 'race' can ignore other potential causes of disease.

Some biomedical and medical researchers believe that gene frequencies cluster by continent, supporting the definition of race based on continental origins. Moreover, they contend that people share a significantly larger number of human leukocyte antigen types with members of their own ethnic or racial group.[8] By contrast, some scientists, such as Raj Bhopal and Liam Donaldson, argue that it is time to abandon race as a variable in public health.[9] A 1998 manifesto by Mindy Fullilove that appeared in the *American Journal of Public Health* recognizes the salience of race in personal identity but calls for the abandonment of race as a formal variable in public health. Another journal, *Archives of Pediatrics and Adolescent Medicine*, announced a new policy in 2001 that required that authors not "use race and ethnicity when there is no biological, scientific or sociological reason for doing so."[10] Similarly, Francis Collins, the director of the National Human Genome Research Institute, wrote:

> Race and ethnicity are poorly defined terms that serve as flawed surrogates for multiple environmental and genetic factors in disease causation. [...] Research must go beyond these weak and imperfect proxy relationships to define the more proximate factors that influence health.[11]

In newly developed bio-bank projects, the concepts of race and ethnicity usually intertwine. Taking the United Kingdom Bio-Bank as a case study, Andrew Smart et al. showed that the ethnic group traits that researchers tend to assume rarely square with the traits that may be relevant to biomedical science (including genotypic, phenotypic, sociocultural, and socioeconomic

---

[8] Esteban González Burchard, "Latino Populations: A Unique Opportunity for the Study of Race, Genetics, and Social Environment in Epidemiological Research," *American Journal of Public Health* 12 (2005): 2161–68.

[9] James W. Buehler, "Abandoning Race as a Variable in Public Health Research," *American Journal of Public Health* 5 (1999): 783.

[10] Steven Epstein, *Inclusion: The Politics of Difference in Medical Research* (Chicago: U of Chicago P, 2007).

[11] Francis S. Collins, "What We Do and Don't Know about Race, Ethnicity, Genetics and Health at the Draw of the Genome Era," *Nature Genetics Supplement* 36 (2004): S13–S15.

characteristics).[12] There is no consensus about what race and ethnicity mean and how they can be operationalized.[13] No one has satisfactorily addressed the operational problems raised by the collection of ethnicity data in bio-bank studies. Indeed, bio-banks could create problems for ethnic groups by emphasizing alleged differences among such groups and thus essentializing them.[14] This important issue remains relatively under-researched.

The second major effect that contemporary research on the human genetic profile has had on identity politics is *the increasing potential that biomedicine has to facilitate the biological expression of social identity and difference.* A number of social scientists have emphasized that novel biomedical techniques make new identity-formations possible.[15] Scholars have begun to speak of a "geneticization of identity."[16] This applies to such existing categories as race, ethnicity, and even national identity. For example, DNA may be regarded as a national resource and a repository of national characteristics; the genomic revolution has furnished potent resources for the expression of nationhood and shared origins. In the same vein, the idiom of genomics can provide a powerful resource for the expression of social differences.[17] Bob Simpson takes Iceland's biogenetic project, DeCODE, as an example of the geneticization of ethnicity.[18] Arnar Arnason and Bob Simpson trace how the images

---

[12] Smart, et al., "Social Inclusivity versus Analytical Acuity?"

[13] George T.H. Ellison, Andrew Smart, Richard Tutton, Simon M. Outram, Richard Ashcroft & Paul Martin, "Racial Categories in Medicine: A Failure of Evidence-Based Practice," *PLoS Medicine* 9 (2007): 1434–436.

[14] Ruth Chadwick, "Genomics, Public Health and Identity," *Acta Bioethica* 2 (2003): 209–18.

[15] Epstein, *Inclusion: The Politics of Difference in Medical Research*; Jenny Reardon, *Race to the Finish: Identity and Governance in an Age of Genomics* (Princeton NJ: Princeton UP, 2005); Margaret Sleeboom–Faulkner, "How to Define a Population: Cultural Politics and Population Genetics in the People's Republic of China and the Republic of China," *BioSocieties* 1 (2006): 399–419; and *New Genetics, New Identites*, ed. Paul Atkinson, Peter Glasner & Helen Greenslade (London: Routledge, 2007).

[16] Deborah Heath, Rayna Rapp, & Karen–Sue Taussig, "Genetic Citizenship," in *A Companion to the Anthropology of Politics*, ed. David Nugent & Joan Vincent (Malden MA: Blackwell, 2004): 153–67.

[17] Atkinson, et al., ed. *New Genetics, New Identities.*

[18] Bob Simpson, "Imagined Genetic Communities: Ethnicity and Essentialism in the Twenty-First Century," *Anthropology Today* 16.3 (June 2000): 3–4.

and metaphors drawn on by Icelanders on all sides of the debate furnish a biogenetic language for conceiving of the link between contemporary identity and the past.[19] DeCODE locates the essence of 'Icelandicness' in the very building blocks of people's bodies, offering novel ways of contemplating and asserting identity. This project has achieved a significant public-relations victory by highlighting features of the past that supposedly make Iceland uniquely appropriate for the development of the database.

Facilitating the biological expression of social identity and difference, biomedicine has significant potential to change identity-politics on the local level. How bio-banks are involved in the process of ethnic categorization, to the extent that they may also be involved in the evolution of identity-politics in different countries, has become a topic of interest.[20] This topic, however, like the racialization of ethnic difference in bio-bank projects, remains under-investigated. Focusing on the case of the TBB, I examine how socially constructed ethnicity is essentialized – if not racialized – and the effect it may have on ethnic politics in Taiwan. I hope to shed light on the complicated and intriguing relationship between genetic discourse and racial/ethnic construction in particular, and between science and politics in general.

3 The Development of the Taiwan Bio-Bank and Ethnic Politics

3.1 The Early Development of Biomedicine in Taiwan: National Economic Transition and Global Influence

In the 1980s, the government of Taiwan began to support the development of biotechnology. Within a decade, genetic research had achieved remarkable vigour, partly as a result of institutional support from the Executive Yuan, the National Science Council, and the Department of Health. In the 1990s, the focus was on basic and clinical research. Beginning in 1996, the Executive Yuan and the National Science Council cooperated on a project called the Advanced Research in Genetic Medicine and Sanitation Plan (ARGMSP). The phrase "advanced research" indicates the specific level of importance as-

[19] Arnar Arnason & Bob Simpson. "Refractions through Culture: The New Genomics in Iceland," *Ethnos* 4 (2003): 533–53.

[20] Arnason & Simpson, "Refractions through Culture: The New Genomics in Iceland."

signed to the plan: a lack of commercial interest kept the government from making it a national project.

But after the Executive Yuan held five meetings on biotech industry strategy in 1998, the National Science Council recommended that the ARGMSP be promoted to a national programme. Four years later, a National Research Program for Genomic Medicine (NRPGM) was inaugurated under the auspices of the Ministry of Economic Affairs, the National Science Council, and the Department of Health. Its goal is the use of human genomic knowledge to give Taiwanese medical research a competitive edge.

The birth of this ambitious programme shows that the state had made a commitment to playing a more active role in facilitating genetic discourse, formulating related policies, and organizing research. Furthermore, with the emergence of a global political economy in which biomedicine promises to play a hugely profitable role, such a programme may reap significant economic returns on investment. The formation of the NRPGM speaks volumes about the changing national interest in an era of globalization.

Science and technology development in Taiwan may be divided into two stages: a labour-intensive stage from 1952 to 1985, and after that a technology-intensive stage.[21] The production of electronics and computer components drove the "Taiwan miracle" over the past two decades. Since the mid-1980s, government support for science and technology and technology-intensive industries has increased. Expectations have been high that biotechnology and biomedicine could become the next engine of Taiwanese economic growth. In 2005, after the NRPGM had decided to establish the TBB, the *Formosan Journal of Medical Humanities* showed its support:

> To ensure progress both economic and technological, Taiwan needs a new direction. The computer industry has brought it much wealth in recent years, yet Taiwan cannot depend solely on that, so a new industry has been targeted – biotechnology. In order to qualitatively fortify Taiwan's bioresearch, a bio-bank seems to be indispensable.[22]

---

[21] See *Economic Development ROC (Taiwan)*, ed. Council for Economic Planning and Development, Executive Yuan (Taipei: Council for Economic Planning and Development Executive Yuan, ROC, 2007).

[22] See the journal editor's introduction to this issue, "The Debate of Establishing a Taiwan Biobank," *Formosan Journal of Medical Humanities* 7.1–2 (2006): 1–2.

The formation of the NRPGM, and the later establishment of the TBB, show the government's commitment to repositioning the national economy during a time of intense international scientific, technological, and economic competition.

3.2 From Disease Research to the National Human Gene Bank

On 3 July 2000, Academia Sinica, Taiwan's leading research institute, convened its 24th academicians' meeting, and the academician Zhuang Ming–Zhe proposed the establishment of a "genetic database for Taiwan" based on cooperation among Academia Sinica, concerned national agencies, schools, and local and foreign experts. With strong support from Academia Sinica's biology division, the motion was revised and passed by majority vote.[23]

Academia Sinica has long played a pivotal role in the formulation of national technology policy and scientific research. In particular, Lee Yuan–Tseh, who was awarded the Nobel Prize in Chemistry in 1986 while working in the USA, subsequently returned to Taiwan to head Academia Sinica; he has become the most influential figure in the advancement of the island's biotechnological aspirations.

The concepts of race and ethnicity were discussed in the initial stages of developing a Taiwanese bio-bank. Zhuang, the driving force behind the project, explained that genetic data collection would be a large inter-agency project and that samples for the bank would be taken only from willing participants. He also made it clear that *a certain number of samples would be taken from each of Taiwan's ethnic groups, proportionate to its share of the population.*[24]

---

[23] The proposed work included: (1) a conference of Taiwanese scientists, physicians, and ethicists to plan; (2) a genetic population database comprising permanent DNA samples from all Taiwanese who agreed to participate; (3) the database was to be available to qualified researchers to study (a) the effects of gene variation on Taiwanese health, (b) the effects of gene variation on response to medication, and (c) the genetic relationship between Taiwanese and other groups; (4) project members were to educate the Taiwanese public about the genetic contribution to health and well-being (5) and to study ethical issues raised by a population genetic database and its impact. See Zhuang Ming–Zhe 楊正敏, "Taiwan People's Genetic Bank, Great Initial Controversy," 台灣人群基因庫,籌備爭議大 [Taiwan Renqun Jiyin Ku, Choubei Zhengyi da] *United Daily News* *聯合報* [*Lianhe Bao*] (5 July 2000): A4.

[24] Yang, "Taiwan People's Genetic Bank, Great Initial Controversy."

This announcement provoked mixed reactions. At the 24th academicians' meeting, there was concern about the potential for "racial discrimination." Some academicians pointed out that genetic data might be used to argue that one racial group had a biological edge over another. The academician Jacqueline Whang–Peng reminded her colleagues that the term 'race' was itself contentious; she suggested "individual differences" as an alternative.[25] Yang Chen–Ning, a 1957 Nobel laureate in physics, brought attention to the sensitive nature of discussions of alleged racial differences based on genes, and he asked the members of the biology division to explain how large the genetic differences among human populations were. In response, Wu Kun–Yu, who was to serve as the chief of the ARGMSP, insisted that the genetic database should be used only for disease research, not for ethnic studies, in order to steer clear of the thorny issue of race and political dispute.[26]

On 4 July 2000, the day after that meeting, Taiwan's major newspapers ran prominent stories about the proposed genetic database; the headline in the *China Times* read "Whose Genes Are Representative of the Taiwanese People?" Some journalists had evidently sensed the danger of applying ethnic categories to the study of Taiwan's genetic profile.[27] In spite of some concerns, the proposed genetic database had been approved during the meeting. Since that date, the project of building a genetic database, with the support of government, has moved ahead and a particular, though untenable, way of imagining Taiwan's ethnicity has shaped the TBB research design.

3.3 A Project in Progress: Identity Politics and the Taiwan Bio-Bank

Since 2000, Taiwan's national policy has shifted from emphasizing disease-based research to focusing on population-based research. Many of the scientists working on the TBB have embraced the concept of 'four great ethnic groups', itself a novel idea that dates from the early 1990s.

Perhaps we should not be surprised that the genetic distinctness of the Taiwanese people has become the focus of their discussion, as the popular

---

[25] Li–Wen Zhang, 張瓅文 "Academia Sinica Discusses Making a Citizens' Gene Bank," 中研院研議設置國人基因庫 [Zhongyanyuan Yanyi Shezhi Guoren Jiyin Ku] *China Times 中國時報* [*Zhongguo Shibao*] (5 July 2000): A5.

[26] Yang, "Taiwan People's Genetic Bank, Great Initial Controversy."

[27] Li–Wen Zhang, 張瓅文 "Whose Genes are Representative of the Taiwanese People," 誰的基因能代表台灣人? [Shei de Jiyin Neng Daibiao Taiwanren?], *China Times 中國時報* [*Zhongguo Shibao*] (5 July 2000): A5.

conceptual framework of ethnic groups has the potential to help construct a distinct Taiwanese identity. The database project is connected to a broad set of cultural and political transformations that have accompanied the emergence of Taiwanese identity over the last three decades or so. In order to examine how recent trends in identity-politics have shaped the conceptual framework of the TBB project, it is essential to understand the political changes of the past century.

A part of the Qing Empire until 1895, Taiwan was ceded to Japan after China's defeat in the Sino-Japanese War. The Xinhai Revolution, led by Sun Zhongshan, overthrew the Qing in 1911, and the Republic of China was established in 1912. The Chinese Nationalist Party (also known by its Chinese name, Guomindang), which was founded in 1919 and traced its origins to several political organizations founded by Sun, was led by Jiang Jieshi after Sun's death – it became the ruling party of China. After Japan lost the Second World War in 1945, Taiwan was reclaimed by the Republic of China. In 1949, the government of the Republic of China, controlled by the Nationalists, took refuge in Taiwan after it lost a civil war against the Chinese Communist Party. From then on, the Republic of China presented itself as the legitimate government of China. Taiwan, the Republic of China, and the Nationalist Party, effectively coterminous, were known in the international press as Free China. In 1971, Taiwan was forced to withdraw from the United Nations as the communist People's Republic of China supplanted it as the formal representative of China. Since then, Taiwan has been vexed by the problem of international recognition.

During nearly forty years of single-party rule, from 1949 to 1987, Taiwan experienced the most protracted period of martial law in the world. Political opposition was ruthlessly suppressed. Then, in 1986, the Democratic Progressive Party was established, and the era of multiparty politics began.

Since the 1990s, China has maintained its claim to sovereignty over the territory of Taiwan. At the same time, China has tried to isolate Taiwan's government, keeping it off the international stage. This island country has been marginalized in the global community because of that pressure. Meanwhile, democratic institutions have developed rapidly. In 1996, Taiwan carried out its first direct presidential election. Four years later, the election was won by the Democratic Progressive Party, a feat repeated in 2004. This represented a significant consolidation of the 'indigenization' or 'taiwanization' of politics and culture.[28] Over time, more people have come to identify themselves as Taiwanese rather than as Chinese, and a majority of the island's population now believes Taiwan should have a distinct political status instead of being seen as part of China.[29]

The development of the study on the particular genetic makeup of the Taiwanese people is intimately connected with an attempt to redefine who they are. In the initial proposal Zhuang submitted to the Academia Sinica meeting in July 2000, he emphasized:

> *Here are several potential uses of a Taiwanese Human Genetic Database*. It would help us learn more about the population's genetic structure, the migration history of Taiwan, and *the degree to which Taiwanese are genetically similar to and different from other Asian populations*. [...] Because there is no assurance that markers discovered in Western populations will also show variability among Taiwanese, specifically Taiwanese studies are needed.[30]

Another Academia Sinica scholar, Chen Jian–Ren, had a similar view:

---

[28] John Makeham & A–Chin Hsiau, *Cultural, Ethnic, and Political Nationalism in Contemporary Taiwan* (New York: Palgrave Macmillan, 2005).

[29] According to a series of surveys conducted by the Center for the Study of Elections at National Chengchi University, during the period 1992–2007, Taiwanese national identity underwent a dramatic transformation. The number of those who identified themselves as Chinese decreased while the number of those who identified themselves as purely Taiwanese rose from 17.3% to 43.7%. The survey conducted in June 2007 shows that 45.8% of interviewees identified themselves as both Taiwanese and Chinese. But while 5.5% identified themselves as exclusively Chinese, the above-mentioned 43.7% declared themselves exclusively Taiwanese. Another index, attitudes toward Taiwan's political future, also showed a significant change from 2001 to 2007. In general, there was an increase in the number of those who support Taiwan's political independence: from 28.1% in 2001 to 44.7% in 2007. The number of those who wished to preserve the status quo rose from 32.1% to 34%. By contrast, the number of those who supported the idea that Taiwan should be reunified with China declined from 20% to 14.4%. See the web page about "Changes in the Taiwanese/Chinese Identity of Taiwanese (1992–2007)," in Election Study Center. *Election Study Center, National Chengchi University* (2010), http://esc.nccu.edu.tw (accessed 23 April 2010).

[30] Ming–Tsuang Zhuang 莊明哲, "Research Proposal: A Population Genetic Data Base for Taiwan," *The Minutes of the 24th Academicians' Meeting, Academia Sinica* (3 July 2000): 3. (My emphases.)

> Taiwan's population is diverse and significantly different from Chinese around the world, as well as from Europeans; a national genetic bank would help Taiwan discover these differences and facilitate linguistic, cultural, and 'national' research.[31]

When the draft proposal for the TBB was being drawn up, Chen Yuan–Tsong and Shen Chen–Yang, two leading members of the TBB from the Institute of Biomedical Sciences, Academia Sinica, sent a letter to the editor of a newspaper. They tried to convince the public by setting out the merits of a national genetic database:

> *The genetic inheritance of the Taiwanese people is unique*; lifestyles and risk factors for disease differ from country to country. Therefore, we hope to build a biological database for Taiwan to look into the factors causing common chronic disease in Taiwan and to understand the impact of the interaction among genetic and environmental factors on such diseases in order to establish effective treatments and preventive strategies to safeguard the health of Taiwan's people. Because Taiwan's social environment is unique, this major research project will be relevant to our national biomedical development and the well-being of all citizens several decades from now.[32]

The emphasis on the genetic uniqueness of the Taiwanese people and on the potential contribution of the TBB has been repeated in a variety of project drafts and reports. In "The Taiwan Bio-Bank Project: For the Health of Future Generations," an article published in English on *Academia Sinica E-News* in 2007, Shen et al. emphasized the following:

> the Taiwan Bio-Bank plans to use prospective cohort studies based on ethnicity (population-based) that will help determine the effects of the environment or the gene alone and of gene-gene interactions and gene-environmental factor interactions in common diseases. [...] The ethnicity of Taiwan is unique and the Taiwan Bio-Bank can be expected to develop into the supply center for a Chinese database.[33]

---

[31] Yang, "Taiwan People's Genetic Bank, Great Initial Controversy."

[32] Yuan–Tsong Chen 陳垣崇 & Chen–Yang Shen 沈志陽, "Be Careful in Designing the Biological Database," 謹慎規劃生物資料庫 [Jinshen Guihua Shengwu Ziliaoku] *China Times 中國時報 [Zhongguo Shibao]* (2 April 2006): A15. (My emphasis.)

[33] Chia–Hao Ou 歐家豪 & Chen–Yang Shen 沈志陽, "The Taiwan Biobank Project: For the health of future generations," *Academia Sinica E-news* (19 April 2007). http://newsletter.sinica.edu.tw/en/file/file/1/125.pdf (accessed 19 April 2007).

It would be a mistake to ascribe the support for the TBB exclusively to the desire to reconstruct Taiwanese identity and to carve a niche for Taiwan in global politics, but the long-standing problem of international recognition has affected how social agents think about their existence, identity, and actions. Claims made by Zhuang, Chen, and Shen suggest that one of the main impetuses behind the call for establishing the TBB project is the desire of an internationally marginalized country to claim its own identity and niche, both domestically and internationally.

3.4 An Island of Biomedical Technology and the Taiwan Bio-Bank

In 2005, Taiwan's government announced the goal of transforming the country into "the island of biomedical technology" – at least NT$15 billion (US$ 480 million) would be provided. The plan has three parts. The first, a national health information project, will digitize and integrate all health records kept by the Bureau of National Health Insurance and other national programmes. The second part is a clinical medical research project. The third part is devoted to the Taiwan Bio-Bank.

To establish a national human genetic database, the Institute of Biomedical Sciences at Academia Sinica would collect and store blood samples and personal information from 200,000 participants aged thirty to seventy. Participants would be chosen to represent the 'four great ethnic groups' and would be drawn from three geographical areas. The medical records for the sample group, all kept at the Bureau of National Health Insurance, would be made available to the database team.[34]

This project was divided into three stages. From September to December 2003, a preliminary feasibility study was executed by the Institute of Biomedical Sciences. From August 2005 to July 2007, a second feasibility study was conducted; it was sponsored by the National Science Council and carried out by the Institute of Biomedical Sciences. The plan was to collect blood samples from 1,032 subjects in Jiayi, a city of 75,000 people in the island's southwest. Since December 2005, scientists have been carrying out the Preliminary Plan for Establishing the Taiwan Biological Sample Bank; work

[34] See the web page about "The Preliminary Project of the Establishment of Taiwan Biological Sample Bank," in Anon., *The Preliminary Project of the Establishment of Taiwan Biological Sample Bank* (1996–2011), http://www.twbiobank.org.tw (accessed 23 April 2010).

scheduled to conclude in October 2010 is still ongoing. They hope to compile 15,000 samples from three areas, relying on the usual method of encouraging subjects to participate by combining the data collection with health checkups and screenings for seasonal illnesses. The Institute of Biomedical Sciences anticipates that the larger study to follow will be funded by the Department of Health.

From December 2005 to January 2006, the National Research Program for Genomic Medicine, the National Science Council, Academia Sinica, and other groups organized a series of conferences where scientists from the humanities, the social sciences, and medicine discussed the scientific, legal, and ethical aspects of genetic research. All of the panel sessions were recorded on video for public access and free download. On 10 January 2006, the session on "The Human Genetic Database: Collecting and Studying Ethnic Blood Samples" had three keynote speakers – the history professor Fu Dai–Wie, the legal scholar Liu Hung–En, and the biomedical scientist Shen Chen–Yang. Shen asserted that ethnic background was the most important factor that influenced the genetic attributes and gene-related diseases of Taiwan's population:

> If we want to understand how genes influence the diseases of the Taiwanese people, [we must know that] the most important factor that influences the genes of Taiwan's population is clan [*shizu* 氏族], that is, [the distinction between] the Hoklo, the Hakka, the Mainlanders, and the aboriginal people. Different clans probably have different genes. Therefore, if you want to see how genes influence diseases, you have to take a look at people from different clan backgrounds. Now we hope that we can collect 15,000 examples [i.e. blood samples] from Taiwan's Miaoli County, Jiayi City, and Hualian County within two years, beginning in 2007.[35]

Although Shen, in a more or less casual way, used the concept of clan, a kinship term which in anthropology typically denotes a group of people with genealogical links (real or alleged) to a common ancestor, he definitely re-

---

[35] See the web page about "The Human Genetic Database: Collecting and Studying Ethnic Blood Samples," in Anon, "The Human Genetic Database: Collecting and Studying Ethnic Blood Samples," *'Gene Technology and Humanism' Lecture Series 「基因科技與人文議題」系列演講* (2005), http://elsi.nccu.edu.tw/lecture2005/index.htm (accessed 12 December 2008).

vealed that his conceptual framework was the popular 'four great ethnic groups'.

The design and development of the TBB show that there is an elective affinity between the development of biomedicine and ethnic politics which has been leading to a geneticized conception of ethnicity. As shown above, this affinity is contingent on domestic cultural and political changes and the globalization of biomedicine. A distinct form of Taiwanese identity may emerge as a result of the interaction between ethnic politics and economic interests, either in the local or the global arena. However, the potential negative social effects of the ideas about ethnicity presumed in preliminary discussions of the TBB have sparked ethical concerns. In fact, the TBB scientists are themselves aware of such effects. In the section of the 2008 ethical code of practice for the Preliminary Plan for Establishing the Taiwan Biological Sample Bank (Second Edition) devoted to "informed consent," the genetic study of ethnicity is said to run the risk of "stigmatizing specific human subjects and the groups they represent."[36] In spite of this admission, the TBB scientists have shown little awareness of the possibility that their own work might benefit from the insights of the social sciences. Moreover, when the Institutional Review Board of Academia Sinica concluded that the TBB postulated ethnicity as a significant variable, it pointedly questioned the conceptual validity of ethnicity and demanded that the project give full attention to the cultural, social, and political implications and effects of its work.[37] Still, the TBB scientists have shown little interest in thinking differently.

For the last decade or so, the pilot project has repeatedly been criticized by citizens' groups, legal scholars, indigenous peoples, and the media, who have expressed concerns about genetic privacy, informed consent, human rights, matters of law, technology policy-making, and the like.[38] However, the ad-

---

[36] See the website about "The Ethical Code of Practice for the Preliminary Project of Establishing Taiwan Biological Sample Bank (Second Edition)," in Anon., "The Ethical Code of Practice for the Preliminary Project of Establishing Taiwan Biological Sample Bank (Second Edition)," *Taiwan Biobank* (16 October 2008), http://www .twbiobank .org.tw/documents/081016/Doc00.pdf (accessed 12 December 2008).

[37] See the web page about "The Preliminary Project of the Establishment of Taiwan Biological Sample Bank," in Anon., *The Preliminary Project of the Establishment of Taiwan Biological Sample Bank* (1996–2011), http://www.twbiobank.org.tw (accessed 12 December 2008).

[38] Winona Liao 廖文孜, "A Preliminary Examination of the Collective Rights Issues in Human Genetic Resources, with a Discussion of Special Considerations for Abori-

verse effects the project could have as a result of drawing on the socially constructed ethnic categories have been overlooked.

4 Problems of Methodology: The Taiwan Bio-Bank and Ethnicity

In this section, I examine three epistemological and methodological assumptions about the 'four great ethnic groups': (1) Taiwan is ethnically homogeneous; (2) the classification of 'four great ethnic groups' is biologically valid; and (3) representative numbers of samples should be based on county boundaries. I hope to show that in defining ethnic groups, recent socio-political classifications used by the state have played an important role in the perception and formulation of genetic target groups in population genetics. However, historically, relations among various groupings in Taiwan have been subject to drastic change: the groups are relatively porous and inter-ethnic marriages are common. In a settler or immigrant society like Taiwan, which has a high degree of intermarriage and blood mixing, it is difficult to find 'pure' blood samples for each of the assumed four great ethnic groups.

> Assumption 1: Iceland is an isolated island and comparatively ethnically homogenous. Taiwan is similar to Iceland. If Iceland can establish its DeCODE biobank, so can Taiwan.
> Fact 1: Taiwan is an immigrant and hybrid society. It is different from Iceland.

---

gines," 人類基因資源的集體權利問題初探—兼論關於原住民族群的特殊考量 [Renlei Jiyin Ziyuan de Jiti Quanli Wenti Chtan – Jianlun Guanyu Yuanzhumin Zuqun de Teshu Kaoliang] *Newsletter of Biotechnology and Law 生物科技與法律研究通訊* [*Shengwu Keji Yu Falyu Yanjiu Tongxun*] 6 (2000): 29–37; Shu–Juo Chen 陳叔倬, "Ethical and Legal Dimensions of Aboriginal Genetic Research," 台灣原住民遺傳基因研究之倫理爭議與立法保障 [Yuanzhumin Renti Jiyin Yanjiu Zhi Lunli Zhengyi Yu Lifa Baohu] *Newsletter of Biotechnology and Law 生物科技與法律研究通訊* [*Shengwu Keji Yu Falyu Yanjiu Tongxun*] 6 (2000): 7–28; Hung–En Liu 劉宏恩, "Biobank Legalization Issues: International Development and the State of the Field in Taiwan," 人群基因資料庫法制問題之研究—國際上發展與台灣現況之評析 [Renqun Jiyin Ziliaoku Fazhi Wenti zhi Yanjiu – Guoji Shang Fazhan yu Taiwan Xiankuang zhi Pingxi] *Taipei Bar Journal 律師雜誌* [*Lüshi Zazhi*] 303 (2004): 71–94; and Kuei–Tien Chou 周桂田, "Conflicts of Technology Policy and Governance Paradigm in a Knowledge-Based Economy: A Case Analysis of the Construction of the Taiwan Biobank," *Issues and Studies* 3 (2007): 97–130.

From the earliest days of planning the TBB, people assumed a series of resemblances between Taiwan and Iceland. At the Academia Sinica meeting of July 2002, for instance, Zhuang Ming–Zhe argued that the Icelandic experience could be applied to Taiwan.[39] In 2003, a national newspaper reported:

> Academia Sinica is planning to establish a national "Super Control Genomic Database," making Taiwan the second country after Iceland to collect samples for a large-scale national gene bank. [...] Academia Sinica's Chen [Yuan–Tsong] said that many countries around the world have research programs of this nature, but so far only Iceland has carried out a national referendum to approve the establishment of a national gene bank. He stressed that Iceland's small population gives it an advantage in this respect, an advantage Taiwan shares, in contrast to the United States, where ethnic diversity augments the difficulty of such a program.[40]

Iceland attracted considerable international attention because of plans to use its country's population as a resource for establishing the first national genetic database. The project began in 1996. After complex and protracted negotiations, the government awarded the contract for creating this database to DeCODE, a biopharmaceutical company based in Reykjavik that is financed primarily by Americans. The company was granted a twelve-year exclusive monopoly license.[41]

Today Iceland's population is roughly 318,000. The country was settled by a relatively homogeneous group of people that has, until recently, lived in relative isolation. So its population is presumably more ethnically homogeneous. Furthermore, the Icelandic National Health Service possesses medical records for all its patients dating back to 1915. In introducing the idea of a genetic database, DeCODE emphasized that the Icelandic population was uniquely suited to the venture.

A patchwork of different peoples drawn from a wide geographical area, Taiwan differs markedly from Iceland. Bio-bank scientists have neglected –

---

[39] Yang, "Taiwan People's Genetic Bank, Great Initial Controversy."

[40] Hao–Lin Song 宋豪麟 & Ming–Yang Li 李明揚, "Academia Sinica Initiates National Genetic Material Bank," 國人基因資料庫, 中研院籌建 [Guoren Jiyin Ziliao Ku, Zhongyan Yuan Choujian] *United Daily News 聯合報* [*Lianhe Bao*] (22 July 2003): A9.

[41] Arnason & Simpson, "Refractions through Culture: The New Genomics in Iceland."

or misunderstood? – Taiwan's complicated history of ethnicity, which I shall now sketch.

Taiwan Before Han Chinese Rule

The ancient history of Taiwan is still largely unknown. Approximately 6,000 years ago, the island was inhabited by aboriginal people whose descendants are still there. By the time the Dutch colonized Taiwan in the 1620s, they found Han Chinese living there who had crossed from two mainland provinces, Fujian (福建) and Guangdong (廣東). Those who came from Fujian identified with their native prefecture – they called themselves 'Zhangzhou people (漳州人)' and 'Quanzhou people (泉州人)'. These two groups were lumped together and called Hoklo (福佬) or Minnan (閩南), although they spoke different forms of the Southern Min language. Immigrants from Guangdong were primarily Hakka (客家); they too spoke a distinctive language.[42]

The population in the southwestern core, under Dutch control, was ethnically diverse. In the 1620s, aborigines made up the majority. There is historical evidence of Han-aborigine intermarriage early in the Dutch period that could have produced mixed ancestry.[43]

From the Zheng Dynasty to Manchu Rule (1661–1895)

The ethnic balance of power in Taiwan changed dramatically in 1661, when a force of 30,000 invaded. This largely Han Chinese group was led by Zheng Chenggong (also known as Koxinga), a Ming loyalist who defeated the Dutch and founded the Zheng dynasty in opposition to the Qing dynasty on the mainland.

In 1683, the Qing sent a naval force to Taiwan and the Zheng dynasty collapsed. After that, Taiwan remained under Qing control until 1895. Early in that period, Han Chinese settlers became the preponderant group, and they have become ever more dominant since.

---

42 A-Chin Hsiau 蕭阿勤, *Contemporary Taiwanese Cultural Nationalism* (London: Routledge, 2000).

43 Melissa J. Brown, *Is Taiwan Chinese? The Impact of Culture, Power, and Migration on Changing Identities* (Berkeley: U of California P, 2004), and William Campbell, *Formosa Under the Dutch* (London: Kegan Paul, 1903).

Taiwan Under Japanese Colonialism (1895–1945)

In 1895, after losing the Sino-Japanese War, Qing China ceded Taiwan to Japan. Japan ruled the island from 1895 to 1945. The official creation of ethnic groups under the Japanese was a political decision. Under the household registration system instituted by the colonial government, people were classified according to 'race' – two categories of Han and two categories of aborigines, those from the plains and those from the mountains.[44]

The Japanese colonial government conducted seven censuses, in 1905, 1915, 1920, 1925, 1930, 1935, and 1940. The distinction between people from Fujian and Guangdong, aborigines from the plains and the mountains was made in all seven, though it was not until 1935 that the terms 'plains aborigines' and 'mountain aborigines' came to replace the old classifications used by the Qing – 'cooked savages' and 'raw savages'.[45] The Japanese colonial government distinguished 'Fujianese people', 'Guangdongese people', 'plains aborigines (平埔族)', and 'mountain aborigines (高山族)' on the basis of their linguistic, cultural, and other differences. However, this does not imply a biological foundation for the classifications.

Taiwan Under Guomindang Rule and After (1945–Present)

The Japanese authorities offered an estimate of the total number of plains aborigines, which grew, they reckoned, from 58,219 in 1936 to 62,199 in 1943. When the Guomindang government conducted its first census in 1956, it dropped the 'plains aborigines' category; those originally so categorized were presumably treated as Hoklo or Hakka as a result of being assimilated into mainstream Han society.[46]

---

[44] Fu–Chang Wang 王甫昌, "From Chinese Natal Place to Taiwanese Ethnicity: an Analysis of Census Category Transformation in Taiwan You," 由「中國省籍」到「台灣族群」:戶口普查籍別類屬轉變之分析 ['Zhongguo ShengJi' dao 'Taiwan Zuqun': Hukou Pucha JibieLeishu Zhuanbian Zhi Fenxi] *Taiwanese Sociology* 台灣 社會學 [*Taiwan Shehuixue*] 9 (2005): 59–117.

[45] Wang, "From Chinese Natal Place to Taiwanese Ethnicity: An Analysis of Census Category Transformation in Taiwan You."

[46] Fu–Chang Wang 王甫昌, "The Evolution of Attitudes toward Ethnic Categories and Assimilation in Taiwan," 台灣族群分類概念與內涵的演變 [Taiwan Zuqun Fenlei Gainian yu Neihan de Yanbian] in *Annual Conference of the Taiwanese Sociological Association* (Taiwan: Academia Sinica, 2008).

Between 1948 and 1950 an estimated 950,000 Mainlanders, most of them soldiers, left mainland China for Taiwan.[47] As political refugees, mainlanders became a new ethnic group. According to Wang, among those born before 1950 ethnic intermarriage was mainly the result of a skewed sex ratio: older mainland males marrying younger Taiwanese females.[48] But soon cultural (including linguistic) assimilation had come into play, residential and educational segregation had begun to break down, the consequences of intermarriages had begun to make their presence felt, and the sex ratio became more balanced.

Xu investigated inter-ethnic marriages for three generations of the 'four great ethnic groups'.[49] The rate of inter-ethnic marriages among those married before 1961 was 12.8%; for those who married between 1961 and 1981, the figure was 21.5%; it grew to 28.2% among those married after 1981 (see Table 1). As the following tables show, the rate of inter-ethnic marriages in the third generation of Hoklo was 15%, 63.4% in Hakka, 82% in Mainlanders, and 38.2% in aborigines (Tables 2, 3, 4, and 5). Although the rate of inter-ethnic marriages among Hoklo remained low, those of the other three groups were very high.

Changing ethnic classifications and the increasing incidence of inter-ethnic marriage have made Taiwan a hybrid society. In fact, during the past two decades or so, among Taiwan's Han Chinese, the idea of a pure Han ethnicity has been weakening, and some have started to acknowledge the possibility that their own forefathers did not always mate within their group.[50] Obviously,

---

[47] Dong–Ming Li 李棟明, "An Analysis of Population Growth in Postwar Taiwan," 光復後台灣人口社會增加之探討 [Guangfu Hou Taiwan Renkou Hehui Zengjia Zhi Tansuo] *Taipei Historical Documents Quarterly* *台北文獻* [*Taipei Wenxian*] 10 (1969): 215–49.

[48] Fu–Chang Wang 王甫昌, "Causes and Patterns of Ethnic Intermarriage among the Hokkien, Hakka, and Mainlanders in Postwar Taiwan: A Preliminary Examination," 光復後台灣漢人族群通婚的原因與形式初探 [Guangfu Hou Taiwan Hanren Zuqun Tong Hun de Yuanyin yu Xingshi Chutan] *Bulletin of the Institute of Ethnology, Academia Sinica* *中央研究院民族學研究所集刊* [*Zhongyang Yanjiuyuan Minzuxue Ynjiusuo Jikan*] 76 (1993): 43–96.

[49] Z–Min Xu 許咨民, "Taiwan and Fujian 3 Generation Intermarriage Survey Analysis," *Newsletter of the Chinese Statistical Association* *中國統計通訊* 13.11 (2002): 13–16.

[50] Sleeboom–Faulkner, "How to Define a Population: Cultural Politics and Population Genetics in the People's Republic of China and the Republic of China," 7.

Taiwan is different from Iceland. The generalizability of Iceland's DeCODE to Taiwan's context is limited by the differing social and cultural context, especially the specificity of Taiwan's historical process of ethnogenesis. Intermarriage makes it a challenge to classify ethnic groups according to genetic data. More important are the doubts cast on the systems of ethnic classification made up in different historical periods: just consider what happened in 1956, when a group previously represented in several censuses was simply erased.

> Assumption 2: TBB scientists take the "four great ethnic groups" for granted.
> Fact 2: The division of Taiwan's population into 'four great ethnic groups' is a recent social construction.

As mentioned earlier, when they presented the TBB project, Chen Yuan–Tsong and Shen Chen–Yang insisted that the most important influence on the genetic attributes and gene-related diseases in Taiwan was the ethnic group.[51] Each of the 'four great ethnic groups' has particular genetic attributes which in turn influence health; they have been targeted for bio-sample collection (see Fig. 1).

FIGURE 1. A page of the online pamphlet introducing the TBB[52]

[51] Chen & Shen, "Be Careful in Designing the Biological Database."

[52] Anon. *The Preliminary Project of the Establishment of Taiwan Biological Sample Bank* (1996–2011), http://www.twbiobank.org.tw.

Moreover, the Preliminary Plan for Establishing the Taiwan Biological Sample Bank refers to collecting blood samples and personal information from members of different *shizu*. The *shizu* are the four ethnic groups.

Spouse's ethnicity \ Time span	before 1961	1961–1981	after 1981
Total	100.0	100.0	100.0
Same	87.2	78.5	71.8
Different	12.8	21.5	28.2

TABLE 1. Intermarriage over three generations (%)

Spouse's ethnicity \ Time span	before 1961	1961–1981	after 1981
Total	100.0	100.0	100.0
Hoklo	95.9	90.4	85.0
Hakka	2.5	5.0	6.1
Aborigines	0.1	0.2	0.5
Mainlanders	1.4	3.5	5.9
Foreigners	0.1	0.7	1.5
unsure	0.1	0.2	1.0

TABLE 2. Hoklo intermarriage (%)
The "before 1961" column of this table in Xu (2002) adds up to 100.1

Spouse's ethnicity \ Time span	before 1961	1961–1981	after 1981
Total	100.0	100.0	100.0
Hoklo	18.1	33.7	51.4
Hakka	78.8	55.0	36.6
Aborigines	-	0.9	0.5
Mainlanders	2.7	9.5	8.7
Foreigners	-	0.7	2.7
unsure	0.4	0.2	-

TABLE 3. Hakka intermarriage (%)
The "after 1981" column of this table in Xu[53] adds up to 99.9

[53] Z-Min Xu 許咨民, "Taiwan and Fujian 3 Generation Intermarriage Survey Analysis."

Spouse's ethnicity \ Time span	before 1961	1961–1981	after 1981
Total	100.0	100.0	100.0
Hoklo	43.9	61.4	68.9
Hakka	5.9	6.1	6.6
Aborigines	0.4	1.4	-
Mainlanders	47.6	28.7	18.0
Foreigners	2.2	1.7	2.7
unsure	-	0.7	3.8

TABLE 4. Mainlander intermarriage (%)

Spouse's ethnicity \ Time span	before 1961	1961–1981	after 1981
Total	100.0	100.0	100.0
Hoklo	6.7	8.9	32.4
Hakka	-	1.8	-
Aborigines	93.3	76.8	61.8
Mainlanders	-	12.5	5.9
Foreigners	-	-	-
unsure	-	-	-

TABLE 5. Aborigine intermarriage (%)
The "after 1981" column of this table in Xu[54] adds up to 100.1.

The concept of four ethnic groups first appeared in a proposal made by Ye Ju–Lan, a legislator from the Democratic Progressive Party, in 1993. The fact is that the current classification of 'four great ethnic groups' is just a 1990s social construct. Drawing on a wide range of historical and contemporary texts, the sociologist Wang convincingly demonstrated the socially fabricated nature of ethnicity in Taiwan.[55] Based on Wang's work, Fig. 2 shows the complex historical changes to Taiwan's ethnic categories over the past few centuries.[56] Changes in these categories have accompanied other transformations in politics and culture.

[54] "Taiwan and Fujian 3 Generation Intermarriage Survey Analysis."

[55] Fu–Chang Wang 王甫昌, *Ethnic Iimagination in Contemporary Taiwan 當代台灣社會的族群想像* [*Dangdai Taiwan Shehui de Zzuqun Xiangxiang*] (Taipei: Qunxue 群學, 2003); Wang, "From Chinese Natal Place to Taiwanese Ethnicity: An Analysis of Census Category Transformation in Taiwan You"; and Wang, "The Evolution of Attitudes toward Ethnic Categories and Assimilation in Taiwan."

[56] Wang, "The Evolution of Attitudes toward Ethnic Categories and Assimilation in Taiwan."

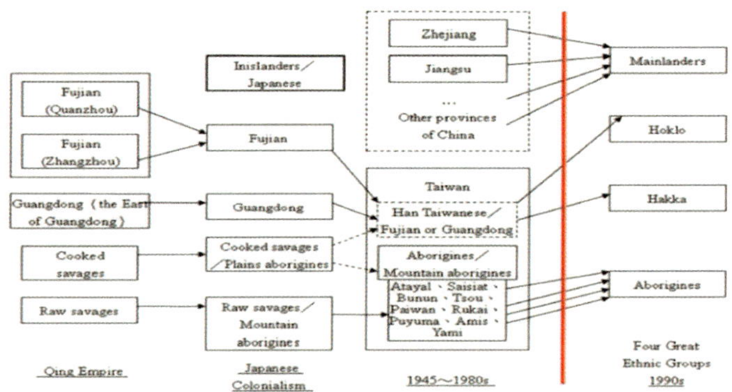

FIGURE 2. The complex historical changes in Taiwan's ethnic categories[57]

4.1 The Hoklo and the Hakka

As mentioned above, the Hoklo and the Hakka arrived in Taiwan in the late Ming and early Qing dynasties from Fujian and Guangdong. The Hoklo were from Zhangzhou and Quanzhou – both in Fujian province. The Japanese noted linguistic and cultural differences between Hakka and Hoklo, and divided the Taiwanese on the basis of province of origin, Guangdong and Fujian, ignoring the fact that some immigrants from the Ding River basin (汀洲流域) in northern Fujian were mostly Hakka, or that people living in Chaozhou (潮州) in Guangdong were mostly Hoklo.

In the first postwar census conducted by the Guomindang government (in 1956), not only were those formerly labelled 'plains aborigines' shunted into other categories, but many Hakka people were wrongly classified as Hoklo because they had lost the ability to speak their ancestral language.[58]

4.2 The Mainlanders

The Mainlanders are those who came from thirty-five provinces in China after the Second World War, and their offspring. As an ethnic group, they do not share the same ancestry, and there are, naturally, significant intra-group differences in language and culture. Owing to the confrontations between local

[57] Wang, "The Evolution of Attitudes toward Ethnic Categories and Assimilation in Taiwan."

[58] "The Evolution of Attitudes toward Ethnic Categories and Assimilation in Taiwan." 5.

Taiwanese and Mainlanders after 1945, the latter began to be regarded as a discrete ethnic group.

4.3 The Aboriginal Peoples

The earliest attempt to categorize Taiwan's aboriginal peoples divided them into 'cooked savages' and 'raw savages'. In 1935, by which time Japan had occupied Taiwan for forty years, this crude distinction was abolished. The colonized were classified into seven ethnic groups, including two sub-categories of Han Chinese – Fujianese and Guangdongese – and two sub-categories of the aboriginal people, plains aborigines and mountain aborigines. Then, in 1956, the postwar Guomindang government dropped the category of plains aborigines because it was believed that they had been assimilated into Han Chinese society.

After 1945, the Republic of China inherited the seven-group classification employed by the Japanese colonial regime. On 14 March 1954, the Ministry of the Interior changed the classification to nine groups.[59] Today, there are fourteen officially recognized ethnic groups in Taiwan, but they do not share the same ancestry and culture any more than the Mainlanders do. This official system used by the government is based mainly on cultural differences such as language, marriage customs, and, to a lesser extent, biological characteristics.[60]

---

[59] Mau–Thai Chen 陳茂泰, *A Study of the Aboriginal Ethnic Groups and Their Distribution in Taiwan 台灣原住民族族群與分佈之研究* [*Taiwan Yuanzhuminzu Zuqun Yu Fenbu Zhi Yanjiu*] (Taipei: The Ministry of Interior 內政部, 1994).

[60] Since the 1990s, taking blood from the aboriginal people for scientific research purposes has raised a host of ethical issues. For example, it is not unusual for scientists to ignore the informed consent procedure by, say, avoiding explaining their research goal. A socially engaged geography professor of aboriginal background, Ming–hui Wang, once argued that because any project that involved taking blood from the aboriginal people, like the TBB, would keep ethnic records, it should obey the Indigenous Peoples Basic Law (原住民基本法), especially its Article 21. Wang asserts that any research based on the collection of the blood samples from the aboriginal people should fulfil ethical requirements by obtaining the informed consent from not only individuals but also CRBs (community review boards). Article 21 of the Indigenous Peoples Basic Law stipulates that "The government or private party shall consult indigenous peoples and obtain their consent or participation, and share with indigenous peoples benefits generated from land development, resource utilization, ecology conservation and academic research in indigenous people's regions. In the event that the government, laws

But the Taiwan Bio-Bank takes these ethnic categories for granted and analyses the genetic data of the Taiwanese people in terms of the four ethnic categories. Given the prevalence of mixed marriages, scientists will have difficulty in finding a 'pure' ethnic group if they adhere to the principle that both parents of group members must have been group-endogamous for three generations.[61] But the ultimate determinant of ethnicity, according to the TBB, is the donor's subjective belief. Unfortunately, when ethnic identity is a matter of choice, people frequently have several ways of categorizing themselves. In fact, during the past few centuries, many people in Taiwan have changed their ethnic identities for one reason or another. Take a recent case. A survey conducted by the Council of Hakka Affairs in 2004 showed that when questionnaires about ethnic identity were provided with multiple-choice answers, subjects tended to disclose their Hakka identity more easily, increasing the number of those who identify themselves as Hakka.[62]

There is a danger of using socially defined groups as proxies for genetic differences. Discussing the difference among the four great ethnic groups, some sociologists state that none of the groups is a concrete reality. Systems of ethnic categorization amount to ideologies. What we should be asking is when, why, and how this classification became so important.[63] The social construction of the four great ethnic groups and the state of ethnic politics in Taiwan raise challenging questions for the TBB. Questions about the biological meaning of ethnic groups obviously cannot be disentangled from questions about their social meaning and political identity.

---

or regulations impose restrictions on indigenous peoples' utilization of their land and natural resources, the government shall first consult with indigenous peoples or indigenous persons and obtain their consent. A fixed proportion of revenues generated in accordance with the preceding two paragraphs shall be allocated to the indigenous peoples' development fund to serve as returns or compensations."

[61] Chen, "Ethical and Legal Dimensions of Aboriginal Genetic Research."

[62] Council for Hakka Affairs Executive Yuan 行政院客家委員會, *97 Niandu Quanguo Kejia Renkou Jichu Ziliao Diaocha Yanjiu 97 年度全國客家人口基礎資料調查* (Taipei: Council for Hakka Affairs Executive Yuan 行政院客家委員會, 2008).

[63] Wang, "*Ethnic Imagination in Contemporary Taiwan*"; Wang, "The Evolution of Attitudes toward Ethnic Categories and Assimilation in Taiwan"; and A–Chin Hsiau 蕭阿勤, *Return to Reality: Political and Cultural Change in 1970s Taiwan and the Postwar Generation 回歸現實:臺灣1970年代的戰後世代與文化政治變遷* [*Huigui Xianshi: Taiwan 1970 Niandai de Zhan Hou Shidai yu Wenhua Zhengzhi Bianqian*] (Taipei: Institute of Sociology, Academia Sinica 中央研究院社會學研究所, 2008).

> Assumption 3: Genetic samples representative of each of the great ethnic groups can be collected in three different areas in Taiwan.
> Fact 3: The areas selected will not provide samples that can represent pure ethnic groups.

Early scientific accounts of the geographical history of *Homo sapiens* were connected with core hypotheses about whether racial differences are the evolutionary effect of migrations from one original location, or the result of different geographical origins for different races. That is, geography is commonly assumed to account for racial differences "as a result of evolution." The main physical index associated with race – skin colour – was until recently believed to be the result of natural selection within populations living under greater or lesser sun exposure.[64]

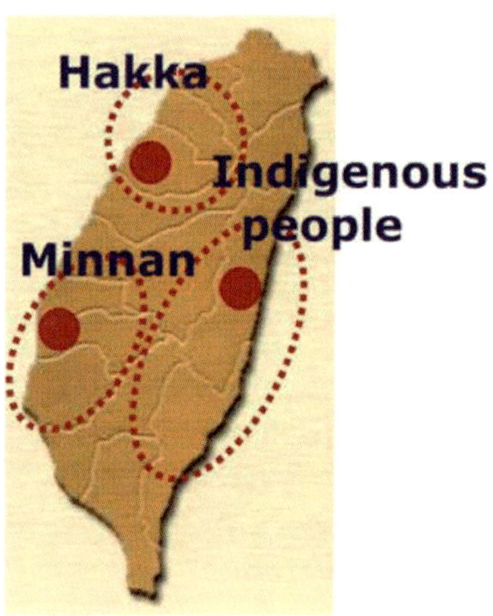

FIGURE 3. The three areas from which blood samples from the 'four great ethnic groups' will be drawn[65]

The TBB has been planning to collect blood samples of the "four great ethnic groups" from three areas (see Fig. 3). The argument is that the majority of the residents of Taoyuan, Hsinchu, and Miaoli counties in northern Taiwan are of Hakka descent; Hoklo make up the majority in Jiayi, Yunlin, and Tainan counties in southern Taiwan; and aboriginal peoples are common in

---

[64] Naomi Zack, *Philosophy of Science and Race* (London: Routledge, 2002).

[65] Anon. *The Preliminary Project of the Establishment of Taiwan Biological Sample Bank* (1996–2011), http://www.twbiobank.org.tw.

Hualien and Taitung counties in eastern Taiwan. The mainlanders, who migrated to Taiwan after 1945, are a minority in all three locations, but they are of a sufficient size in each location to be included as a target population.[66]

However, the idea that there exists a clear relationship between ethnicity and geography is doubtful when applied to so small an area as Taiwan. Government survey data show that 60 percent of the population in Miaoli County is Hakka, 33 percent Hoklo, 3 percent Mainlander, and 0.9 percent aboriginal. In Jiayi, 2.7 percent of the population is Hakka, 81 percent Hoklo, 8.5 percent Mainlander, and 0.6 percent aboriginal. And 22.1 percent of the population of Hualian County is Hakka, 46.2 percent Hoklo, 10.2 percent Mainlander, and 18.4 percent aboriginal (see Table 6). Moreover, the traditional tribal communities of such aboriginal groups as Saisiat, Bunun, Tsou, and Rukai fall outside of Hualien and Taitung counties. The blood samples collected from these two counties of eastern Taiwan would be limited, failing to represent the diversity of aboriginal genetic attributes.

Ethics groups County/city	Hakka	Hoklo	Mainlander	Aboriginal	other	unknown
Taoyuan County	32.2	51.7	9.0	1.8	1.2	2.0
Hsinchu County	63.0	25.9	3.8	3.2	0.8	0.9
Miaoli County	60.6	33.6	3.0	0.9	0.3	0.1
Yunlin County	2.7	92.1	2.8	0.3	0.8	1.1
Jiayi County	3.4	85.1	3.8	1.5	2.5	3.0
Tainan County	1.5	91.8	3.4	0.6	1.0	1.3
Taitung County	12.7	48.8	7.2	28.2	0.1	1.4
Hualian County	22.1	46.2	10.2	18.4	0.3	0.9
Jiayi City	2.7	81.9	8.5	0.6	3.4	2.4
Tainan City	2.8	86.4	5.3	0.1	1.8	3.3

TABLE 6. Ethnic groups in Taiwan

Briefly, there is no correspondence between the boundary of a county and the boundary of an ethnic group. Each of the three counties includes diverse ethnic groups. Studying only geographically stable samples does not take into account the effects of intermarriage and other forms of population mobility. Obviously, the blood samples collected in this way should not be used to pin down the genetic attributes of these three ethnic groups.

---

[66] Council for Hakka Affairs Executive Yuan 行政院客家委員會, *97 Niandu Quanguo Kejia Renkou Jichu Ziliao Diaocha Yanjiu 97 年度全國客家人口基礎資料調查*, 6.

5 Conclusion: An Alarm Bell for Science and Ethnic Politics

Due to a faulty understanding of ethnicity, the Taiwan Bio-Bank has essential problems it has to clear up. My analysis shows how the categories that defined ethnicity in Taiwan have changed many times over the past few centuries. The examination of the TBB's policy narratives shows how, in turn, the TBB's production of scientific knowledge has been informed by concepts emerging from a specific social-political context. Situated at a particular moment in Taiwanese history, the ambitious plan to map out the genetic attributes of ethnicity is contingent on the interactive dynamics of state governance, genetic discourse, the interests of a varied scientific community, and identity-politics. The historicity of the project and its unintended cultural and political effects are a matter subject to investigation.

By relying on the conceptual framework of the 'four great ethnic groups', the TBB provokes deep misgivings about the methodological validity of the project. In this framework, the TBB is trying to posit a gene-environment interaction specific to an ethnic group. It is true that many biomedical researchers focus not only on ethnic genetic affinity but on environmental influences as well. However, this does not justify conflating the social construction of ethnicity and the biological essence of ethnicity, which is what the TBB does. Such an act should be investigated, not legitimated in the name of science. As the American Anthropological Association argues, human populations are not clearly demarcated, and the differences among what one might call biologically distinct groups are genetically minute. This is particularly true of such a comparatively small area with a long history of mixed marriage as Taiwan. The TBB has to face a crucial challenge: why should a socially constructed and politically and culturally loaded conceptual framework be accepted as a valid biological entity? Neglecting the complicated history of Taiwanese ethnic politics renders the project unsatisfactory and leaves room for essentializing the 'four great ethnic groups'.

The TBB is not the only project that faces methodological problems regarding ethnic conceptualization. Using insights from science and technology studies about classification and standardization, Smart et al. showed how classifications of race and ethnicity as boundary objects are used across a range of social settings.[67] The conceptualization of race and ethnicity, the

---

[67] Andrew Smart, Richard Tutton, Richard Ashcroft, Paul Martin, Andrew Balmer, Richard Elliot & George T.H. Ellison, "The Standardization of Race and Ethnicity in

authors pointed out, was ambiguous in the samples they collected – biomedical science editorials and UK bio-bank studies.

The multiple and conflicting constructions of meaning, measurement, and utility of race and ethnicity, they argued, indicated that the concepts were contingent, heterogeneous, and locally situated. Furthermore, the routinized or unreflective adoption of the census classification "threatens to erode the epistemological status of its categories as socio-political constructs."[68]

By emphasizing the alleged biological differences among the 'four great ethnic groups', TBB scientists would encourage the belief that ethnicity has a genetic essence and forms an objective reality. This would facilitate the ethnicization of biomedicine and the racialization of ethnic difference. As the long history of racial science has shown, a national project like the TBB could transform sociopolitical constructs into objective biological categories that are simply taken for granted. It could also help to construct an imagined community, ethnically divided but nationally associated. Although plans for a Taiwanese genetic database have not progressed beyond a pilot study, my study should, I hope, ring some early alarm bells by drawing attention to the potentially adverse implications of the manner in which the social sciences treat ethnicity.

Works Cited

Anon. "The Ethical Code of Practice for the Preliminary Project of Establishing Taiwan Biological Sample Bank (Second Edition)," *Taiwan Biobank* (16 October 2008): http://www.twbiobank.org.tw/documents/081016/Doc00.pdf (accessed 12 December 2008).

——. "The Human Genetic Database: Collecting and Studying Ethnic Blood Samples," *'Gene Technology and Humanism' Lecture Series* 「*基因科技與人文議 題系列演講* (2005): http://elsi.nccu.edu.tw/lecture2005/index.htm (accessed 12 December 2008).

——. *The Preliminary Project of the Establishment of Taiwan Biological Sample Bank* (1996–2011): http://www.twbiobank.org.tw (accessed 12 December 2008).

Biomedical Science Editorials and UK Biobanks," *Social Studies of Science* 38 (2008): 407–23, 416–18.

[68] Smart et al., "The Standardization of Race and Ethnicity in Biomedical Science Editorials and UK Biobanks," 416–18.

American Anthropological Association. "American Anthropological Association Statement on 'Race'" (17 May 1998), *American Anthropologist* 100.3 (1999): 712–13.

Arnason, Arnar, & Bob Simpson. "Refractions through Culture: The New Genomics in Iceland," *Ethnos* 4 (2003): 533–53.

Ashcroft, Bill, Gareth Griffiths, & Helen Tiffin. *Key Concepts in Post-Colonial Studies* (London: Routledge, 1998).

Atkinson, Paul, Peter Glasner & Helen Greenslade, ed. *New Genetics, New Identities* (London: Routledge, 2007).

Brown, Melissa J. *Is Taiwan Chinese? The Impact of Culture, Power, and Migration on Changing Identities* (Berkeley: U of California P, 2004).

Buehler, James W. "Abandoning Race as a Variable in Public Health Research," *American Journal of Public Health* 5 (1999): 783.

Campbell, William. *Formosa Under the Dutch* (London: Kegan Paul, 1903).

Chadwick, Ruth. "Genomics, Public Health and Identity," *Aeta Bioethica* 2 (2003): 209–18.

Chen, Mau–Thai 陳茂泰. *A Study of the Aboriginal Ethnic Groups and Their Distribution in Taiwan 台灣原住民族族群與分佈之研究* [*Taiwan Yuanzhuminzu Zuqun Yu Fenbu Zhi Yanjiu*] (Taipei: The Ministry of Interior 內政部, 1994).

Chen, Shu–Juo 陳叔倬. "Ethical and Legal Dimensions of Aboriginal Genetic Research," 台灣原住民遺傳基因研究之倫理爭議與立法保障 (Yuanzhumin Renti Jiyin Yanjiu Zhi Lunli Zhengyi Yu Lifa Baohu, *Shengwu Keji Yu Falü Yanjiu Tongxun, 生物科技與法律研究通訊 Newsletter of Biotechnology and Law* 6 (2000): 7–28).

Chen, Yuan–Tsong 陳垣崇 & Chen–Yang Shen 沈志陽 "Be Careful in Designing the Biological Database," 謹慎規劃生物資料庫 [Jinshen Guihua Shengwu Ziliaoku] *China Times 中國時報* [*Zhongguo Shibao*] (2 April 2006): A15.

Chou, Kuei–Tien 周桂田. "Conflicts of Technology Policy and Governance Paradigm in a Knowledge-Based Economy: A Case Analysis of the Construction of the Taiwan Biobank," *Issues and Studies* 3 (2007): 97–130.

Collins, Francis S. "What We Do and Don't Know about Race, Ethnicity, Genetics and Health at the Draw of the Genome Era," *Nature: Genetics Supplement* 36 (2004): S13–15.

Council for Economic Planning and Development Executive Yuan, ed. *Economic Development ROC (Taiwan)* (Taipei: Council for Economic Planning and Development Executive Yuan, ROC, 2007).

Council for Hakka Affairs Executive Yuan 行政院客家委員會, *97 Niandu Quanguo Kejia Renkou Jichu Ziliao Diaocha Yanjiu 97 年度全國客家人口基礎資料調查* (Taipei: Council for Hakka Affairs Executive Yuan 行政院客家委員會, 2008).

Duster, Troy. "Selective Arrests, an Ever-Expanding DNA Forensic Database, and the Specter of an Early Twenty-First-Century Equivalent of Phrenology," in *DNA and the Criminal Justice System: The Technology of Justice*, ed. David Lazer (Cambridge, MA: MIT Press, 2004): 315–34.

——. "Policy Forum: Race and Reification in Science," *Science* 307 (2005): 1050–1051.

Ellison, George T.H., Andrew Smart, Richard Tutton, Simon M. Outram, Richard Ashcroft & Paul Martin. "Racial Categories in Medicine: A Failure of Evidence-Based Practice," *PLoS Medicine* 9 (2007): 1434–436.

Epstein, Steven. *Inclusion: The Politics of Difference in Medical Research* (Chicago: U of Chicago P, 2007).

González Burchard, Esteban. "Latino Populations: A Unique Opportunity for the Study of Race, Genetics, and Social Environment in Epidemiological Research," *American Journal of Public Health* 12 (2005): 2161–168.

Goodman, Alan H. "Why Genes Don't Count (For Racial Differences in Health)," *American Journal of Public Health* 11 (2000): 1699–1702.

Guo, Yi–Chun 郭怡君. "Taiwan Plans Han Chinese Infectious Materials Center," 我將設華人遺傳資訊中心 [Wo Jiang She Huaren Yichuan Zixun Zhongxin] *The Liberty Times 自由時報* [*Ziyou Shibao*] (19 March 2005): 19.

Heath, Deborah, Rayna Rapp & Karen–Sue Taussig. "Genetic Citizenship," in *A Companion to the Anthropology of Politics*, ed. David Nugent & Joan Vincent (Malden MA: Blackwell, 2004): 153–67.

Hsiau, A–Chin 蕭阿勤. *Contemporary Taiwanese Cultural Nationalism* (London: Routledge, 2000).

——. *Return to Reality: Political and Cultural Change in 1970s Taiwan and the Postwar Generation 回歸現實:臺灣1970年代的戰後世代與文化政治變遷* [*Huigui Xianshi: Taiwan 1970 Niandai de Zhan Hou Shidai yu Wenhua Zhengzhi Bianqian*] (Taipei: Institute of Sociology, Academia Sinica 中央研究院社會學研究所, 2008).

Li, Dong–Ming 李棟明. "An Analysis of Population Growth in Postwar Taiwan," 光復後台灣人口社會增加之探討 [Guangfu Hou Taiwan Renkou Hehui Zengjia Zhi Tansuo) *Taipei Historical 台北文獻* [*Documents QuarterlyTaipei Wenxian*] 10 (1969): 215–49.

Liao, Winona 廖文孜. "A Preliminary Examination of the Collective Rights Issues in Human Genetic Resources, with a Discussion of Special Considerations for Aborigines," 人類基因資源的集體權利問題初探—兼論關於原住民族群的特殊考量 [Renlei Jiyin Ziyuan de Jiti Quanli Wenti Chtan – Jianlun Guanyu Yuanzhumin Zuqun de Teshu Kaoliang], *Newsletter of Biotechnology and Law 生物科技與法律研究通訊* [*Shengwu Keji Yu Falyu Yanjiu Tongxun*] 6 (2000): 29–37.

Liu, Hung–En 劉宏恩. "Biobank Legalization Issues: International Development and the State of the Field in Taiwan," 人群基因資料庫法制問題之研究—國際上發展與台灣現況之評析 [Renqun Jiyin Ziliaoku Fazhi Wenti zhi Yanjiu – Guoji Shang Fazhan yu Taiwan Xiankuang zhi Pingxi] *Taipei Bar Journal 律師雜誌* [*Lüshi Zazhi*] 303 (2004): 71–94.

——. "Public Trust, Commercialization, and Benefit Sharing in Biobanking," 基因資料庫研究中的公眾信賴、商業介入與利益共享 [Jiyin Ziliaoku Yanjiu Zhong de Gongzhong Xinlai, Shangye Jieru yu Liyi Gongxiang] *Taipei University Law Review 台北大學法學論叢* [*Taipei Daxue Faxue Luncong*] 57 (2005): 367–93.

Makeham, John, & A–Chin Hsiau. *Cultural, Ethnic, and Political Nationalism in Contemporary Taiwan* (New York: Palgrave Macmillan, 2005).

Ou, Chia–Hao 歐家豪, & Chen–Yang Shen 沈志陽. "The Taiwan Biobank Project: For the health of future generations," *Academia Sinica E-news* (19 April 2007), http://newsletter.sinica.edu.tw/en/file/file/1/125.pdf (accessed 19 April 2007).

Reardon, Jenny. *Race to the Finish: Identity and Governance in an Age of Genomics* (Princeton NJ: Princeton UP, 2005).

Schwartz, Robert S. "Racial Profiling in Medical Research," *New England Journal of Medicine* 344 (2001): 1392–93.

Shen, Zh–Yang 沈志陽, Yuan–Tsong Chen 陳垣崇, Der–Tsai Lee 李德財 & Chien–Te Fan 范建得. "Final Report of the National Science Council's Research Grant Proposal: The Completion of the Taiwan Biological Sample Bank," *National Science Council*, NSC 94-3112-B-001-017 (2007).

Simpson, Bob. "Imagined Genetic Communities: Ethnicity and Essentialism in the Twenty-First Century," *Anthropology Today* 16.3 (June 2000): 3–4.

Sleeboom–Faulkner, Margaret. "How to Define a Population: Cultural Politics and Population Genetics in the People's Republic of China and the Republic of China," *BioSocieties* 1 (2006): 399–419.

Smart, Andrew, Richard Tutton, Richard Ashcroft, Paul Martin, Andrew Balmer, Richard Elliot & George T.H. Ellison. "Social Inclusivity versus Analytical Acuity? An Aualitative Study of UK Researchers Regarding the Inclusion of Minority Racial/Ethnic Groups in Biobanks," *Medical Law International* 9.2 (2008): 169–90.

——. "The Standardization of Race and Ethnicity in Biomedical Science Editorials and UK Biobanks," *Social Studies of Science* 38 (2008): 407–23.

Song, Hao–Lin 宋豪麟, & Ming–Yang Li 李明揚. "Academia Sinica Initiates National Genetic Material Bank," 國人基因資料庫, 中研院籌建 [Guoren Jiyin Ziliao Ku, Zhongyan Yuan Choujian] *United Daily News 聯合報* [*Lianhe Bao*] (22 July 2003): A9.

Wang, Fu–Chang 王甫昌. "Causes and Patterns of Ethnic Intermarriage among the Hokkien, Hakka, and Mainlanders in Postwar Taiwan: a Preliminary Examination," 光復後台灣漢人族群通婚的原因與形式初探 [Guangfu Hou Taiwan Hanren Zuqun Tong Hun de Yuanyin yu Xingshi Chutan] *Bulletin of the Institute of Ethnology, Academia Sinica 中央研究院民族學研究所集刊* [*Zhongyang Yanjiuyuan Minzuxue Ynjiusuo Jikan*] 76 (1993): 43–96.

——. *Ethnic Imagination in Contemporary Taiwan 當代台灣社會的族群想像* [*Dang-dai Taiwan Shehui de Zzuqun Xiangxiang*] (Taipei: Qunxue 群學, 2003).

——. “From Chinese Natal Place to Taiwanese Ethnicity: an Analysis of Census Category Transformation in Taiwan You,” 由「中國省籍」到「台灣族群」:戶口普查籍別類屬轉變之分析 [‘Zhongguo ShengJi’ dao ‘Taiwan Zuqun’: Hukou Pucha Jibie Leishu Zhuanbian Zhi Fenxi] *Taiwanese Sociology 台灣社會學* [*Taiwan Shehuixue*] 9 (2005): 59–117.

——. “The Evolution of Attitudes toward Ethnic Categories and Assimilation in Taiwan,” 台灣族群分類概念與內涵的演變 [Taiwan Zuqun Fenlei Gainian yu Neihan de Yanbian], in *Annual Conference of the Taiwanese Sociological Association* (Taiwan: Academia Sinica, 2008).

Xu, Z–Min 許咨民. “Taiwan and Fujian 3 Generation Intermarriage Survey Analysis,” *Newsletter of the Chinese Statistical Association 中國統計通訊* 13.11 (2002): 13–16.

Yang, Zheng–Min 楊正敏. “Taiwan People’s Genetic Bank, Great Initial Controversy,” 台灣人群基因庫,籌備爭議大 [Taiwan Renqun Jiyin Ku, Choubei Zhengyi da] *United Daily News 聯合報* [*Lianhe Bao*] (5 July 2000): A4.

Zack, Naomi. *Philosophy of Science and Race* (London: Routledge, 2002).

Zhang, Li–Wen 張璨文. “Academia Sinica Discusses Making a Citizens’ Gene Bank,” 中研院研議設置國人基因庫[Zhongyanyuan Yanyi Shezhi Guoren Jiyin Ku], *China Times 中國時報* [*Zhongguo Shibao*] (5 July 2000): A5.

——. “Whose Genes are Representative of the Taiwanese People?” 誰的基因能代表台灣人? [Shei de Jiyin Neng Daibiao Taiwanren?] *China Times 中國時報* [*Zhongguo Shibao*] (5 July 2000): A5.

Zhuang, Ming–Tsuang 莊明哲. “Research Proposal: A Population Genetic Data Base for Taiwan,” in *The Minutes of the 24th Academicians’ Meeting, Academia Sinica* (3 July 2000).

⌘

Part III
Surviving and Resisting

Genome Survivance

GERALD VIZENOR

> Reason was to be given priority as an instrument of knowledge, not as a motive for human conduct; it was opposed to faith, not to passions. The emancipation of knowledge paved the way for the development of science.[1]

CHARLES DARWIN, THE "GENTLEMAN GENIUS" OF NATURAL SELECTION, observed in *The Descent of Man* more than a century ago that "sympathy and cooperation" were factors of evolution. His original theories of evolution have survived the counter sentiments and refutations of pious monotheists and fundamentalists.

Native American Indians have likewise survived the churchy politics of monotheism, cultures of dubious science, phrenology, blood types, arithmetic levels, and the conquest caricatures and ideologies of natural selection, only to be measured and compared once more by recent genetic codes and haplotype signatures to determine, contest, and separate origins, names, identities, and cultures.

The US government established an arithmetic blood quantum, a pernicious racial computation to register and determine Native American Indian eligibility for federal reservation membership, or citizenship, and to determine and regulate federal health, education, and other services. This obscure arithmetic and bureaucratic system was named the Chart to Establish Degree of Indian Blood.

Recent genetic science and notions of race were considered in my preparation of a new constitution for the White Earth Reservation in northern Minne-

---

[1] Tzvetan Todorov, *In Defence of the Enlightenment* (London: Atlantic, 2009): 7.

sota. I was appointed the principal writer of the Constitution of the White Earth Nation. I was also a sworn delegate to serve at four Constitutional Conventions. The forty delegates struggled over the appropriate language to define a citizen of the White Earth Nation. The federal scheme of blood quantum was favoured by many of the delegates who worried that they might lose certain federal entitlements that were based on arithmetic blood levels or quantification.

I wrote two specific articles in the new Constitution of the White Earth Nation that satisfied the serious interests of those delegates who favoured the blood-quantum concession, and those delegates who insisted that genealogy or direct family descent and identity determine the actual meaning of citizenship in the White Earth Nation.

> Article 1
> Citizens of the White Earth Nation shall be descendants of Anishinaabeg families and related by linear descent to enrolled members of the White Earth Reservation and Nation, according to genealogical documents, treaties and other agreements with the government of the United States.
>
> Article 2
> Services and entitlements provided by government agencies to citizens, otherwise designated members of the White Earth Nation, shall be defined according to treaties, trusts, and diplomatic agreements, state and federal laws, rules and regulations, and in policies and procedures established by the government of the White Earth Nation.

The Constitution of the White Earth Nation was duly ratified, after a detailed discussion of each article, by a secret vote of the official delegates on 4 April 2009, at the Shooting Star Casino in Mahnomen, Minnesota.[2]

Genetic ancestry has never been the natural reason for native survivance but, rather, a scientific and systematic means to resolve and disconnect tribal, totemic associations, and the familiar histories of native families. Natives have mingled with adventurers, consorted with various colonial missionaries, and many settlers as a course of assurance, education, and survivance, and many woodland natives eagerly participated in that premier union of the fur

---

[2] Constitution of the White Earth Nation. Ratified text of the Constitution published in *Anishinaabeg Today* and available on the website of the White Earth Reservation, http://www.whiteearth.com/home.html.

trade for centuries. The French, English, Spanish, Russian, and many other nations are in the bloodline, reminiscence, and history of the Northern Hemisphere. In other words, diverse and determined native families trace their surnames and ancestry to these colony settlements, dynamic continental unions of culture and chance, the natural traces of a sense of presence, history, hard done by, and survivance.

Cultural and totemic associations inspire the humane tropes of native visionary stories, traditions, and memoirs. The genetic tests and traces of maternal mitochondrial DNA and the signature paternal Y chromosomes, separate creation stories of families, and the actual visionary narratives of a community, by genetic markers, linear counts, and other scientific measures.

Genetic ancestry is not a family. We are animals, but there are no genomes of visionary totemic associations. The chance unions of humans, animals, and native families are connections by creation and trickster stories, by transmotion, or by the visionary sense of natural motion, sacred and secular imagination, but not by the genetic science of haplotypes or the abstract counts and codes of ancestry. Clearly, natives have created a sense of presence by emotive and ironic stories, associations, and memories, a distinct sense of presence in the natural world. Genetic ancestry creates an absence, not a humane sense of presence, in historical narratives.

The Biblical descent of Adam and Eve, or the genome traces and tropes of mitochondrial DNA and Y chromosomes, leaves out many significant branches of native associations and ancestors.

Genetic ancestry is the most recent intrusion of scientific modernism into the diverse associations of native identity, stories of direct descent, totemic associations, and tribal relations. The primary motivation of this modernist intrusion is more cynical than racialist, more curious than chauvinistic. Yet, some reservation governments might consider genetic tests as devious evidence to determine enrolment or citizenship, and other reservations could use genetic codes of ancestry to ascertain the entitlement of per-capita royalty payments by reservation casinos.

The current considerations of genetic tests to determine native ancestry, however, are not as fierce or as destructive as the dubious scientific debates between the polygenists and the monogenists.

Robert Bieder observed in *Science Encounters the Indian*:

> By the end of the first two decades of the nineteenth century, when philanthropy and the churches could show few positive results from their efforts to lift the Indians, doubts were raised about whether they

> could really be civilized. Many such critics began to question the monogenetic assumptions, set forth in the Bible, that all mankind shared the same origin." The racial detractors at the time "began to explain Indians' recalcitrant nature in terms of polygenism. To polygenists Indians were separately created and were an inferior species of man.[3]

The ostensible scientists of polygenism, the theory that there were many creations, and natives were separate from other human creations, manoeuvred to dominate the professional organizations and institutions of the time. The monogenists derived their persuasive authority from monotheism, the fundamental creation of humans, and reasoned that natives were merely disadvantaged by nature and culture, and could be educated and assimilated into the wider culture and dominion.

Samuel Morton, a medical doctor, studied the distinctive "crania of the five races of man." *Crania Americana*, his comparative study of the skulls of aboriginals, was published in 1839, at the rise of the scientific debate between the polygenists and monogenists. Bieder pointed out that many phrenologists at the time argued that the "brain was the organ of the mind; that the brain was composed of individual faculties that controlled personality, thought, and moral action," and that these "faculties could be determined by protuberances on the skull; and that each race manifested its cultural traits through the shape of the cranium. Thus, phrenologists believed, each race possessed a typical, or national, cranium."

Morton sealed and filled the skull "cavity with mustard seed," and weighed the "seed to determine cranial volume." He concluded that "Indian crania were smaller in volume" than the skulls of Caucasians. The publication of *Crania Americana* "provided a scientific foundation for a polygenetic racial history of man." Morton was convinced that "Indians never could be 'civilized' because they lacked the necessary brains," writes Bieder, "and what brains they had were more animal than human."[4]

Stephen Jay Gould noted that Morton had actually averaged the smaller native crania from South America with the larger native crania from North America. Morton leveraged the comparative study of crania to show larger crania for Caucasians.

---

[3] Robert Bieder, *Science Encounters the Indian, 1820–1880* (Norman: U of Oklahoma P, 1986): 12.

[4] Bieder, *Science Encounters the Indian*, 59, 69, 79.

Bieder argued that the

> issue of polygenism was a crucial one for a society that was splitting itself over the question of slavery, eager to expand its boundaries westward and south into Mexico, and proclaiming that none but whites should rule. The question of the capacity of the 'inferior' dark races for progress had tremendous political and social implications.[5]

The notions of this dubious science of racial categories have prevailed in popular culture, including native perceptions of human differences. Marge Anderson, for instance, formerly the elected chief executive of the Mille Lacs Band of Ojibwe Indians in Minnesota, declared that Ojibwe or Anishinaabe "tradition speaks of peace among the four colors – red, white, yellow, and black – which represents the four races on earth. Sadly, some people in this world don't show respect for all races, and they hurt others though their lack of understanding." The notion of race and colour, of course, is the primary misconception.[6]

Sadly, indeed, rather than begetting a sense of peace, the notion of four racial colours perpetuates a crude separation of humans, natives, and cultures. The four-colours notion, and dubious native traditions, insinuates the cynical pseudoscience of polygenism, or the separate creation of humans, that has been advanced in the past two centuries.

William Warren, the Anishinaabe historian, observed more than a century ago that some native creation stories were told as a courtesy to missionaries. "These tales, though made up for the occasion by the Indian sages, are taken by his white hearers as their bona fide belief, and, as such, many have been made public, and accepted by the civilized world," he wrote in the *History of the Ojibway Nation*, first published in 1885 by the Minnesota Historical Society.

The Great Spirit, in one of these stories, created three races: white, black, and red. "To the first he gave a book, denoting wisdom," noted Warren. To the "second a hoe, denoting servitude and labour; to the third, or red race, he gave the bow and arrow, denoting the hunter state." These tricky creation stories were translated and treasured as representations of native culture by missionaries at the time. Warren continued:

---

[5] Bieder, *Science Encounters the Indian*, 83.

[6] Gerald Vizenor, *Native Liberty: Natural Reason and Cultural Survivance* (Lincoln: U of Nebraska P, 2009): 172.

> We have every reason to believe that America has not been peopled from one nation or tribe of the human family, for there are differences amongst its inhabitants and contrarieties as marked and fully developed as are to be found between European and Asiatic nations – wide differences in language, beliefs, and customs.

Warren, in his history, and the Anishinaabe in their stories and teases of creation, sustained a sense of native diversity, survivance, literary irony, and modernity.[7]

The scientific theories of cardinal colours were advanced more than two centuries ago. Johann Friedrich Blumenbach studied crania and named "five basic races" – "Mongolian, American, Caucasian, Malayan, and Ethiopian." Moreover, "through his observations of skulls he identified the Caucasian as the original type from which other races have degenerated," wrote Bieder. Blumenbach, in his dissertation *On the Natural History of Mankind*, published in 1775, declared that the three races of colour were degenerate.[8] Adolf Hitler and the Nazi Party declared a similar racial superiority. The Germans proclaimed with a gruesome vengeance their racial purity as Aryans. The Jews – including their literature and modern art – were degenerate.

Carolus Linnaeus, the eighteenth-century naturalist, created a binomial nomenclature for plants and animals, and he also classified humans. Barbara Katz Rothman points out that Linnaeus "found five natural categories of humans, four of which were geographically based." He named and characterized five human categories: *Homo sapiens Americanus*, red harsh face, wide nostrils, scanty beard, obstinate, free, and ruled by customs; *Homo sapiens Asiaticus*, yellow, melancholy, greedy, haughty, severe, and ruled by opinions; *Homo sapiens Afer*, black, lazy, flat nose, silky skin, ruled by caprice; *Homo sapiens Europaeus*, white, serious, strong, blond hair, blue eyes, active, very smart, inventive, ruled by laws; *Homo sapiens Monstrosus*, those curious and strange humans.

Linnaeus represented three categories of the modern human species with racial detractions and caricatures, ruled by custom, opinion, and caprice. *Sapiens* whites, ruled by laws, he vested with care and fair countenance.[9]

---

[7] William Warren, *History of the Ojibway Nation* (Minneapolis MN: Ross & Haines, 1957): 58.

[8] Bieder, *Science Encounters the Indian*, 61.

[9] Barbara Katz Rothman, *Genetic Maps and Human Imaginations* (New York: W.W. Norton, 1998): 46–47.

Russell Means, the brave dancer, radical spiritualist, errant reservation politician, presidential candidate, 'post-Indian' movie actor, and crux of Indian simulations, has his own notions of race, reason, and culture. "Humans are able to survive only through the exercise of rationality since they lack the abilities of other creations to gain food through the use of fang and claw," he declared in his autobiography *Where White Men Fear to Tread*. "But rationality is a curse since it can cause humans to forget the natural order of things in ways other creatures do not. A wolf never forgets his or her place in the natural order. American Indians can. Europeans almost always do."

Means argued that he does not care about skin colour, and at the same time he noted that white is a race and "one of the sacred colors of the Lakota people – red, yellow, white and black. The four directions. The four seasons" and the "four races of humanity." He described the brew of four colours as a fifth race. "Mix red, yellow, white and black together and you got brown, the color of the fifth race. This is a natural order of things."[10]

Race is a racial simulation, a pseudoscience, and the duplicitous 'science' of race is political not biological; human differences are genetic, of course, but the simulations of four or more races are *faux*-traditions, mistaken and detractive notions.

"We are not separate species," declared Rothman. "Race was a liquid concept," but the once common metaphor of blood, as blood would show and tell blue, red, hot, cold, thicker than water, has lost practice and significance.

> No longer visible, no longer divisible, race has moved inward from body to blood to genes; from solid to liquid to a new crystallization. Blood no longer tells. Race is now a code to read; the science of race is the science of decoding.[11]

Families, names, visionary associations, experiences, and imagic memories, not genetic codes or racial colours, or the size of crania, are the sources of native identities. Stories create the names, and a sense of presence in the world; actions, ancestors, and memories are the sources and contingencies of recognition in native communities.

Native stories are wise, tricky, and ironic. The stories of native nicknames, for instance, are imagic moments, and names are memorable experiences in

---

[10] Russell Means, *Where White Men Fear to Tread: The Autobiography of Russell Means* (New York: St Martin's, 1995): 551, 553.

[11] Rothman, *Genetic Maps and Human Imaginations*, 64, 65.

many communities. Customarily, the tease of nicknames is an inclusive notice, the sanction of a presence not an absence; likewise, the ironic turn of nickname stories is an act of native imagination, survivance, and literary modernity. There are no genetic codes or trickster haplotypes of irony.

"Races, as natural divisions of the human species, are thus rather like angels," proclaimed the molecular anthropologist Jonathan Marks in his essay "The Realities of Races." Race "now becomes the simple facts of ancestry and appearance, not something to be diagnosed or identified." He critiqued the companies that market genetic tests to determine racial identity. The technological and statistical business is sophisticated, but the epistemology is "very primitive." Marks concludes: "In other words, this business has far more to do with the modern culture of science than with the production of reliable knowledge."[12]

The Human Genome Diversity Project, for instance, has collected biological and genetic material from populations around the world. Luigi Luca Cavalli–Sforza, Emeritus Professor at Stanford University, initiated the systematic research on the scientific classification of human populations. Mitchell Leslie noted more than a decade ago in the *Stanford Magazine*:

> The [genome project] proposal won support from geneticists and some anthropologists, who saw it as a logical way to pull together irreplaceable data. But it also drew sharp criticism. Project planners, most of them white academics, were denounced as gene pirates, neocolonialists and racists by some who believe the project would backfire on minority groups. One Australian aboriginal group came up with the name 'vampire project' to describe the plan.[13]

There were many contentious issues raised by the collection of genetic material. For instance, who would profit from the commercial patents of generic discoveries? "Critics of the new project say it would allow comparable 'biopiracy' in the human genetic realm. For example, companies might trawl the data for patentable genes that could lead to new medicines." Cavalli–Sforza

---

[12] Jonathan Marks, "The Realities of Race," Social Science Research Council, 7 June 2006, revised version, "Race: Past, Present, and Future" in *Revisiting Race in a Genomic Age,* ed. Sandra Lee, Barbara Koenig & Sarah Richardson (Piscataway NJ: Rutgers UP, 2008): 21–38. Online: http://raceandgenomics.ssrc.org/Marks/

[13] Mitchell Leslie, "The History of Everyone and Everything," *Stanford Magazine* (May–June 1999), http://www.stanfordalumni.org/news/magazine/1999/mayjun/articles/cavalli_sforza.html

responded that worries about "economic exploitation are baseless [...] the results of misunderstood intentions. We are very much against patenting DNA." Cavalli–Sforza's insistence that his "studies can serve as an antidote to racism," have, Leslie points out, led to him receiving "stacks of hate mail from white supremacists."[14]

GenEthics News has been critical of the practices of the project and for issues of practical and situational ethics:

> Apart from the reactions of indigenous people, the project has also been heavily criticized by other geneticists and anthropologists. Many anthropologists, like psychologists, are alarmed by what they see as the growing dominance of their field by genetics, and the growth of genetic determinism. [...] Jonathan Marks [...] points out that "different genetic studies produce different 'family trees" for populations."[15]

According to the same issue of *GenEthic News*, the Human Genome Diversity Project "tends to assume that indigenous groups are genetically pure and unaffected by the massive population movements that have taken place over the last five hundred years and [would] help to provide a picture of what humanity looked like genetically before migrations." As Marks points out, "there is extensive evidence from ethnohistory of intermarriage between, for example, Native American groups, evidence which geneticists ignore. On the whole, there are very few 'pure' population groups which have not intermarried as the result of migration," colony settlement, or "military conquest."[16]

The Morrison Institute for Population and Resource Studies describes the Human Genome Diversity Project as a research programme that "seeks to understand the diversity and unity of the entire human species." The Project overview of ethical issues provides three principles for consideration in research. These recent principles, informed consent, respect for the participating populations and cultures, and adherence to international standards of human rights, were established to ensure the rights of participating communities.

"The Project categorically rejects the idea of 'bleed and run' collecting, "done by researchers who disappear without a trace," according to the conclusion of the published outline of ethical principles.

---

[14] Leslie, "The History of Everyone and Everything."

[15] "The Human Genome Diversity Project," Human Genetics Alert, *GenEthics News* 10 (2000), http://www.hgalert.org/topics/personalInfo/hgdp.htm

[16] "The Human Genome Diversity Project."

> Collecting must be done only with the full consent, cooperation, and engagement of those sampled. Although this will require close and expert knowledge about the populations and may take a long time, respect for the populations as partners in the scientific enterprise – rather than as objects of it – requires no less.[17]

The recent establishment of genetic testing companies that promote genetic traces of ancestry presents new political, racial, genealogical, and cultural issues for Native American Indians. Kimberly TallBear, an assistant professor of Science, Technology, and Environmental Policy at the University of California at Berkeley, pointed out in a recent essay that "tribal governments should be wary of trying to 'solve' contentious enrollment processes with science. Western scientific and cultural values about kinship lie behind genetic testing technologies." Genetic testing "privileges the cultural values that inhere in those technologies over American Indian cultural values about kinship, ancestry, and citizenship."[18]

WORKS CITED

Constitution of the White Earth Nation. Ratified text of the Constitution published in Anishinaabeg Today and available on the website of the White Earth Reservation, http://www.whiteearth.com/home.html.

Bieder, Robert. *Science Encounters the Indian, 1820–1880* (Norman: U of Oklahoma P, 1986).

"The Human Genome Diversity Project," Human Genetics Alert, *GenEthics News* 10, 2000, http://www.hgalert.org/topics/personalInfo/hgdp.htm

[17] "Modal Ethical Protocol for Collecting DNA Samples," Morrison Institute for Population and Resource Studies, Stanford University, Human Genome Project. Guidelines subsequently published in the *Houston Law Review* 33.5 (1997), http://www.stanford.edu/group/morrinst/hgdp/protocol.html

[18] Kimberly TallBear, "'Native American DNA': Implications for Citizenship and Identity," American Indian Policy Center and Leadership Development Center, Arizona State University (2008), http://www.ncaiprc.org/files/Native%20American%20DNA %20Implications%20for%20Citizenship%20and%20Identity.pdf; revised version, "Native-American-DNA: In Search of Native American Race and Tribe" in *Revisiting Race in a Genomic Age*, ed. Sandra Lee, Barbara Koenig & Sarah Richardson (Piscataway NJ: Rutgers UP, 2008): 235–52.

Leslie, Mitchell. "The History of Everyone and Everything," *Stanford Magazine* (May–June 1999), http://www.stanfordalumni.org/news/magazine/1999/mayjun /articles /cavalli_sforza.html

Marks, Jonathan. "The Realities of Race," *Social Science Research Council* (7 June 2006), http://raceandgenomics.ssrc.org/Marks/ Revised version, "Race: Past, Present, and Future" in *Revisiting Race in a Genomic Age,* ed. Sandra Lee, Barbara Koenig & Sarah Richardson (Piscataway NJ: Rutgers UP, 2008).

Means, Russell. *Where White Men Fear to Tread: The Autobiography of Russell Means* (New York: St Martin's, 1995).

"Modal Ethical Protocol for Collecting DNA Samples," Morrison Institute for Population and Resource Studies, Stanford University, Human Genome Project. Guidelines subsequently published in the *Houston Law Review* 33.5 (1997), http://www .stanford.edu/group/morrinst/hgdp/protocol.html

Rothman, Barbara Katz. *Genetic Maps and Human Imaginations* (New York: W.W. Norton, 1998).

TallBear, Kimberly. "'Native American DNA': Implications for Citizenship and Identity," American Indian Policy Center and Leadership Development Center, Arizona State University (2008), http://www.ncaiprc.org/files/Native%20DNA %20 Implications %20for%20Citizenship%20and%20Identity.pdf. Revised version: "Native-American-DNA: In Search of Native American Race and Tribe," in *Revisiting Race in a Genomic Age*, ed. Sandra Lee, Barbara Koenig & Sarah Richardson (Piscataway NJ: Rutgers UP, 2008): 235–52.

Todorov, Tzvetan. *In Defence of the Enlightenment* (London: Atlantic, 2009).

Vizenor, Gerald. *Native Liberty: Natural Reason and Cultural Survivance* (Lincoln: U of Nebraska P, 2009).

Warren, William. *History of the Ojibway Nation* (Minneapolis MN: Ross & Haines, 1957).

⌘

The Edge of Extinction

Ethnic Survival Among the Yukaghirs of Northern Yakutia

JAROSŁAW DERLICKI

THE YUKAGHIRS ARE ONE OF THE SMALLEST AND THE MOST ANCIENT indigenous peoples of Northern Siberia. Out of 1,500 Yukaghirs living in the Russian Federation, almost a thousand live in the Republic of Sakha (Yakutia). The main Yukaghir settlements in Yakutia are located in the Lower Kolyma and Upper Kolyma districts.[1] This division is not only geographical but also cultural and linguistic. The two groups are considered to be quite different from each other. The Lower Kolyma Yukaghirs (also called the Tundra Yukaghirs) were traditionally reindeer breeders, just like their neighbours, the Chukchi and the Eveny. The Upper Kolyma Yukaghirs (called the Taiga Yukaghirs) are hunter–gatherer. Both of their dialects, which are practically dead, are as different from each other as Polish and Russian.

Waldemar Jochelson was the first researcher to conduct ethnographic fieldwork among the Yukaghirs, at the end of the nineteenth century. His book *The Yukaghir and Yukaghirized Tungus*[2] is still the most comprehensive work on the Yukaghirs and their past. In it, the author states that the Yukaghirs are on the edge of extinction. It could be reasonably presumed, then, that by the beginning of the twenty-first century, all the Yukaghirs had already died out. In this article, I would like to discuss the discourse of extinction and try to answer why, contrary to that expectation, the Yukaghirs have not vanished yet. The case of the Yukaghirs shows that it is difficult to list, or even name,

---

[1] In Russian, the *nizhnekolymskii* and *verkhnekolymskii raion*.

[2] Waldemar Jochelson, *The Yukaghir and Yukaghirized Tungus* (New York: E.J. Brill & G.E. Stechert, 1926).

all the factors responsible for ethnic survival. It also shows that ethnicity is not a primordial phenomenon. Anthony Smith underlines the role of demographic and cultural continuity. According to him, the extinction of an ethnic group does not mean the extinction of its population, but of its characteristics – culture, ways of living or the sense of community. Ethnic survival does not assume the survival of all of the group features (purity of blood, language, culture, etc.). The most important are the so-called specific ethnic elements, which may vary from one group to another.[3] It seems that the Yukaghirs, who have lost to a significant degree historical and cultural continuity with the former Yukaghir tribes, have nevertheless saved some specific ethnic elements and survived as an ethnic or local group.

James Clifford writes that existence among ruins has always been considered as a collapse or the end of the group, yet he does not agree with that view. He argues that every assimilated group brings in something new, leaves its trace, or establishes a new group. As an example, he cites the Mashpee Indians, who, despite losing most of their cultural traits, have established a specific local culture.[4] Similarly, Smith writes that an ethnic group may survive even if it is assimilated into the other culture.[5] An example of this is one of the Yukaghir tribes called the Chuvantsy: in the eighteenth century they assimilated with the Koryaks and became a distinct ethnic group. Over the next decades, the Chuvantsy were russianized, but they are still a separate *ethnie*.

The issue of cultural traits which are the base of ethnic development brings the discussion to the core of humanity. From the genetic point of view, people are almost identical on the DNA level. There are, however, some parts of the genome that reveal information about our ancestry, and these can be traced to the common 'father' and 'mother' of all humans.[6] Culture is somewhat similar to DNA – everyone has it.[7] Yet it seems as though it is much more diverse than DNA. This diversity developed over the course of thousands of years,

---

[3] Anthony D. Smith, *The Ethnic Origins of Nations* (Oxford: Blackwell, 1996): 96.

[4] James Clifford, "Identity in Mashpee," in Clifford, *The Predicament of Culture: Twentieth-Century Ethnography, Literature, and Art* (Cambridge MA: Harvard UP, 1988): 277–348.

[5] Smith, *The Ethnic Origins of Nations,* 96–97.

[6] Spencer Wells, *Deep Ancestry: Inside the Genographic Project* (Washington DC: National Geographic Society, 2007): 68.

[7] We could argue whether culture is something people possess or not, but my idea is only to show that culture is a common trait of humanity, like DNA.

just like the mutations of certain parts of DNA. It is mainly culture that is responsible for the contemporary ethno-cultural diversity of humans in the world. Does our ethnicity depend on our genes? It probably would if there had been no intermixing of genes throughout the history of humanity. However, many contemporary cases show that ethnic identification is not a matter of genetic background but self-identification.

One more aspect to be considered in the discussion of ethnic extinction is ethnicity and nationality. It must be remembered that these concepts were abstract ideas for Indigenous populations before their contact with white people. In northern Siberia, people belonged to particular tribes and clans. They usually called themselves 'true', 'genuine' people, as opposed to the neighbours who spoke a different language or possessed 'strange' features. Yet from the very dawn of human culture, the processes of assimilation, acculturation, and creolization have always existed. For centuries Yukaghir tribes were yukaghirizing their neighbours and being assimilated by them.[8] It is hard to specify when this exchange of genotype, culture, and ways of living started, since people have always migrated, traded and fought.[9] In some regions, such as the Lower Kolyma, the population was so inter-mixed it would call itself by one name – Khangai, the Tundra people – even though its members were of Yukaghir, Eveny or Chukchi descent. The only difference between them was the language used at home, and archive materials show how particular clans who, in the past, had considered themselves Chukchi or Eveny were becoming Yukaghir and vice-versa. Only after the arrival of the Russians, and especially after the October Revolution, were they divided into official ethnicities (or nationalities), since everyone was legally obliged to have a passport and a definite nationality. Ethnic extinction did not matter to those peoples until they were told it mattered.

---

[8] There are also genetic facts to prove it. At the European Human Genetics Conference in 2009, I.O. Mazunin, R.I. Sukernik and E.B. Starikovskaya presented a paper in which they argued that the mtDNA of Eveny/Evenki living in the middle of ancient Yukaghir territory shows the recent amalgamation of the Yukaghir remnants with Tungusic population ("Mitochondrial Genome Diversity in Tungusic-speaking Populations (Even and Evenki) and Resettlement of Arctic Siberia After the Last Glacial Maximum" (25 May 2009), http://www.eshg.org/eshg2009/abstracts.htm).

[9] Genographic data shows how and when people started to migrate from Africa into other parts of the world. It also seems that the scientists tracing the mutations of the DNA are able to identify the common ancestors of all contemporary humans. See Wells, *Deep Ancestry: Inside the Genographic Project*.

The history of Yukaghir tribes is so far an unsolved mystery; however, the traces of proto-Yukaghir tribes have been found in various parts of Yakutia.[10] Tungus tribes began to invade Yukaghir lands in the thirteenth century. When the Russians and Yakuts appeared in these territories in the seventeenth century, the Yukaghirs lived in the northern basins of the Yana, Indygirka, Kolyma, and Lena rivers. Yakut folktales describe the Yukaghirs as a relatively populous group. It has been estimated that, at the beginning of the seventeenth century, there were 5,000 Yukaghirs. Then – in under two centuries – several tribes vanished due to assimilation, invasions, and epidemics.[11] By the end of the nineteenth century, the Yukaghir population had dropped to several hundred.[12]

The Russian conquest changed the life of the Yukaghirs in many ways; on the other hand, the Russians did not really interfere with their internal affairs. They were only interested in the collection of a sort of Indigenous tax called *yasak*, economic exploitation, and keeping up the peace in the district. Those Yukaghirs (or members of their families) who could not fulfil their duties would become slaves. Sometimes clan leaders (or, again, members of their families) were kept by the Russian administration as hostages in order to make sure the Yukaghirs would obey the authorities. Even though intensive christianization was introduced in the 1860s, the Yukaghirs still believed in the forces of nature. Their traditional ways of living did not change much. They were still wandering along the rivers and lakes during summers and spending the coldest season in winter camps. Innovations and products introduced by the Russians made Yukaghir life easier – or, in the case of alcohol, miserable.[13]

---

[10] I wonder what genographic data could reveal about Yukaghir past? Some genetic information can be found in N.V. Volodko et al., "Analysis of the Mitochondrial DNA Diversity in Yukaghirs in the Evolutionary Context," *Russian Journal of Genetics* 45.7 (July 2009): 870–74.

[11] More in Jochelson, *The Yukaghir and Yukaghirized Tungus*, A.P. Okladnikov, *Iukagiry: Istoriko-etnograficheskii ocherk* (Novosibirsk: Nauka, 1975), V.A. Tugolukov, *Kto vy, Iukagiry* (Moscow: Nauka, 1979).

[12] Different sources give varying numbers. According to Jochelson, there were 1,455 Yukaghirs. Soviet sources mention 700–800 Yukaghirs.

[13] M.I. Kolesov, *Istoriia kolymskogo kraia* (Yakutsk: Iakutskoe Knizhnoe Izdatelstvo, 1991): 6–25 , Tugolukov, *Kto vy, Iukagiry*, 14–39.

The October Revolution made Yukaghir life even more miserable. Several waves of famine and epidemics, as well as thefts of animals and equipment, not to mention war contributions brought Yukaghir population to the 'edge of extinction'.[14] The situation improved in the 1920s, when the Soviet authorities introduced a rather liberal economic policy. For the first time, the Soviet state became interested in the matters of its aboriginal citizens. A special state agency, known as the Committee of the North, was organized to conduct ethnographic and linguistic studies, as well as to improve the condition of the tribal peoples of the north. Another policy, the so-called *korenisatsia*, was introduced to 'produce' the first indigenous Soviet elites, who would then become true Soviet leaders for their kinsmen.[15]

By the beginning of the twentieth century, basically two Yukaghir groups survived: the Lower and the Upper Kolyma ones.[16] First, permanent settlements were built in the 1930s. The Upper Kolyma (the Taiga) group was settled in the village of Nelemnoe[17] on the Yasachna river, while Tustah–Sen on the Chukochia river became a home for the Lower Kolyma (the Tundra) Yukaghirs. At the same time, the first cooperative enterprises were organized among the Yukaghirs. These were based on private property, but their members were obliged to share a part of their income with the organization. It was also the time when the first Russian teachers and medical doctors, as well as Soviet propagandists, arrived in Yukaghir settlements.[18] Until the 1950s, Nelemnoe was located in the Magadan district (not Yakutia) and from the

---

[14] Fewer than 400 Yukaghirs in 1926. See E.N. Fedorova, *Naselenie Iakutii: Proshloe i nastoiashchee* (Novosibirsk: Nauka, 1998): 84.

[15] James Forsyth, "The Indigenous Peoples of Siberia in the Twentieth Century," in *The Development of Siberia: People and Resources*, ed. Alan Wood & Alfred R. French (London: U of London P, 1989): 78–83; Okladnikov, *Iukagiry: Istoriko–etnograficheskii ocherk*; Nikolai Vakhtin, "Native Peoples of the Russian Far North," in *Polar Peoples: Self Determination and Development* (London: Minority Rights Group Report 5, 1992): 6–37.

[16] There are some yakutized and evenized Yukaghirs living in the northern basins of the Yana, Lena, and Indygirka rivers. All these groups number around a hundred persons altogether.

[17] Only in the 1950s were their kinsmen living on the Korkodon River forced to move to Nelemnoe, which became a Yukaghir showcase settlement in the Yakut Autonomous Soviet Socialist Republic (YASSR).

[18] Okladnikov, *Iukagiry: Istoriko-etnograficheskii ocherk*; Vakhtin, "Native Peoples of the Russian Far North."

very beginning was exposed to the influence of the Russian language, whereas the Yukaghirs living in the Lower Kolyma district treated the Yakut language as a local lingua franca.[19]

This rather liberal policy changed in the mid-1930s when forced collectivization was introduced and the Committee of the North was dissolved. Everything became state property and former cooperative enterprises were turned into *kolkhozes*.[20] At the same time, the state started a campaign against shamans and clan elites[21] (considered to be wealthy *kulaks*). Due to the outbreak of the Second World War, the state was unable to fully implement russification and industrialization, but shortly after the war, giant industrial companies started to exploit tribal territories and Russian became the language of education. Centralization, another Soviet policy, introduced in the 1950s, involved the liquidation of smaller villages and the re-location of local populations in larger settlements. Smaller *kolkhozes* were united into larger *sovkhozes*.[22] The Tundra Yukaghirs were forced to move to the multi-ethnic village of Andrushkino, where they became a minority among the Eveny and Yakut.[23]

The Soviet regime considered Indigenous peoples to be backward and handicapped savages who needed Russian help to become civilized. New labour and educational systems aimed at producing Soviet citizens out of Siberian natives. People were divided into *kolkhoz* and *sovkhoz* brigades and had to fulfil Soviet economic plans. Children had to attend boarding schools

---

[19] This situation is still, to a certain extent, noticeable today. The Upper Kolyma Yukaghirs are russianized and they usually do not speak the Yakut language. Among the Lower Kolyma Yukaghirs, only the younger generations are russianized, while the elder and middle generations speak Yakut better than Russian.

[20] *Kolkhoz* was a form of collective work introduced by Soviet state. *Kolkhoz* basically meant 'collective farm' (*kolektivnoe khoziaistvo*).

[21] Both categories were literally liquidated. In 1938, Teki Odulok, the only Yukaghir to obtain higher education, was accused of being a Japanese spy and executed.

[22] *Sovkhoz* was somewhat similar to *kolkhoz*, but it was a more advanced version of collective farming. The word itself means 'Soviet farm' (*sovetskoe khoziasistvo*).

[23] Forsyth, "The Indigenous Peoples of Siberia in the Twentieth Century," and Okladnikov, *Iukagiry: Istoriko–etnograficheskii ocherk*. Some Tundra Yukaghirs lived with Chukchi in Kolymskoe, in Cherskii (the district capital), and in other settlements at the mouth of the Kolyma river.

dominated by Russians and the Russian language.[24] According to the policy 'ethnic in form, but Soviet in content', traditional culture was supposed to be locked up in a museum and replaced by the modern Soviet one. The state introduced a system of small local museums and the so-called 'houses of culture', where traditional culture could survive in the form of folksongs, dances, and costumes.[25]

It seems that the Yukaghir were among the Indigenous peoples least resistant to Russian assimilation. They were also one of the poorest, and had one of the highest mortality rates, as well as the lowest average life expectancy. In the 1950s the Soviet authorities came up with the idea of 'the Soviet people' – they believed various nations living in the Soviet Union would unite into one homogeneous Soviet society. To show that ethnic (or national) boundaries were vanishing, they excluded several ethnicities from national censuses. The Yukaghirs from Andrushkino were all considered to be Eveny and, due to this fact, the Yukaghir population dropped to 295 individuals in 1959. In the next censuses (1979, 1989), the Yukaghir population in Yakut ASSR could be seen to be steadily increasing.[26] Traditional culture, however, was vanishing and being replaced by Soviet culture. Since the Yukaghirs were convinced their culture had no value, they tried to become Soviet as fast as possible. The only areas where traditional culture was still alive were their traditional occupations of living off the taiga and tundra.[27]

The Yukaghir territories became the property of the *kolkhozes* and then the *sovkhozes*, but the Yukaghirs continued to exploit the land. Before the October Revolution, every Yukaghir family or clan had its traditional lands

---

[24] People who attended such schools very often refer to traumatic memories. They remember unpleasant smells, claustrophobic rooms, being punished by teachers for speaking languages other than Russian, etc. Many children tried to escape from the school and go back to their village. Many people who acquired higher education argued that they were Yukaghirs no longer, because a genuine Yukaghir lives in the taiga or tundra. They considered themselves a lost generation, living between the Yukaghir and Russian cultures, which quite often led them to alcohol abuse.

[25] Debora L. Schindler, "Theory, Policy, and the *Narody Severa*," *Anthropological Quarterly* 64.2 (April 1991): 68–79, and Vakhtin, "Native Peoples of the Russian Far North," 14–17.

[26] Fedorova, *Naselenie Iakutii: Proshloe i nastoiashchee*, and Kolesov, *Istoriya kolymskogo kraya*, 29–33.

[27] Jarosław Derlicki, "Jukagirzy: Psi ludzie," in *Pierwsze narody: Społeczności rdzenne i idea tubylczości we ws* (Warsaw: DIG, 2002): 249–50.

(usually along a particular river or lake). This traditional system of land distribution survived even under communism. Yukaghir hunters and fishermen were re-organized into *kolkhoz* brigades, but were allowed to hunt or fish in their clan territories. It was also a time when two hunting/fishing categories appeared: licensed hunters, who were *sovkhoz* workers, and 'hobbyists' (Rus. *lubitel*), who did something else for a living, but were allowed to hunt and fish. Licensed hunters could hunt not only in their traditional territories but also on additional lands assigned by the *kolkhoz* or *sovkhoz*. In fact, if *sovkhoz* authorities wanted to exploit someone's traditional territories, they had to ask for permission. Hunting territories were formally state property, but in fact people treated them like their own.[28]

Life in the taiga and tundra did not change much in the Soviet period. Obviously, hunters and reindeer herders acquired modern technologies and equipment, but they continued to do what their ancestors had done for centuries, preserving the 'Yukaghir way of life'. Even though the USSR was an atheistic state, the Yukaghir, as well as the other ethnic groups, still negotiated with the spirits of the land, asking them for prey or success and bringing small offerings like a drop of vodka, some bread or cigarettes. The whole environment of taiga and tundra was perceived as consisting of supernatural beings. Every plant, animal, lake, and river had its own spirit. Far from being anonymous, the vast area was divided into well-known places (Rus. *mestnost*) which had their guardian spirits called the Owners (Rus. *khozain*). All in all, even though the Yukaghirs had lost much of the essential contents of their traditional culture, they continued to call themselves 'Yukaghir', for they still occupied the territories of their ancestors and did the same things for a living.[29]

At the beginning of the 1990s, the Yukaghirs, just like other tribal groups in Russia, had to face two crucial issues: economic survival and ethnic revival. The first was related to the collapse of *sovkhozes* and the fall of the Soviet economic system. The state could no longer support its citizens in the way it had done in the past. It was the end of the governmental subsidies and reimbursements which kept the USSR in one piece. It was also a time when 'ethnic' elites in particular provinces of Russia became sufficiently aware of the specificity of their identity to demand the right to save and develop their culture and language. A lot of effort was put into creating ethnic mythologies,

[28] Jaroslaw Derlicki, "Spirits of the Land," *Academia* 16 (2007): 12–15.

[29] Derlicki, "Spirits of the Land."

inventing traditions and resurrecting the 'traditional' holidays and celebrations.[30]

The former *sovkhozes* and their real estate as well as all movable goods were divided among the newly created clan communities.[31] The Upper Kolyma Yukaghirs from Nelemnoe established a clan community called Teki Odulok. The situation in Andrushkino was much more complicated, since it was a multi-ethnic village. Every ethnic group founded its own clan community, but they all went bankrupt and were subsequently united as one Yukaghir–Eveny[32] community named Chaila. Both communities (in Nelemnoe and Andrushkino) faced the same problems. At the beginning, it was hard to divide the *sovkhoz*'s property among all the communities of the district on an equal basis. Then this property vanished in mysterious circumstances. The *obshchina* in Nelemnoe consists mainly of hunters/fishermen and several workers involved in cattle- and horse-breeding. In Andrushkino, there are several reindeer-herding tribes as well as hunters and fishermen. Unfortunately, neither community can exist without governmental support, and for both of them the economic results are very poor.[33]

In 1992, there were congresses organized in Yakutia for every ethnic group living in the republic. A congress for the Yukaghirs living in Russia was organized in Nelemnoe. During that congress, the Yukaghir elite established the Yukaghir Council of Elders (Rus. *Sovet Stareyshin*) to advise the government on Yukaghir issues and help to solve current problems. There were,

---

[30] Jarosław Derlicki & Wojciech Lipiński, "Na krańcach Jakucji," *Sprawy Narodowościowe* 6.2 (1997): 329–40.

[31] The Yakutian parliament (*Il Tumen*) passed a law reorganizing *kolkhoz*es into "clan communities" (Rus. *rodovaya obshchina*), which were administrative units rather than true clans or kin structures. Before the October Revolution the tsarist administration used to create artificial administrative clans based not on the kin system but on the inhabited area or taxation system. The new law perceived the clan in a similar way: i.e. as an economic unit with legal status. These units were supposed to replace the former *sovkhozes* and *kolkhozes* as well as to focus on ethnic and cultural revival. In fact, they became small enterprises.

[32] Yukaghirs from Andrushkino had utopian ideas of establishing their own community, leaving Andrushkino and moving back to the village of Tustah–Sen. After the bankruptcy of the community, they became more realistic and accepted the fact they would never move back to Tustah–Sen, since there were no funds for such an enterprise and almost all Yukaghirs lived in ethnically mixed families anyway.

[33] Derlicki & Lipiński, "Na krańcach Jakucji."

naturally, a lot of speeches and lectures on how to save Yukaghir culture and language. Even though the congress did not bring any measurable or significant changes, it showed for the first time that Yukaghir culture was interesting and of value and should therefore be saved. It was also a time when the Yukaghirs from different parts of Russia (if mainly from Yakutia) gathered together and could experience the differences that existed among them as well as the common features that bound them together. They started to perceive themselves as one group having a common past. During the congress, the traditional Yukaghir holiday of *Shakhadibe* was introduced as part of the ethnic revival. Although it continued to be celebrated for several years afterwards, it had lost all its meaning by the end of the 1990s.[34]

After the congress, the newly established Council of Elders put a great deal of effort into building Yukaghir nationalism. With the support of Russian and Yakut authorities, its members initiated several ethnic projects. A Teki Odulok's Yukaghir School was created in Nelemnoe, where students learned their native language, national culture, and the literature of the northern peoples, as well as traditional occupations. In Andrushkino, children were divided into ethnic classes (Eveny, Yukaghir, and Yakut) on the basis of their background, but the idea was basically the same – to teach students their native language and culture. This ethnic education has produced some positive results: the children have become aware of their native heritage, for instance, and especially enjoy such subjects as traditional occupations, which are very practical and useful in everyday life. The problem is, however, that several years after graduation they can barely pronounce a handful of Yukaghir words or say anything about traditional Yukaghir culture.[35]

Another important idea from the point of view of Yukaghir nationalism was the project of Yukaghir autonomy called *Suuktul*. Since the United Nations considered the Yukaghirs and their language endangered, the intelligentsia tried to use this fact to establish something resembling an Indian reservation. It was supposed to include both of the largest Yukaghir settlements, Nelemnoe and Andrushkino, and their native territories. The project postu-

[34] Jarosław Derlicki, "The New People: The Yukaghir in the Process of Transformation," in *Between Tradition and Postmodernity*, ed. Lech Mróz & Zofia Sokolewicz (Warsaw: DIG, 2003): 124–25.

[35] Jarosław Derlicki, "Ethno-pedagogy – the Curse or the Cure? The Role of the School Among Youth in Nelemnoe (Yakutia)," *Sibirica: Journal of Siberian Studies* 4.1 (April 2004): 63–73.

lated the creation of native authorities with a Supreme Chief, a Council of Elders, and a Great Assembly. It took several years to solve all the practical issues, as well as the legislative and economic problems. It was also not clear which territories were to be included in *Suuktul*. Nobody questioned the layout of the Yukaghir lands around their former village of Tustah–Sen, but when the idea of moving back to Tustah–Sen was abandoned, new questions arose. It was impossible to separate the Yukaghirs from the rest of the village (due to mixed marriages, etc.) and it was clear the *Suuktul* must encompass all villagers.[36] The territories of the Upper Kolyma Yukaghirs comprised two-thirds of the entire district area, and, additionally, there is a gold mine on that land. District authorities called that territory 'a special land fund of the district' and only allowed the Yukaghirs to exploit it, not own it. After a long legal battle, the Yukaghirs regained severely trimmed lands. Finally, the *Suuktul* plan passed through the Russian and Yakutian parliaments and was accepted for implementation. Unfortunately, at present there are no funds to bring *Suuktul* into being, but it has become a sort of Yukaghir nationalist idea.

Yet the intelligentsia's efforts have very little meaning in the life of the ordinary people who have to earn their living. At the beginning of the 1990s there was a spirit of enthusiasm; people actually felt and believed that things would get better. Nobody knew how hard it would be to revive the culture and the language. The Yukaghir elites live mostly in the capital, Yakutsk, and do not fully understand local village problems. Basically, the village people, or at least the majority of them, are concerned mostly with economic issues.

Since the intelligentsia has no influence over the economic situation in its native villages and can do very little to improve this, it concentrates on cultural issues. The elites understand that they must attract people to the Yukaghir culture and language. They are also aware of the devastation left by the Soviet period, when native cultures were presented as backward, savage, and a thing to be ashamed of. Using different tools and mechanisms, the Yukaghir leaders have built a new image of the Yukaghirs and their culture, trying to convince people it has value and must be saved.[37]

The discourse of 'dying out' or being on 'the edge of extinction' plays a quite significant role in this process. Press headlines such as "Perishing

---

[36] Derlicki, "The New People: The Yukaghir in the Process of Transformation," 127–28.

[37] "The New People: The Yukaghir in the Process of Transformation," 128.

Cultures"[38] or "The Last Mohicans"[39] suggest that a part of human heritage is being irrevocably lost. Many Indigenous peoples and many native languages are considered to be endangered. Since the 1970s, Indigenous peoples have started to organize themselves into international associations and societies[40] as well as to present the problems of their survival to the United Nations Organization.[41] In most cases, the issue of extinction has enabled international help to be given to these groups. The discourse of 'dying out' is a very popular argument used by Indigenous peoples to attract international attention.[42]

This argument is also used by the Yukaghir intelligentsia, which tries to convince its kinsmen, as well as public opinion, that Yukaghir culture ought to be saved. Titles such as *Problems of the Revival of Vanishing Yukaghirs*[43] or *When Will the Last Descendant of the Odul Tribe Be Born?*[44] are so well-known that, if any child in Nelemnoe or Andrushkino is asked who the Yukaghirs are, s/he answers: 'A dying people'. In both villages, certain individuals quite often introduce themselves as 'the last true Yukaghir'. It must be also remembered that the 'extinction' argument helps in getting financial support from the government and international organizations.

The image of the Yukaghirs as a vanishing/dying people consists of several elements. The Yukaghir leaders, referring to the concepts of Lev Gumilev,[45]

---

[38] Wade Davis, "Vanishing Cultures," *National Geographic* 196.2 (August 1999): 64–89.

[39] Olgierd Budrewicz, "Ostatni Mohikanie," *Wprost* 31 (2001): 90–92.

[40] Endangered Peoples Trust, National Indigenous Organization of Colombia, RAIPON, Inuit Circumpolar Conference and many, many others.

[41] UN Permanent Forum on Indigenous Issues.

[42] *BBC News* (26 January 2011), http://news.bbc.co.uk/2/hi/europe/2238333.stm; *Living in Ecuador Blog* (26 January 2011), http://blog.pro-ecuador.com/?p=1171; *Galdu: Research Centre for the Rights of Indigenous Peoples* (26 January 2011), http://www.galdu.org/web/index.php?odas =2749&giella1=eng Australian Law Reform Commission (26 January 2011), http://www.alrc.gov.au/publications/23.%20General%20Issues%20of%20Evidence%20and%20Procedure/aboriginal-dying-declarations

[43] I.E. Tomskii, *Problemy vozrozhdeniia ischezaiushchikh iukagirov* (Yakutsk: Severoved, 1996).

[44] V. Khristoforov, "Kogda roditsia poslednii potomok plemenii odul?" *Gazeta Iakutiia* (6 July 2000).

[45] Lev N. Gumilev, *Ethnogenesis and the Biosphere* (*Etnogenez i biosfera zemli*; Moscow: Progress, 1999), and *Ethnosphere: Human History and Natural History* (*Etnosfera: Istoriia liudei i istoriia prirody*; Moscow: Progress, 1993).

state that the nation is a living unit, which is born, develops, disintegrates and finally dies out. This argument allows the Yukaghir past to be reconstructed in a very flexible way. Using scientific facts and their own assumptions, authors describe how the Yukaghir nation was built, how it colonized vast parts of Siberia, how it civilized neighbouring tribes and cultures, and, finally, how it started to turn inert and to die out after Russian colonization. In this way, they explain how one mighty Yukaghir nation and language disintegrated into numerous smaller Yukaghir tribes and dialects.[46] In my view, this is the intelligentsia's way of overcoming the differences between the Upper and Lower Kolyma Yukaghirs, to show that they were united in the past. It also explains the contemporary situation of Yukaghirs, showing that the 'golden age' has irrevocably passed.

Even though the Yukaghirs have 'almost' died out, they have left a heritage of which every Yukaghir can be proud. Nikolay Kurilov, one of the Yukaghir leaders, tells of how the Yukaghirs migrated from Yakutia to Chukotka, giving rise to Chukchi and Eskimo. A part of the Yukaghirs migrated to the south and participated in the formation of the Mongol and Tungus–Manchurian peoples. In the same way, the Khanty, Mansi, and Nenets owe much to the Yukaghirs. According to Kurilov, the Yukaghirs were the ancestors of the ancient Scythians. The author also argues that the Yukaghirs crossed the Bering Strait and populated Northern America. His argument is that by the end of the 1980s, the Yukaghirs from Andrushkino lacked precisely the same blood groups as American Indians do.[47] The Yukaghir version of the Indians' origins has more supporters; for example, one Yukaghir linguist believes that "according to linguistic data, there are relations with western California, far away. And this is quite strange […] we should expect similarities with northern Indians, but not Californians."[48] It is clear that the Yukaghir elites try to mobilize ethnic pride by showing how much everyone

---

[46] Derlicki, "The New People: The Yukaghir in the Process of Transformation," 130. Some linguists believe that it was a group of distinct languages somewhat related to the Finno-Ugric group.

[47] Nikolai Kurilov, *Iukagiry: Neraskrytaia zagadka chelovechestva (Razmyshleniia iukagira)* (Yakutsk: Severoved, 1999).

[48] Male, aged 65, Yakutsk. This is one of a series of ethnographic interviews that I carried out in 1999–2006. All further quotations emanating from the interviews conducted during that period will be referenced in subsequent footnotes as 'DI' (Derlicki Interviews).

owes to the Yukaghirs. These arguments are also followed by some Messianic premisses – everyone is in debt to the Yukaghirs, who are now dying out, having sacrificed themselves for the good of their neighbours.

What is unique in the Yukaghir heritage? According to Yukaghir ideologists, the Yukaghirs were able to develop a unique relationship with nature. It is constantly emphasized that the Yukaghirs depend on nature. Living off the taiga and tundra is identified with the Yukaghir way of life:

> The Yukaghirs are the ultimate taiga people. Our traditions were adopted by other nations, even by the Russians. Taiga traditions and customs – everything was taken from the Yukaghirs. Nobody hunted like we did.[49]

This exceptional approach has survived only due to traditional ways of living, such as hunting, fishing, and reindeer-breeding. Since all of nature is thought to be made up of spiritual beings, there exists an entire set of rules and laws defining the relationship between humans and nature. Luck in pursuit of a traditional occupation is secured by small offerings to the spirits of a place or animal. Even though many other Siberian peoples practise such animistic rituals, they are perceived as genuine Yukaghir traditions, which are now being re-invented and applied to modern celebrations (at school, in houses of culture, etc.).

Since the Yukaghirs have developed such a unique relationship with nature, they perceive themselves as the most ecological people in the world. They take from nature only what they need, because nature is their mother, who knows best what their needs are.

> If you take the traditional outlook, you're not allowed to take more from nature than you need at the moment. You're not allowed to gather more – what you need, you take, what you'll need tomorrow, you'll take tomorrow. That is why others say the Yukaghirs are light-hearted, we don't care about stocks, because our mother nature is going to take care of us.[50]

The intelligentsia believe that only respect for nature allowed the Yukaghirs to survive for centuries. Contemporary problems are quite often associated with the changing relationship with nature, with pragmatism and over-

---

[49] Male, aged 65, Nelemnoe, DI.

[50] Male, aged 35, Yakutsk, DI.

exploitation. Very often this is perceived as a threat to the Yukaghirs: "When there are no reindeer, I don't know what we'll do for a living. We are nomads, we cannot sit in one place."[51]

There is a whole set of psychological traits assigned to Yukaghirs by their leaders. Living in extremely hard conditions resulted in the development of hospitality and mutual help. The refusal of help or hospitality could mean a death sentence and everyone was aware they might need such help one day; as one interviewee put it, "I don't sell meat, I give it to people. I know that tomorrow I might not have the meat and will have to ask others for it."[52] Shyness, modesty, and reticence are considered to be positive traits, but are often perceived as a threat, too, since they make the Yukaghirs unable to fight for their rights and needs.

The Yukaghir intelligentsia sketch an image of a good and naive 'Man Friday', the last true child of nature, still unspoiled by civilization. Unfortunately, his advantages are also a curse, since he is unable to fight for his rights and to defend his values. It is quite interesting how Yukaghir leaders attempt to reverse the Soviet stereotype which presented the Yukaghirs as uncivilized, backward, and savage. They try to convince their kinsmen that the Yukaghir way of living is the best. It proves that Anthony D. Smith was right in saying that even if a particular group is truly backward and 'savage', it is a duty of its elite to present the native culture as pure, primordial, and unique.[53]

The elements of traditional culture that have survived are basically those which are related to traditional occupations, animistic religiosity, respect for nature, etc. People still use traditional winter clothing, which is much better than contemporary garments. The problem is that it is hard to say whether those articles of clothing are of Yukaghir or Eveny origin. Most authors underline Eveny influence on Yukaghir costume.[54] The Yukaghir intelligentsia argue that it was the Eveny who adapted Yukaghir clothing to their needs, not the other way round. After all, the argument goes, the Eveny migrated from the South, while the Yukaghirs lived in the North for centuries and developed clothing best suited to harsh polar conditions. One Yukaghir teacher from Andrushkino argued:

---

[51] Male, aged 35, Andrushkino, DI.

[52] Male, aged 60, Nelemnoe, DI.

[53] Smith, *The Ethnic Origins of Nations*.

[54] Liudmila N. Zhukova, *Odezhda iukagirov* (Yakutsk: Iakutskii krai, 1996).

> Whose are the clothes? It depends on who wrote more books about clothes. Since most books about Yukaghir clothes were written by the Eveny, it seems like our clothes were borrowed from the Eveny, but we know the truth […].”[55]

The Yukaghirs still build traditional *urasa*,[56] where they smoke and dry fish and meat. In the tundra, where people breed reindeer, many traditional tools and items have been preserved, but just as in the case of clothes, it is hard to say whether they are of Eveny, Chukchi, or Yukaghir origin. The Yukaghir argument is always the same, however: “we were here first, so it must be ours.”

Many authors insist that ethnic survival does not necessarily assume the survival of traditional culture. According to Anthony D. Smith, it is specific ethnic elements that are responsible for the survival of the group.[57] James Clifford writes that culture and identity are not stable phenomena which have a long genealogy, but live thanks to pollination and transplantation. This means that culture is flexible and open to critical and creative re-arrangements of both new and old elements.[58] The Yukaghir elites understand that traditional culture is vanishing. They are trying to save as much as possible, keeping in mind that traditional culture must be modernized and re-arranged.

> I support the idea we must modernize culture. We won’t be able to save pure patriarchal culture […] we have to take some elements and direct our children. Let’s take traditional clothing – it will survive in the museum, but we could inspire our kids to somehow rearrange it and make something new using traditional elements. Our girls like fashion and they like sewing, maybe they could create a new, neo-Yukaghir style.[59]

This kind of invention of tradition has been described by Eric Hobsbawm as an important part of nationalism.[60]

---

[55] Female, aged 55, Andrushkino, DI.

[56] *Urasa* was a kind of tent, similar to the Indian tepee, but made out of larch bark. Yukaghirs used to live in such tents from spring to fall.

[57] Smith, *The Ethnic Origins of Nations*, 96.

[58] Clifford, “On Ethnographic Authority,” in Clifford, *The Predicament of Culture*, 22–23.

[59] Male, aged 35, Yakutsk, DI.

[60] Eric Hobsbawm, “Introduction: Inventing Traditions,” in Eric Hobsbawm & Terence Ranger, *The Invention of Tradition* (Cambridge: Cambridge UP, 1983): 1–15.

At the end of the 1980s, genetic research was conducted among the Yukaghirs. According to the results, only 7.7 percent of the population were direct descendants of the Yukaghir tribes, and about 80 percent were of mixed origins. The prediction was that "within two generations the Yukaghir genotype will be lost."[61] Today, twenty years after this research, it is probably even more difficult to find a 'pure-blooded' Yukaghir.[62] In Nelemnoe, there are people who look Russian yet call themselves Yukaghir. In Andrushkino, the influence of European genes is less visible, since most people living in the village are of the Asiatic type. However, I think that from the genetic point of view, Jochelson's Yukaghirs have already died out. This, however, does not mean that this ethnic group is extinct.

Official statistics and censuses say otherwise. From 1989 to 2002, the number of Yukaghirs rose from 1,142 to more than 1,500 persons. In Nelemnoe, about 92 percent of children from mixed marriages are registered as Yukaghirs, while in Andrushkino the figure is about 76 percent.[63] It seems that Yukaghir nationality is preferred over other nationalities. Is this the result of the intelligentsia's activities or maybe of ethnic mobilization? Probably not. In the Soviet era, the so-called 'Small Peoples of the North' were granted many state privileges, discounts, and the like. There were some reductions in income tax, there were stipends for university study, free hunting licenses, special equipment for hunting and reindeer-herding brigades. Since Eveny are also one of the Small Peoples of the North, the tendency to assign Yukaghir nationality in Andrushkino is weaker than in Nelemnoe, where the Yukaghirs are the only small-numbered natives. It must be remembered that some of the former privileges exist nowadays as well. The Small Peoples of the North can obtain a non-repayable credit or lease; there are places at universities reserved only for them; they get a discount for hunting licenses and are exempt from military service. There are also many international and state projects devoted to improving the situation of the Small Peoples of the North. Since the Yukaghirs are considered an endangered people, there is a strong tendency to

---

[61] Tomskii, *Problemy vozrozhdeniia ischezaiushchikh iukagirov*, 31.

[62] To me as a cultural anthropologist, the idea of the purity of genes is questionable, but I presume geneticists are able to find and define the Yukaghir genotype. I just wonder what the specificity of the Yukaghir genome is and what data it reveals.

[63] Derlicki, "The New People: The Yukaghir in the Process of Transformation," 123.

choose Yukaghir nationality for children.[64] Paradoxically, 'extinction discourse' helps to increase population numbers.

It also shows that while the Human Genome Diversity Project[65] might reveal some information about the genetic history of particular groups, it tells us nothing about ethnic survival. It is not genes that make us a member of our ethnic or cultural group. The Yukaghir example shows one more thing – in the case of human groups or ethnicities it is hard to speak of endogamy. Genome diversity shows how people have been migrating and intermixing with other human groups for thousands of years.[66] If human groups had been truly endogamic, we would probably all have remained in one kinship system or ethno-cultural group.

It has already been said that from the genetic point of view the Yukaghirs have died out. If other objective factors or features of the ethnic group, such as native language, common ancestry, traditional power structure or common territory, are taken into consideration, it is also obvious that the Yukaghirs have died out. The similar case of the Mashpee Indians was described by James Clifford. In order to have self-government and land title, the Mashpee had to prove before the courts that they were indeed an Indian tribe. All objective features failed – they did not have a common native language, beliefs, culture, history, etc. The court, even after applying a very broad definition of what constituted a tribe, was unable to determine tribal continuity among the Mashpee. Indian activists, taking what remained of their traditional culture and past, attempted to sketch an integral picture of an Indian tribe and culture.[67] The court's judgment was unfavourable towards the local Indians.[68]

---

[64] Nikolai Vakhtin, *The Yukaghir Language in Socio-Linguistic Perspective* (Poznán: IIEOS, 1991): 8–12.

[65] Wells, *Deep Ancestry: Inside the Genographic Project.*

[66] Natalya V. Volodko et al. wrote that in the seventeenth century Yukaghirs met the criteria of genetic isolates; she didn't, however, provide any proof or arguments. On the other hand, her genetic findings show that Yukaghir genetic prehistory is a mystery and there is a wide range of subhaplogroups (C and D) of which some are very rare. See Volodko et al., "Mitochondrial Genome Diversity in Arctic Siberians, with Particular Reference to the Evolutionary History of Beringia and Pleistocenic Peopling of the Americas," *American Journal of Human Genetics* 82.5 (May 2008): 1084–1100.

[67] As Anthony D. Smith says, they used specific ethnic elements (*The Ethnic Origins of Nations*, 96).

[68] Clifford, "Identity in Mashpee," in Clifford, *The Predicament of Culture*, 277–348.

The cases of the Mashpee and the Yukaghirs show that if a group loses its ethnic characteristics, it still may be distinguished as a specific local culture.

The Yukaghirs survive only in their traditional occupations and everything that goes along with them – animistic beliefs, respect for nature, hunting and reindeer-breeding rituals and ceremonies. In Nelemnoe, where the Yukaghirs are the only natives, all elements of traditional culture are considered Yukaghir. In Andrushkino, the Yukaghirs argue with the Eveny over what is Yukaghir and what is Eveny. It seems that ethnicity started to matter to these peoples only after the October Revolution, when the Soviet authorities began to assign a specific nationality to specific peoples. Before that, they all called themselves *Khangai* – the Tundra People. Archive materials show how particular clans were being yukaghirized or evenized. Ancestors of one of the Yukaghir leaders used to belong to a clan whose name comes from the Chukchi language. Among the Eveny, after Russian conquest there was a clan called Yukaghir.[69] Assimilation, adaptation, migration, acculturation, and other processes have accompanied human beings from their very beginnings. Peoples were migrating, trading or invading; some have vanished, others have emerged. But vanishing peoples and cultures always leave their traces and become part of a local past and local culture.[70] What matters in ethnic survival is not the physical existence of human beings, but the continuing existence of their cultural traits. Contemporary Yukaghirs are very different from Jochelson's Yukaghirs. But they still live where their ancestors used to live, and do basically the same things for a living. In this sense, it is not the objective or scientific criteria that constitute Yukaghirs, but the self-definition and willingness to be Yukaghir. This justifies treating the notions of 'edge of extinction' or 'dying out' as somewhat relative phenomena.[71]

---

[69] It is also worth remembering the title of Jochelson's book: *The Yukaghir and Yukaghirized Tungus*.

[70] Just as DNA can reveal a great deal of information but is also limited, since there is no trace of male lineages which existed more than 60,000 years ago. More in Wells, *Deep Ancestry*, 228–29.

[71] The research on Yukaghir identity, including that presented in this essay, is being conducted as a part of the research project "Landscape and Identity: The Perception and Role of Taiga and Tundra Among Indigenous Communities in Kolyma Basin," supported by the Polish Ministry of Science and Higher Education (grant no. IP 2010 007070).

WORKS CITED

Budrewicz, Olgierd. "Ostatni Mohikanie," *Wprost* 31 (2001): 90–92.

Clifford, James. *The Predicament of Culture: Twentieth-Century Ethnography, Literature, and Art* (Cambridge MA: Cambridge UP, 1988).

Davis, Wade. "Vanishing Cultures," *National Geographic* 196.2 (August 1999): 64–89.

Derlicki, Jarosław. "Ethno-pedagogy – the Curse or the Cure? The Role of the School Among Youth in Nelemnoe (Yakutia)," *Sibirica: Journal of Siberian Studies* 4.1 (April 2004): 63–73.

——. "Jukagirzy: Psi ludzie," in *Pierwsze narody: Społeczności rdzenne i idea tubylczości we współczesnym świecie*, ed. Jarosław Derlicki & Wojciech Lipiński (Warsaw: DIG, 2002): 247–57.

——. "The New People: The Yukaghir in the Process of Transformation," in *Between Tradition and Postmodernity*, ed. Lech Mróz & Zofia Sokolewicz (Warsaw: DIG, 2003): 121–36.

——. "Spirits of the Land," *Academia* 16 (2007): 12–15.

——, & Wojciech Lipiński. "Na krańcach Jakucji," *Sprawy Narodowościowe* 6.2 (1997): 329–40.

Fedorova, E.N. *Naselenie Iakutii: Proshloe i nastoiashchee* (Novosibirsk: Nauka, 1998).

Forsyth, James. "The Indigenous Peoples of Siberia in the Twentieth Century," in *The Development of Siberia: People and Resources*, ed. Alan Wood & Alfred R. French (London: U of London, 1989): 72–96.

Gumilev, Lev N. *Ethnogenesis and the Biosphere* (*Etnogenez i biosfera zemli*; Moscow: Progress, 1999).

——. *Ethnosphere: Human History and Natural History* (*Etnosfera: Istoriia liudei i istoriia prirody*; Moscow: Progress, 1993).

Hobsbawm, Eric. "Introduction: Inventing Traditions," in Eric Hobsbawm & Terence Ranger, *The Invention of Tradition* (Cambridge: Cambridge UP, 1983): 1–15.

Okladnikov, A.P. *Iukagiry. Istoriko-etnograficheskii ocherk* (Novosibirsk: Nauka, 1975).

Jochelson, Waldemar. *The Yukaghir and Yukaghirized Tungus* (New York: E.J. Brill & G.E. Stechert, 1926).

Khristoforov, V. "Kogda roditsia poslednii potomok plemenii odul?" *Gazeta Iakutiia* (6 July 2000).

Kolesov, M.I. *Istoriia kolymskogo kraia* (Yakutsk: Iakutskoe Knizhnoe Izdatelstvo, 1991).

Kurilov, Nikolai. *Iukagiry: Neraskrytaia zagadka chelovechestva (Razmyshleniia iukagira)* (Yakutsk: Severoved, 1999).

Schindler, Debora L. "Theory, Policy, and the *Narody Severa*," *Anthropological Quarterly* 64.2 (April 1991): 68–79.

Smith, Anthony D. *The Ethnic Origins of Nations* (Oxford: Blackwell, 1996).

Tomskii, I.A. *Problemy vozrozhdeniia ischezaiushchikh iukagirov* (Yakutsk: Severoved, 1996).

Tugolukov, V.A. *Kto vy, Iukagiry* (Moscow: Nauka, 1979).

Vakhtin, Nikolai. "Native Peoples of the Russian Far North," in *Polar Peoples: Self Determination and Development* (London: Minority Rights Group Report 5, 1992): 6–37.

——. *The Yukaghir Language in Socio-Linguistic Perspective* (Poznán: IIEOS, 1991).

Volodko, Natalya V. et al. "Analysis of the Mitochondrial DNA Diversity in Yukaghirs in the Evolutionary Context," *Russian Journal of Genetics* 45.7 (July 2009): 970–874.

——. "Mitochondrial Genome Diversity in Arctic Siberians, with Particular Reference to the Evolutionary History of Beringia and Pleistocenic Peopling of the Americas," *American Journal of Human Genetics* 82.5 (May 2008): 1084–1100.

Wells, Spencer. *Deep Ancestry: Inside the Genographic Project* (Washington DC: National Geographic Society, 2007).

Zhukova, Liudmila N. *Odezhda iukagirov* (Yakutsk: Iakutskii krai, 1996).

⌘

Genetic Signatures of Australia's First Peoples Survive Recent History*

SHEILA VAN HOLST PELLEKAAN

Introduction

THE STORY OF COLONIZATION IS A FAMILIAR ONE for Indigenous peoples whose countries were taken by colonizers, whose culture, language, and traditional life-styles were regarded as inferior, and who were forced to move aside to make room for large-scale expansion dominated by technological, political, and economic power. In Australia the process began in earnest in 1788 and, despite the intentions of some colonizers to treat the inhabitants with respect, most failed to recognize the richness of cultural diversity and the long history of human habitation of the continent that was handed down through language, stories, art, and music. The prevailing attitude was that the process of British colonization would gradually and inevitably see the demise of Australia's First Peoples. Conflict over land and stock lead to punitive killing expeditions, and introduced disease wiped out many of the people who came in contact with white settlers. Much of the earlier ethnographic and anthropological work reflected the prevailing social attitude that Aboriginal people would 'vanish' within a few generations. The 'vanishing' concept is not only insulting but deeply hurtful to survivors of this period. It is only relatively recently that, thanks to the struggles of the survivors, enlightened academic research, and improved media reporting, the

* My special thanks go to the Aboriginal people from the Paakintji (Barkindje), Ngiyambaa and associated language groups, to the Walbiri people from Yuendumu, and to all Aboriginal and Torres Strait Islanders with whom I have worked. The genesis and motivation of the present essay are indebted to them all.

Australian public is beginning to appreciate the value of the heritage that has been compromised since 1788.

Aboriginal and Torres Strait Islander Australians have gained some influence over their own affairs during the past thirty years but control is still too rare, being subject to political decision-making that affects education and employment opportunities. Much racist social baggage remains, socio-economic disadvantage is extreme for most Aboriginal Australians, and, against a background of insensitive research,[1] it is not surprising that the response to new research proposals is often negative.[2] In the area of genetic research, resistance from many groups is very strong due to the legacy of distrust generated by Australia's history, coupled with concerns that are universal. In this era, genetic technology has reached extraordinary levels of power to probe living organisms in minute detail and has the potential to generate information that can be used in many ways, meaning that likely benefits and risks require careful consideration and explanation to potential individual and organizational research participants.

This article will focus on the experiences of the author, a researcher who has been doing genetic research in Australia for eighteen years. Mixed responses from Aboriginal groups to proposals have been received during that period. Nevertheless, progress has been made slowly by working closely with Aboriginal participants in the Darling River region of western New South Wales. For some, there remain issues that touch on protection of identity, privacy, and the perceived threat to 'Aboriginality'. On the broader level of state and national Aboriginal and Torres Strait Islander organizations, scientists and educators need to foster a better understanding of the potential benefits of the research if effective partnerships are to be forged.

The Colonization Experience for the People of the Darling River Region

The rich diversity of language and culture practised by the descendants of Australia's First Peoples was not appreciated by the European explorers and colonizers who later settled in the continent after 1788. Experiences of contact

---

[1] Sheila van Holst Pellekaan, "Human Genome Diversity: Ethics and Practice in Australia," *Human Evolution* 19.2 (2004): 131–44.

[2] Michael Dodson & Robert Williamson, "Indigenous peoples and the morality of the Human Genome Diversity Project," *Journal of Medical Ethics* 25.2 (April 1999): 204–208.

between incoming European colonizers and Australian inhabitants varied. Those in the areas that were first settled were confronted suddenly, others in more remote areas were contacted more sporadically over a longer period. The Darling River runs south-west from present day Queensland through north-western New South Wales and joins the Murray River at Wentworth, from where it flows south to the sea. The river country has been a rich habitat throughout human history, capable of supporting many people. As shown by studies in anthropology[3] and archaeology,[4] Aboriginal people have lived in the Darling River area for more than 40,000 years. The strong language group called Paakintji (Barkindje) derives its name from 'Paaka' meaning 'river' in the local language.[5]

The earliest documented reports of the inhabitants were from explorers' notes, dating from the 1830s through to around the 1850s,[6] and these were added to by reports from various ethnographers.[7] Exploration of the region was deemed essential by the colonizers to open up opportunities for pastoralism and to provide a route for the transport of livestock between Sydney and South Australia. A period of unofficial pastoralization began during the period of exploration, so that by the time leaseholds along the Darling and its Great Anabranch were documented in 1858,[8] there were many white settlers on what had been traditional Aboriginal land.

---

[3] Norman B. Tindale, "Eagle and Crow Myths of the Maraura Tribe, Lower Darling River, New South Wales," *Records of the South Australian Museum* 6.3 (1939): 243–61.

[4] James M. Bowler et al., "New ages for human occupation and climatic change at Lake Mungo, Australia," *Nature* 421 (2003): 837–40.

[5] Luise A. Hercus, *Paakintyi Dictionary* (Canberra: Australian Institute of Aboriginal and Torres Strait Islander Studies, 1993).

[6] Charles Sturt, *Two Expeditions into the Interior of Southern Australia* (1833; North Adelaide SA: Corkwood, 1999), http://freeread.com.au/ebooks/e00058.html (accessed 30 October 2010); Thomas Mitchell, *Three Expeditions into the Interior of Eastern Australia*, 2 vols. (London: T. & W. Boone, 1839), http://freeread.com.au/ebooks/e00035.html (accessed 30 October 2010).

[7] See, for example, Frederick Bonney, unpublished MS (Sydney: Mitchell Library, [1881]).

[8] Jeanette Hope, *Lake Victoria: Finding the Balance*, Cultural Heritage Report, Background Report No. 1: Lake Victoria EIS (Canberra: Murray–Darling Basin Commission, 1998).

Relationships between the newcomers and the Aboriginal people were mixed, often quite harmonious, but there were also incidents of serious conflict over stock, water and resource access, and disrespect for important traditional places. Numbers were severely reduced by punitive and deliberate killing and by introduced disease. However, in many cases, the surviving Aboriginal people remained on or near their own country, working on properties in a range of roles including child-care and domestic work, and as station hands and postal workers, and were regarded by many white settlers as great assets. A short period of reasonable, albeit paternalistic, coexistence was possible until telegraph communications and river transport opened the area to intensified white settlement from the 1860s to the 1880s. Aboriginal people were no longer regarded as assets, but as being in the way of white settlement. Fewer records of this period seem to have survived, perhaps because there was less interest from white settlers and, in many cases, downright hostility.

The Aboriginal Protection Board was established in 1881 and effectively marked the end of traditional life, as access to rightful country became almost impossible.[9] The purported 'protective' role of these Boards really meant shifting people into reserves for the convenience of white settlers and in the belief that, eventually, traces of the First Australians would disappear. Not all lived in reserves, some managing to escape the Protector and existing in small groups moving within the region of the river country. The 1882 census noted that 81.2% of the people in the Darling region were 'self sufficient', meaning they were either employed or had seasonal work combined with subsistence activities.

The establishment of reserves saw a decline in respect for the original inhabitants of the area. Under assimilation policies, children of mixed partnerships were removed. By the 1930s, Aboriginal people began to lose their kinship links, were forced into groups not necessarily appropriate to their traditional partnership rules, and the authority of the Elders was compromised. A sad period of decline of traditional values, engendered by the life-style forced upon people, resulted in loss of dignity for many. Yet many of those who survived retained strong family ties that remain today.[10]

---

[9] Heather Goodall, *Invasion to Embassy: Land in Aboriginal Politics in New South Wales, 1770–1972* (St. Leonards, NSW: Allen & Unwin/Black Books, 1996).

[10] Sarah Martin, *Aboriginal Cultural Heritage of Menindee Lakes Area, Part 1. Aboriginal Ties to the Land.* A Report to the Menindee Lakes Ecologically Sustainable

As a result of recent history, the population of today consists of people whose ancestry derives from partnerships between traditional language groups (mainly Paakintji and Ngiyambaa and associated dialectal groups), and incoming groups, such as Scottish, Irish, English, and some Afghan immigrants. Renewed identity and pride in traditions is evident, and many have worked hard to gain some control over their heritage and services, living with dignity in the world of the dominant society, despite continued disadvantage in education, employment opportunities, and health.[11] Elders and community members from the traditional groups have significant roles in managing such important places as the Willandra World Heritage Area, Lake Mungo (site of the oldest known skeletal remains in Australia and the oldest known cremation site in the world),[12] and Mutawintji National Park.

Government recognition of the rights of Australia's First Peoples has been slow, but is occurring through recognition of Native Title, acknowledgement of misguided assimilation policies through the important Human Rights and Equal Opportunity Commission Report, 'Bringing Them Home,'[13] and culminating in an apology from the then Prime Minister, Kevin Rudd, in 2008.[14]

A Lingering Legacy Hinders Research

Despite the gains made in recent years, the legacy of the past translates into distrust for genetic research. There remains a feeling among many Indigenous people that scientists are still steeped in attitudes of Western superiority and are studying them for outcomes that have little to do with Indigenous interests. Some responsibility for the lasting distrust of motives behind genetic

---

Development Project Steering Committee (Buronga, NSW: Department of Land and Water Conservation, DLWC, 2001).

[11] Australian Bureau of Statistics, http://www.abs.gov.au/ausstats/abs@.nsf/mf/4704.0/ (accessed 1 November 2010).

[12] Alan Thorne et al., "Australia's Oldest Remains: Age of the Lake Mungo 3 Skeleton," *Journal of Human Evolution* 36.6 (1999): 591–612.

[13] Human Rights and Equal Opportunity Commission, *Bringing them Home*, National inquiry into the removal of Aboriginal and Torres Strait Islanders from their families (Canberra: Commonwealth of Australia, 1997), http://www.hreoc.gov.au/social_Justice/bth_report/report/index.html (accessed 1 November 2010).

[14] Kevin Rudd, "Apology to Australia's Indigenous Peoples," Australian Parliament House, Canberra (13 February 2008), http://www.aph.gov.au/house/rudd_speech.pdf (accessed 1 November 2010).

studies arose from the presentation of the Human Genome Diversity Project (HGDP) in the 1990s to some Aboriginal Australian organizations in central and northern Australia.[15] It is so well remembered today that the experience and publicity that it generated are still frequently referred to and have set back the efforts of many Australian scientists who have worked hard with Aboriginal organizations and individuals to present the benefits of genetic research, to involve participants in projects, and to deal with the negative concerns together.

The particular experience of the HGDP in Australia, and the urgency with which it was presented, is not unique, and the issues raised elsewhere serve as lessons for us all. I do not intend to repeat what has already been extensively discussed and reviewed by others,[16] but to reflect on my own experience during the very same period. I began working cooperatively with Aboriginal Australians in western New South Wales in 1992. Together with my colleague, the late Dr June Roberts–Thomson, we both also generated results from DNA samples given by the Walbiri people of Yuendumu in central Australia. A long and ongoing period of friendship and negotiation with the Walbiri people was initiated by Professor Barry Boettcher of Newcastle University during the 1980s, continued for years by June Roberts–Thomson, and currently continues through the custodian, Professor Rodney Scott, also from Newcastle University. My own experience of consultation and negotiation has been discussed in open forums and described in publications.[17]

It is abundantly clear that, for consultation and negotiation to succeed, researchers need to demonstrate willingness and ability to meet Aboriginal people and organizations frequently, in their own place, on their own terms, and with appropriate explanatory documents. Information that states clearly the intention and possible outcomes of the research needs to be available in lan-

---

[15] Dodson & Williamson, "Indigenous peoples and the morality of the Human Genome Diversity Project."

[16] Matthew Rimmer, "The Genographic Project: Traditional Knowledge and Population Genetics," *Australian Indigenous Law Review* 11.2 (2007): 34–54, and Indigenous Peoples Council on Biocolonialism, http://www.ipcb.org/ (accessed 27 October 2010).

[17] Sheila M. van Holst Pellekaan, "Genetic research: what does this mean for Indigenous Australian communities?" *Journal of Australian Aboriginal Studies* 1–2 (2000): 65–75, and van Holst Pellekaan, "Human Genome Diversity: Ethics and Practice in Australia."

guage suited to the needs of people with different educational experiences, as well as conforming to the expectations and demands of institutional or regional ethics committees. This may require sets of information on different levels of language that at the same time give the essential accurate information, and it may mean repeating verbal explanations. It is a long-term commitment, as ongoing progress reports to the communities and individual participants are necessary to maintain trust between parties. In turn, I have received respect and friendship from the communities and assurance that the work has been appreciated. Some negative comments have come from external organizations or individuals not involved in the research that appear to disrespect the intelligent decisions made by participants in consenting to take part and deal with outcomes.

For this researcher, eighteen years of slow but successful research has eventuated, initially through the University of Sydney and, from 2003, through the University of New South Wales. It is now useful and timely to ask two questions: first, have the outcomes been positive and, second, have concerns been dealt with adequately?

Genetic Research Outcomes, Signatures of the Past and Implications for Health

Ancestry-Focused Research

Against an already powerful background of a Pleistocene presence in the Darling River region as shown by archaeology,[18] the project that was presented to people, beginning in 1992, was to study (maternally inherited) mitochondrial DNA (mtDNA). The aim was to explore genetic evidence of maternal ancestry and ancient shared connections with other human groups as they dispersed into Sahul, the land mass that included present-day Australia and New Guinea, and then throughout the Australian continent. The results of that study have been well-documented and discussed,[19] providing evidence of strong

---

[18] Bowler et al., "New ages for human occupation and climatic change at Lake Mungo, Australia."

[19] Sheila M. van Holst Pellekaan, Max Ingman, June Roberts–Thomson & Rosalind M. Harding, "Mitochondrial genomics identifies major haplogroups in Aboriginal Australians," *American Journal of Physical Anthropology* 131.2 (October 2006): 282–94, Sheila M. van Holst Pellekaan, "Origins of the Australian and New Guinea Aborigines," in *Encyclopedia of Life Sciences (ELS)* (Chichester: John Wiley, 2008), and Georgi Hudjashov et al., "Revealing the prehistoric settlement of Australia by Y

signatures of the past. They have confirmed the presence of ancient Australian maternal haplogroups, some of which have ancestral connections with people from Papua New Guinea; others are Australia-specific, indicating that the continent was settled very early in the dispersal of modern humans. People may have entered at different parts of the northern shores around similar times, then dispersed. A period of relative genetic isolation followed when sea levels rose and Sahul became Australia and New Guinea, resulting in the diversity observed today.

This type of research is based on an evolutionary paradigm assuming that humans evolved over millions of years and that our single species, *Homo sapiens,* now populates the globe after dispersing from small founding populations that left Africa relatively late in the story of life on planet Earth. This is clearly in conflict with views of *in-situ* creation held by many Indigenous populations, but in the process of this research, it has not caused difficulty. I do not intend to debate the conflicting models here but to comment on how the ancestry-related research has been received by Aboriginal participants. The following excerpts are taken from a community report that was distributed in 1997 to explain the mtDNA work:

> Each Aboriginal Australian group has their own stories about how they belong to the land. Some groups say that their ancestors came from over the sea, others say that they have always been in their country and were created from the land. I would like you to know that I think these stories are very important in explaining the special relationship that all Aboriginal groups have with their country. As a scientist, I believe that humans came from other parts of the world and spread into Australia at least 40,000 years ago, maybe longer. [...] Many scientists believe that humans came from Africa and spread into all other parts of the world. They base these beliefs on archaeological remains and also on genetic information, some of it like the work I have done. IT IS IMPORTANT THAT YOU KNOW THAT MY WORK DOES NOT PROVE THAT THIS IS TRUE OR UNTRUE [...] it is another sort of information about the past.[20]

chromosome and mtDNA analysis," *Proceedings of the National Academy of Sciences of the United States of America* 104 (2007): 8726–30.

[20] Sheila van Holst Pellekaan, "Patterns in DNA passed by our mothers: What does this mean for the Aboriginal people of the Darling River region?" a report to the Aboriginal communities of Wilcannia, Menindee & Dareton, 1997.

Conversations with community members produce a variety of responses. Some find the DNA work interesting but simply ignore the implications for creation. Others are fascinated with the results, and discuss how it sits with spiritual beliefs. I try to emphasize that, whatever belief-system one has, it does not take away from the rich story of human history that their ancestors began and continued in this continent, at least 40,000 years ago, well before my ancestors lived in Scotland, Ireland, and England. Since 1788, the teaching of Christianity has constituted a threat to traditional cosmology, yet, as in other parts of the world, Indigenous peoples have shown a great ability to embrace elements of different faiths. Many Aboriginal Australians accommodate both systems into their world-view, and the ancestry-focused DNA work is another piece of information that can be added to the story for participants, their families, and their community.

The perception that genetic research threatens the notion of 'Aboriginality' is in several ways more important to address. Loss of traditional identity has resulted from events and policies of recent colonization. Over the last four decades in Australia there has been a surge of 'cultural regeneration', so that individuals and families deeply affected by the events of the past have been able to re-connect with their traditions. For many of these, whose parentage was mixed due to exploitative miscegenation or by chosen partnerships, Aboriginal identity was hidden, so it may also require a difficult search for those wishing to reconnect with Aboriginal family. Because these personal journeys are often painful, there is a fear expressed by some (usually external to the sample group) that by participating in genetic studies the admixture will be exposed, their Aboriginal genotype 'lost', and that a demonstration of this in the form of scientifically based data 'plays right into the hands' of those with racist attitudes and may influence rights to country through Native Title, culture, and services. To counterbalance this, I have needed to address two key points.

First, from a scientific perspective it is simply untrue that the genotype passed on by Aboriginal ancestors will be totally lost. Genetic traces of all of our ancestors are contained within our genomes and that remarkable heritability is why geneticists strive to detect patterns of common ancestry that stem from connections beyond the documented or remembered genealogies. Certainly, with each generation there is mixing, because each parent contributes 50% of his/her genetic material to each new child, and identifying specific, ancient genetic signatures becomes harder with each generation. This reminds us of the HGDP and the urgency with which its promotion was made on the basis that the genomes of Indigenous peoples would 'vanish'. The promoters of the HGDP, however, made a valid point, because the task of understanding diversity becomes more difficult (not impossible) with time, but they used inappropriate words and failed to see why promoting the project in the way they did insulted Indigenous peoples. While the genetic information (called haplotype) that is obtainable from mtDNA (direct maternal lineage), or in the paternal lineage by Y-chromosome analysis, does give informative results, it is only useful for showing the Aboriginal parentage if the descent from an Aboriginal ancestor is direct. However, it must also be remembered that, for those whose ancestry includes mixed Aboriginal and non-Aboriginal ancestors, cultural identity cannot be denied on the basis of genetic testing.

Secondly, proceeding from this, 'Aboriginality' is not defined by biological attributes, although it must be acknowledged that this idea was inherent in the attitudes that prevailed during the colonization period and had influence on policies and events. Biologists showed, with sound studies, that the social notion of 'race' linked to biological traits associated with skin colour and morphology has no scientific basis. The history of medicine was strongly influenced by the concept of 'race' but this did not keep pace with accurate biological science.[21] Scientists today do not use the term 'race', because it is neither useful nor accurate, as genetic variation in major population groups is a result of many influences including the vicissitudes of migration, population bottlenecks, culturally influenced mating patterns, and isolation by distance. Social values have taken a long time to change, to influence thinking, but change is now reflected in Australian federal policy. The working definition used by Aboriginal people in Australia claiming identification for various services and rights, such as Native Title claims, and accepted by the Federal and State governments, encompasses three parts:

> Till the late 1950s States regularly legislated all forms of inclusion and exclusion (to and from benefits, rights, places etc.) by reference to degrees of Aboriginal blood. Such legislation produced capricious and inconsistent results based, in practice, on nothing more than an obser-

---

[21] Clarence C. Gravlee & Elizabeth Sweet, "Race, ethnicity and racism in medical anthropology, 1977–2002," *Medical Anthropology Quarterly* 22.1 (March 2008): 27–51, Jeffrey C. Long, Jie Li & Meghan E. Healey, "Human DNA sequences: more variation and less race," *American Journal of Physical Anthropology* 139.1 (May 2009): 23–34, and John H. Relethford, "Race and global patterns of phenotypic variation," *American Journal of Physical Anthropology* 139.1 (May 2009): 16–22.

> vation of skin colour. [...] For the modern anthropologist a 'human tree' can do no more than show the frequency (not exclusiveness) of genetic traits in sample populations and more meaningful divisions of humankind are suggested by region, culture, religion and kinship. [...] An Aboriginal or Torres Strait Islander is a person of Aboriginal or Torres Strait Islander descent who identifies as an Aboriginal or Torres Strait Islander and is accepted as such by the community in which he (she) lives.[22]

In practice, this means that a person needs to have verified knowledge (preferably by documentation, though it is not always possible to find records) that they are descended from an Aboriginal person, that they live or engage with their known traditional group, and that they are accepted by that group. There is no requirement for demonstrating genetic links; furthermore, genetic tests cannot be used to deny 'Aboriginality' if the three criteria are met.

Genetics for Health

During the mid-1990s, in addition to the mtDNA results that were accumulating in the first part of my research, efforts were made to extend the project to investigate genetic markers that might be important for understanding cardiovascular diseases and Type II diabetes, both of which have a high incidence in Aboriginal Australian people and contribute to early morbidity and mortality.[23] Participants were again consulted for their consent to the extra work, and institutional ethical approval was granted. The intention at this stage was to expand the study to include local Aboriginal health personnel and to streamline the sample collection, to include controls, and to collect phenotypic data. Several candidate genetic markers were examined but, owing to limited resources, the study was not fully developed. At this point, awareness of ethical issues for all research resulted in more detailed approval procedures. There is now an expectation that, in addition to institutional ethical clearance, all proposals for research involving Aboriginal people in New South Wales be submitted to the Aboriginal Health and Medical Research

---

[22] John Gardiner–Garden, "The Definition of Aboriginality," Research Note 18 2000-01, Parliament of Australia, Parliamentary Library (5 December 2000): http://www.aph.gov.au/library/pubs/rn/2000-01/01RN18.htm (accessed 1 November 2010).

[23] Australian Bureau of Statistics, http://www.abs.gov.au/ausstats/abs@.nsf/mf/4704.0/ (accessed 1 November 2010).

Council (AH&MRC) of New South Wales.[24] This was established in 1996 to promote the interests of Aboriginal people and to help them regain control over health management that had been lost during the colonial period. Some hard-fought, impressive gains have been made despite uncertainties that arise from State and Federal political and financial constraints. The accelerating impact of genetic technological power, combined with the memory of poor research practices of the past and the HGDP experience, served to intensify resistance from some groups to genetic research. National guidelines for all research practice are in place[25] but extra information and evidence of acceptance of projects by communities and organizations is required by the AH&MRC. This researcher has great respect for the principles motivating the AH&MRC; however, in practice, the length of time taken to receive responses to submission is great, much required material is repetitive, and research has not been able to proceed at a reasonable pace. The process of negotiation with the AH&MRC Ethics Committee has taken far longer than it has taken to deal directly with the participants and communities who have given consent for ongoing work.

Approval for continued research was finally given by the AH&MRC Ethics Committee in 2009 and ratified by the University of New South Wales Human Research Ethics Committee to extend the research in two ways. First, we proposed to better understand the basic genetic diversity of Aboriginal Australians in order to evaluate the relevance of that diversity for health, in particular cardiovascular disease, and, secondly, to develop the project to include Aboriginal personnel in the collection and analysis of phenotypic data. To begin the first stage, we carried out genotyping of 37 existing DNA samples using new technology (Affymetrix 6.0) that allows a genome-wide scan (GWS) to examine ~907,000 single nucleotide polymorphisms (SNPs) at one time.[26] The advantage of this technology is that large amounts of data can be interrogated, but one of the limitations is that the markers are derived

---

[24] Aboriginal Health & Medical Research Council (AH&MRC), http://www.ahmrc.org.au/Ethics%20and%20Research.htm (accessed 1 November 2010)

[25] Australian Government, National Health & Medical Research Council (NHMRC), Health and research ethics guidelines, http://www.nhmrc.gov.au/guidelines/ethics_guidelines.htm (accessed 1 November 2010).

[26] Brian P. McEvoy et al., "Whole-genome genetic diversity in a sample of Australians with deep Aboriginal ancestry," *American Journal of Human Genetics* 87.2 (August 2010): 297–305.

from previous human studies that do not include Aboriginal Australians. Consequently, if there are novel variations in the Australian samples, they may not be detected by this method. What is yielded is an idea of the frequency of the variant alleles, and this enables comparative evaluation of frequencies between population groups. Thus, it produces some baseline data which can be further probed for association with function and may inform us about parts of the genome worth more intensive study.

Critics of the GWSs point out that it is hard for Indigenous people to see the benefit of this basic approach if it does not give specific information about the cause of disease. Yet, until we do this baseline work, preferably with a wider Australian sample, we will not know if there are specific variations in the genomes resulting from a long period of adaptation to conditions and lifestyle that we need to consider for facilitating diagnosis, designing health-care strategies, and formulating appropriate treatments. At present, the large-scale genetic epidemiological studies that are being carried out in other populations are being used as reference studies to design better policies and to improve diagnosis and treatment, and these may not be universally appropriate. This argument has been used for some time by scientists to justify genetic studies, but, although understood by some Aboriginal Australians,[27] it is either not understood by many or not seen as sufficiently beneficial to outweigh risks. In the current situation, for the most part, Aboriginal Australians are simply being left out.

In order to understand the influence of mixed ancestry or admixture on the results we have generated, we estimated the level of admixture facilitated by what is known from the genealogies of participants, freely given under consent and with assurances of confidentiality. In keeping with ethical obligations, this author is the only researcher with access to personalized information. All other collaborators use anonymized identifying labels only for samples and data files. The inclusion of admixture estimates in published reports has prompted comments from the AH&MRC and several critics, external to the study group, who again raise the issue of a potential return to classification of people based on blood quotient, or genetic quotient estimates. This would be a misuse of information and, for the reasons outlined above, I do not believe it would occur today. To reject genetic research on that basis is to close the door on the benefits that new knowledge can bring.

---

[27] Dodson & Williamson, "Indigenous peoples and the morality of the Human Genome Diversity Project."

Benefits, Risks, and Unsolved Issues

The goal of understanding genetic diversity should not be labelled racist, as the concept of social hierarchy is neither useful nor accurate. Dozens of global studies that do not include Aboriginal Australians show that there are clusters of populations that show patterns of genetic similarity, but there are others, definable by geographical region (e.g., Europeans), whose patterns of diversity suggest a more complex demographic history.[28] Australia was isolated from other regions during the period of human habitation and people lived in a range of extreme and changing conditions. People who survived did so with genomes that are the result of the gene-pool of the founders evolving within the influence of the total environment, including life-style. We do not know the detail of that genetic/environmental compatibility for humans generally and we are a long way off from understanding the Australian story, because so few studies have been carried out.

Since colonization, a different life-style was foisted on people, diet changed, and both settlers and introduced animals brought different pathogenic organisms, the adverse effects of which were influenced by social degradation and disadvantage. Fundamental understanding of genetic variation is a baseline step towards working out details of cellular mechanisms that influence wellness and illness. The benefits will flow for all humans because policies and treatments will be developed from better information, but they will only become evident gradually.

There are acknowledged risks for all people in relation to the scale of genetic work now possible and the relative ease with which the public can access direct-to-consumer (DTC) services.[29] Are members of the public who pay for their genome to be sequenced through these services sufficiently informed or advised on how to interpret the information that is returned to them, and could this information be accessed by insurance companies or other services and used inappropriately? Critics may raise these issues in a timely reminder that legal processes in most countries do not deal adequately with the

---

[28] Keith L. Hunley, Meghan E. Healey, & Jeffrey C. Long, "The global pattern of gene identity variation reveals a history of long-range migrations, bottlenecks and local mate exchange: implications for biological race," *American Journal of Physical Anthropology* 139.1 (May 2009): 35–46.

[29] Charmaine D. Royal et al., "Inferring Genetic Ancestry: Opportunities, Challenges, and Implications," *American Journal of Human Genetics* 86.5 (May 2010): 662–63.

pace and potential of the technology. From small amounts of DNA it is possible to examine many parts of the genome, raising the issue of what should be covered in consent. Should researchers seek 'blanket' consent or single use?[30]

The difficulty of assuming that blanket consent is sufficient is exemplified in the case of *Havasupai v Arizona State University*,[31] where original consent was broad, apparently including the study of causes of 'behavioural/medical' conditions, but it would seem that discussions with communities centred on diabetes research which became the part that was well understood. When other unrelated projects used the same samples, the community felt wronged, stigmatized, and offended. It may be reasonable to ask participants to give consent for analysis of a few loci for the purposes of studying one type of disease, but is it fair to ask them to consent to work in the future that might involve other research teams, thereby consenting to the sharing of samples? In Australia, sharing of samples beyond the personnel named in research protocols is no longer acceptable, and for any additional analyses renewed consent must be sought from participants and ethics committees informed of changes in personnel.

Similarly, the use of data generated from projects goes beyond the immediate published reports that emanate from discrete projects. The expectation in the scientific and academic communities that results be made available to other legitimate researchers has produced large databases that are also accessible to the public. There are strong arguments against this policy. Databases include full genome sequences and, as genetic markers become associated with health, wellness or 'desirable' traits, there is also the risk of stigmatization if families or communities are shown to have a high risk of illness or culturally 'offensive' attributes. In patient-focused studies there is also a risk of the re-identification of individuals despite anonymization of data.[32] For this

---

[30] Amy L. McGuire, Timothy Caulfield & Mildred K. Cho, "Research ethics and the challenge of whole-genome sequencing," *Nature Reviews: Genetics* 9.2 (February 2008): 152–56.

[31] Michelle M. Mello & Leslie E. Wolf, "The Havasupai Indian Tribe Case – Lessons for Research Involving Stored Biologic Samples," *New England Journal of Medicine* 363.3 (July 2010): 204–207.

[32] Michael Krawczak, Jürgen W. Goebel & David Cooper, "Is the NIH policy for sharing GWAS data running the risk of being counterproductive?" *Investigative Genetics* 1.3 (September 2010): 1–4.

researcher, anonymized results of mtDNA sequences were made available in the first ancestry-focused study, but in the later SNP study[33] raw data has not been made available, as the question is not resolved in Australia.

Conclusion

In summary, I return to the two questions posed earlier. Have there been positive results of doing genetic research with Aboriginal Australians? Yes – the ancestry-related research has contributed to our knowledge by confirming the great time-depth since the arrival of First Peoples, the genetic relationships with near neighbours, and the distribution of matrilines in Australia. It has been received positively by communities and individuals and, I believe, is slowly alleviating concerns about genetic research expressed by others. Health-related genetic research has not produced direct benefit at this stage and will need to persist if it is to demonstrate positive value.

Secondly, concerns expressed along the way have mostly been dealt with by continued contact with communities, and the continued ability to do research indicates trust and respect between researcher and participants, suggesting that concerns have been dealt with adequately. External to the study group, occasional hostility arises from people who do not spend time discussing their reservations, so some issues still need to be addressed. The general principle of appropriate consultation, negotiation, and respect for individuals and communities is easy to maintain if enough time is allowed for continued liaison. Ethics committee representatives, Indigenous and non-Indigenous social scientists and geneticists need to develop better lines of communication to deal with issues about use and misuse of genetic data in a spirit of mutual respect and trust. This approach was employed in 2010 with a small interdisciplinary workshop hosted by the Lowitja Institute, Victoria,[34] and it is envisaged that it will be continued in order to develop fruitful dialogue between relevant groups.

Concerns that are unresolved include: the broader acceptance of genetic research by other Aboriginal and Torres Strait Islanders, so that results are more representative; the continued ability to build on research using the same samples, and how ethics committees will deal with this; the avoidance of repeti-

---

[33] Brian P. McEvoy et al., "Whole-genome genetic diversity in a sample of Australians with deep Aboriginal ancestry."

[34] Lowitja Institute, Australia's National Institute for Aboriginal and Torres Strait Islander Health Research, http://www.lowitja.org.au/ (accessed 1 November 2010).

tion in dealing with many levels of ethical evaluation; and the agreement to share results in restricted-access data bases to allow collaborative data analysis with other legitimate researchers. An additional shortcoming not yet resolved is the ability to make results of ancestry-related research in the future available to generations not directly participating in the study. This availability is currently constrained by the obligation to maintain strict anonymity in reports and papers. While this is quite understandable in the immediate context of a project, it limits the value of research for historically linked peoples. Why should it not be possible for a member of a participant's family to access results, maybe generations into the future, because they know that their granny or relative was in the study and would like to find out about the work? The results are part of their ancestors' story that they may have missed hearing of through their family. I have not been able to resolve this issue of access, but I would like to find a solution in return for the friendship, trust, and shared experiences that I continue to enjoy.

Works Cited

Aboriginal Health & Medical Research Council (AH&MRC). http://www.ahmrc.org.au/Ethics (accessed 1 November 2010).

Australian Bureau of Statistics. http://www.abs.gov.au/ausstats/abs@.nsf/mf/4704.0/ (accessed 1 November 2010).

Bonney, Frederick. unpublished MS (Sydney: Mitchell Library, [1881]).

Bowler, James M. et al. "New ages for human occupation and climatic change at Lake Mungo, Australia," *Nature* 421 (2003): 837–40.

Dodson, Michael, & Robert Williamson. "Indigenous peoples and the morality of the Human Genome Diversity Project," *Journal of Medical Ethics* 25.2 (April 1999): 204–208.

Gardiner–Garden, John. "The Definition of Aboriginality," Research Note 18 2000-01, Parliament of Australia, Parliamentary Library (5 December 2000): http://www.aph.gov.au/library/pubs/rn/2000-01/01RN18.htm (accessed 1 November 2010).

Goodall, Heather. *Invasion to Embassy: Land in Aboriginal Politics in New South Wales, 1770–1972* (St. Leonards, NSW: Allen & Unwin/Black Books, 1996).

Gravlee, Clarence C., & Elizabeth Sweet. "Race, ethnicity and racism in medical anthropology, 1977–2002," *Medical Anthropology Quarterly* 22.1 (March 2008): 27–51.

Hercus, Luise A. *Paakintyi Dictionary* (Canberra: Australian Institute of Aboriginal and Torres Strait Islander Studies, 1993).

Hope, Jeanette. *Lake Victoria: Finding the Balance*, Cultural Heritage Report. Background Report No.1. Lake Victoria EIS (Canberra: Murray–Darling Basin Commission, 1998).

Hudjashov, Georgi et al. "Revealing the prehistoric settlement of Australia by Y chromosome and mtDNA analysis," *Proceedings of the National Academy of Sciences of the United States of America* 104 (2007): 8726–30.

Human Rights and Equal Opportunity Commission. *Bringing them Home*, National inquiry into the removal of Aboriginal and Torres Strait Islanders from their families (Canberra: Commonwealth of Australia, 1997): http://www.hreoc.gov.au/social_Justice/bth_report/report/index.html (accessed 1 November 2010).

Hunley, Keith L., Meghan E. Healey & Jeffrey C. Long. "The global pattern of gene identity variation reveals a history of long-range migrations, bottlenecks and local mate exchange: implications for biological race," *American Journal of Physical Anthropology* 139.1 (May 2009): 35–46.

Indigenous Peoples Council on Biocolonialism. http://www.ipcb.org/ (accessed 30 October 2010).

Krawczak, Michael, Jürgen W. Goebel & David Cooper. "Is the NIH policy of sharing GWAS data running the risk of being counterproductive?" *Investigative Genetics* 1.3 (September 2010): 1–4.

Long, Jeffrey C., Jie Li, & Meghan E. Healey. "Human DNA sequences: more variation and less race," *American Journal of Physical Anthropology* 139.1 (May 2009): 23–34.

Lowitja Institute. Australia's National Institute for Aboriginal and Torres Strait Islander Health Research, http://www.lowitja.org.au/ (accessed 30 October 2010).

Martin, Sarah. *Aboriginal Cultural Heritage of Menindee Lakes Area, Part 1. Aboriginal Ties to the Land*, A Report to the Menindee Lakes Ecologically Sustainable Development Project Steering Committee (Buronga, NSW: Department of Land and Water Conservation, DLWC, 2001).

McEvoy, Brian P. et al. "Whole-genome genetic diversity in a sample of Australians with deep Aboriginal ancestry," *American Journal of Human Genetics* 87.2 (August 2010): 297–305.

McGuire, Amy L., Timothy Caulfield & Mildred K. Cho. "Research ethics and the challenge of whole-genome sequencing," *Nature Reviews: Genetics* 9.2 (February 2008): 152–56.

Mello, Michelle M., & Leslie E. Wolf. "The Havasupai Indian Tribe Case – Lessons for Research Involving Stored Biologic Samples," *New England Journal of Medicine* 363.3 (July 2010): 204–207.

Mitchell, Thomas. *Three Expeditions into the Interior of Eastern Australia*, 2 vols. (London: T. & W. Boone, 1839): http://freeread.com.au/ebooks/e00035.html (accessed 30 October 2010).

National Health & Medical Research Council. http://www.nhmrc.gov.au/guidelines/ethics_guidelines.htm (accessed 1 November 2010).

Relethford, John H. "Race and global patterns of phenotypic variation," *American Journal of Physical Anthropology* 139.1 (May 2009): 16–22.

Rimmer, Matthew. "The Genographic Project: Traditional Knowledge and Population Genetics," *Australian Indigenous Law Review* 11.2 (2007): 33–54.

Royal, Charmaine D. et al. "Inferring Genetic Ancestry: Opportunities, Challenges, and Implications," *American Journal of Human Genetics* 86.5 (May 2010): 661–73.

Sturt, Charles. *Two Expeditions into the Interior of Southern Australia* (1833; North Adelaide, SA: Corkwood, 1999), http://freeread.com.au/ebooks/e00058.html (accessed 30 October 2010).

Thorne, Alan et al. "Australia's Oldest Remains: Age of the Lake Mungo 3 Skeleton," *Journal of Human Evolution* 36.6 (1999): 591–612.

Tindale, Norman B. "Eagle and Crow Myths of the Maraura Tribe, Lower Darling River, New South Wales," *Records of the South Australian Museum* 6.3 (1939): 243–61.

van Holst Pellekaan, Sheila. "Genetic research: what does this mean for Indigenous Australian communities?" *Journal of Australian Aboriginal Studies* 1–2 (2000): 65–75.

——. "Patterns in DNA passed by our mothers: What does this mean for the Aboriginal people of the Darling River region?" a report to the Aboriginal communities of Wilcannia, Menindee, and Dareton, 1997.

——. "Human Genome Diversity: Ethics and Practice in Australia," *Human Evolution* 19.2 (2004): 131–44.

——. "Origins of the Australian and New Guinea Aborigines," in *Encyclopedia of Life Sciences (ELS)* (Chichester: John Wiley, 2008).

—— et al. "Mitochondrial genomics identifies major haplogroups in Aboriginal Australians," *American Journal of Physical Anthropology* 131.2 (October 2006): 282–94.

⌘

Nutrition and the Indigenous Body

A Genetic Concept of Food

ANDREA ZITTLAU

WHEN LOOKING FOR A CULINARY EXPERIENCE in Washington, D.C., Mitsitam, the café in the National Museum of the American Indian, located directly on the Mall, is no longer exclusively an insider's tip. The secret to its success is a menu inspired by the traditional cuisine of the Indigenous peoples of the Americas. Hungry customers can find anything from corn bread and freshly prepared salads to buffalo meat and prickly-pear ice cream. Reviews have praised the exotic range of the dishes, but have often reproved the café for its 'inauthentic' albeit creative food preparation and combinations. Tom Sietsema, the food critic for the *Washington Post*, found the food selections satisfying, although the menu, he wrote, included "a few silly exceptions":

> The Indian tacos from the Plains heap iceberg lettuce, grated cheddar cheese and chili on saucer-size fry bread, which doesn't bring to mind the prairies so much as it does a TGIF. And I seriously doubt any Indians traditionally finished their meals with fruit tarts or coconut macaroons, both of which are also available here. But the success stories prevail over the lapses.[1]

This perception of Indigenous cuisine is a prevalent one. Indigenous peoples have long been portrayed as belonging to the past, to vanished civilizations that were assumed to be inferior to those of the Western world. The twentieth century revolutionized this view in several arenas, including those of art, music, and the celebration of numerous festivals and rituals. The contempo-

[1] Tom Sietsema, "Mitsitam Café," *Washington Post Magazine* (16 October 2005).

rary expressions of these cultural practices, which have refused to be constricted or constrained by traditional notions of authenticity, contributed to the acknowledgment that today's Indigenous cultures are essential parts of modern society. An opposite move, however, can be observed concerning environmental issues, and food in particular. The concept of an authentic cuisine, which has over the centuries resisted all cultural influences, is a common one. And in the wake of claims of Indigenous identities based on genetics, food discourses have carved out a field all their own. The Indigenous body is perceived (by Natives as well as by non-Natives) as fundamentally different from that of the non-Native, thus requiring a special diet and particular kinds of physical activity. Often portrayed as hunter–gatherers, warriors, and shamans, and as masters of the wilderness, Indigenous people remain closely and stereotypically connected with discourses on food. And although Native communities are making their contemporary voices heard, the long shadow of the past continues to follow them into the future. This is an interesting phenomenon, one that relates to centuries-old debates, arguments that have tried to prove the Indigenous body unfit for Western civilization. The present essay seeks to explore a genetic concept of food in relation to diabetes resulting in a quest for authentic Indigenous ingredients, which then revive colonial constellations of Self and Other.

Decoding Diabetes

Contemporary civil society, as Janet Flammang summarizes it, is a fast-paced consumer culture driven by workaholics.[2] People's daily routines allow little time for diversified meals and food rituals. The market is dominated by processed and flavoured food that has been manufactured, not grown. The success of books like Jonathan Safran Foer's *Eating Animals* and Michael Pollan's *In Defense of Food* reflects a public concern with food production and careless consumption:

> Most of what we are consuming today is not food, and how we are consuming it – in the car, in front of the TV, and increasingly alone – is not really eating. Instead of food, we're consuming 'edible foodlike substances' – no longer the products of nature but of food science.[3]

---

[2] Janet Flammang, *The Taste of Civilization: Food, Politics, and Civil Society* (Chicago: U of Illinois P, 2009): 1.

[3] Michael Pollan, *In Defense of Food: An Eater's Manifesto* (New York: Penguin, 2008): blurb.

This element of artificiality is thought to cause numerous health problems, including allergies and diabetes. Formerly known as non-insulin-dependent diabetes mellitus (NIDDM), or adult-onset diabetes, it is currently one of the world's most serious health threats – some scholars even refer to it as an emerging global epidemic,[4] and *Indian Country Today* has called diabetes "an enemy that must be confronted."[5] Presently, at least 100 million people suffer from Type 2 diabetes worldwide (hereafter referred to as 'diabetes'), rising fastest in developing countries and among ethnic-minority groups, migrant populations, and disadvantaged communities in developed nations.[6] Native Americans have been described as a population particularly at risk for diabetes, the disease having surfaced in many Native communities after World War Two, increasing after the 1970s, and continuing to spread at alarmingly high rates.

Diabetes causes the body to either fail to produce enough insulin through the pancreas or to render it incapable of properly using the insulin it does produce. An essential hormone, insulin transports glucose, the body's basic fuel, to all its cells. When this function breaks down, it can lead to, among other things, infection and gangrene, and may further result in nerve damage, blindness, kidney failure, limb amputation, and ultimately death. There is, as yet, no cure for diabetes, because its causes remain unknown. The disease is controlled as much as possible through a variety of protocols that are known to prevent it from progressing, such as a change in diet and an increase in physical activity.

The supposition is that, in order to manifest itself, diabetes requires a certain genetic precondition, one that is found more routinely in Native Americans than in any other population:[7]

---

[4] See, for example, Paul Zimmet et al., "Epidemiology, Evidence for Prevention: Type 2 Diabetes," in *The Epidemiology of Diabetes Mellitus: An International Perspective*, ed. Jean–Marie Ekoé, Paul Zimmet & Rhys Williams (New York: John Wiley, 2001): 41.

[5] "Confronting Diabetes with Tradition," *Indian Country Today* (20 December 2000) in *America is Indian Country. Opinions and Perspectives from Indian Country Today*, ed. José Barreiro & Tim Johnson (Golden CO: Fulcrum, 2005): 236.

[6] See Daniel C. Benyshek, John F. Martin & Carol S. Johnston, "A Reconsideration of the Origins of the Type 2 Diabetes Epidemic among Native Americans and the Implications for Intervention Policy," *Medical Anthropology* 20.1 (2001): 25–26.

[7] Other sources mention the Australian Aboriginals and the Torres Strait Island people as the populations with the highest risk (Zimmer et al., "Epidemiology,

> There are approximately 100,000 genes packed into 23 pairs of chromosomes in each person. Within a gene, chemicals form individual codes, like words, which tell the cells of the body what to do. It is the code within a gene that directs the body to […] circulate blood and hormones such as adrenalin and insulin.
>
> Some diseases are caused by bacteria or viruses that infect the body and make it sick. Others, such as diabetes, occur because a gene's code causes it to function differently under some circumstances. […] A person can't choose his or her genes.[8]

This genetic argument can be traced back to the 'thrifty genotype' hypothesis, which was most prominently developed by James Neel in an article published in 1962 in the *American Journal of Human Genetics*. According to Neel, Native Americans faced difficult conditions on their way from Asia to North America thousands of years ago. They suffered through cycles of starvation that alternated with periods of an overabundance in food supplies. This situation favoured the survival of those whose bodies could "more readily convert blood sugar into stored fat when food was available."[9] Whereas Neel's 'thrifty genotype' hypothesis applies to all humans who undergo this starvation/ overabundance cycle with food, a "population specific evolutionary model […], the New World Syndrome, was proposed."[10] This syndrome was considered applicable to Indigenous populations and was thought to be related to the traditional ways in which they lived, ways that were believed to have been lost.[11] In contemporary society, the body's ability to effectively use all the energy sources it consumes has become a disadvantage, leading, in many

---

Evidence for Prevention: Type 2 Diabetes," 41). And Benyshek et al. call attention to the fact that urban Fijians of East Indian ancestry, the Chinese population of Mauritius, and Singaporean Malays have some of the highest rates of Type 2 diabetes in the world. Benyshek et al., "A Reconsideration of the Origins of the Type 2 Diabetes Epidemic among Native Americans and the Implications for Intervention Policy," 34–35.

[8] Jane DeMouy, "Genetic Research," *The Pima Indians: Pathfinders for Health*, http://diabetes.niddk.nih.gov/dm/pubs/pima/genetic/genetic.htm (accessed 23 September 2010).

[9] Benyshek et al., "A Reconsideration of the Origins of the Type 2 Diabetes Epidemic among Native Americans and the Implications for Intervention Policy," 30.

[10] Emőke J.E. Szathmáry, "Non-Insulin Dependent Diabetes Mellitus among Aboriginal North Americans," *Annual Review of Anthropology* 23 (1994): 465.

[11] Szathmáry, "Non-Insulin Dependent Diabetes Mellitus among Aboriginal North Americans," 465.

cases, to diabetes and obesity. Since the Indigenous body, according to the assumption of the New World Syndrome, processes food differently, it is more likely to succumb to diseases brought by Western civilization.

As a hypothesis, the genetic argument is now taken for granted, markedly influencing clinical studies on diabetes over the last five decades, with investigations of the disease continuing to focus on Native Americans and their food habits. Titles such as "Diabetes in Relation to Serum Levels of Polychlorinated Biphenyls and Chlorinated Pesticides in Adult Native Americans"[12] and "Mental Health Status and Diabetes among Whites and Native Americans: Is Race an Effect Modifier?"[13] are typical, confirming that Indigenous peoples have yet again become subjects of scientific surveys.[14] Whereas the pesticides study was carried out in a community identified as Mohawk, the examination of mental health in relation to diabetes makes no mention of tribal affiliations. In these two studies, as well as many others, because ethnicity was categorized not as a cultural phenomenon but as a biological factor that contributed significantly to the outbreak of the disease, chemical, biological, medical, and anthropological surveys of diabetes have most often been conducted among Indigenous communities.

One prominent example is the Strong Heart Study (SHS), which has been observing cardiovascular disease and its risk factors among Native Americans since 1988, and has become "the largest epidemiologic study of American Indians ever undertaken."[15] Thirteen American Indian tribes and communities situated in four states (Arizona, North and South Dakota, and Oklahoma) are participating in the long-term study, which in the course of its investigations

---

[12] Neculai Codru et al., "Diabetes in Relation to Serum Levels of Polychlorinated Biphenyls and Chlorinated Pesticides in Adult Native Americans," *Environmental Health Perspectives* 115.10 (October 2007): 1442–47.

[13] Abe E. Sahmoun, Mary J. Markland & Steven D. Helgerson, "Mental Health Status and Diabetes among Whites and Native Americans: Is Race an Effect Modifier?" *Journal of Health Care for the Poor and Underserved* 18.3 (August 2007): 599–608.

[14] The genetic argument has also been taken up in studies of other ethnicities – for example, "A Comparison Between Japanese-Americans Living in Hawaii and Los Angeles and Native Japanese," another study that presupposes genetic origins of diabetes Type 2. The participants were especially chosen because they were "genetically identical to native Japanese" and had not "intermarried with other races." Shigetada Nakanishi et al., *Biomedicine and Pharmacotherapy* 58.10 (December 2004): 571–77.

[15] *Strong Heart Study*, http://strongheart.ouhsc.edu/ (accessed 12 September 2010).

of heart disease, has identified age, parental diabetes, obesity, and "a higher degree of American Indian ancestry" as diabetes risk factors.[16]

Evidence of genetic susceptibility and diabetes, however, has yet to be found, and scholars have challenged the 'thrifty genotype' hypothesis. Benyshek et al., for example, point out that there is no proof that hunter–gatherers suffered significant periods of starvation in the prehistoric past:

> In fact, the vast majority of archaeological and ethnographic evidence suggests that agricultural societies throughout time have been much more susceptible to severe, periodic famine than have hunters and gatherers.[17]

They also call attention to the fact that several Indigenous communities, including the Pueblo, have a long agricultural history and still experience a high rate of diabetes; according to the 'thrifty genotype' hypothesis, the opposite should be the case.[18] Gilbert Velho and Philippe Froguel offer the proposition that diabetes "seems to result from several combined gene defects,"[19] that "many different combinations of gene defects may exist among diabetic patients,"[20] and conclude that "complex interactions between genes and environment complicate the task of identifying any single genetic susceptibilily factor."[21] Still, none of the extensive inquiries into the relationship between diabetes and genetics has met with success so far.[22]

---

[16] Lee et al., quoted in Yvette Roubideaux & Kelly Acton, "Diabetes in American Indians," in *Promises to Keep*, ed. Mim Dixon & Yvette Roubideaux (Washington DC: American Public Health Association, 2001): 195.

[17] Benyshek et al., "A Reconsideration of the Origins of the Type 2 Diabetes Epidemic among Native Americans and the Implications for Intervention Policy," 34.

[18] Benyshek et al. suggest a connection between undernutrition *in utero* and the appearance of Type 2 diabetes in later life. They point out that many Native Americans share a history of deprivation and forced relocations, which resulted in such conditions ("A Reconsideration of the Origins of the Type 2 Diabetes Epidemic among Native Americans and the Implications for Intervention Policy," 35–37).

[19] Gilbert Velho & Philippe Froguel, "Type 2 Diabetes: Genetic Factors," in *The Epidemiology of Diabetes Mellitus: An International Perspective*, ed. Jean–Marie Ekoé, Paul Zimmet & Rhys Williams (New York: John Wiley, 2001): 141.

[20] Velho & Froguel, "Type 2 Diabetes: Genetic Factors," 141.

[21] "Type 2 Diabetes: Genetic Factors," 141.

[22] However, it was proven that genetic factors play a significant role in some rare forms of diabetes, such as in the Maturity Onset Diabetes of the Young (MODY) from

These studies do suggest, however, that because of supposed genetic differences, Indigenous people process modern food differently from non-Native people. Less tolerant of processed food, according to this argument, Native populations should go back to traditional diets. This assertion has been made most prominently by Kerin O'Dea, who suggests a return to an archaic way of life. O'Dea himself supervised a project in 1980 in which "13 full-blood Aboriginal men and women"[23] participated. Together, they simulated the conditions on which Neel based his 'thrifty genotype' hypothesis: living as hunter–gatherers. The thirteen men and women spent three months "living a relatively traditional lifestyle"[24] which consisted mainly of hunting and gathering the food essential for their survival. Among the outcomes was the fact that all "Aboriginals […] lost weight during the study,"[25] especially those overweight prior to the study, and their state of health improved significantly. O'Dea confirmed, as have many others, that a healthy diet minimizes the symptoms of diabetes and slows the progress of the disease. Healthy eating habits are also believed to reduce the risk of developing the disease in the first place. At the same time, his investigation illustrates the significance of the assumption that the Indigenous body is genetically different from the non-Native body. The simulation project considered no parameters (e.g., social, environmental, educational) other than ethnicity (perceived to be biologically determined), and it did not test other populations for their tolerance of culturally specific foods. Furthermore, the fabrication of an historical setting demonstrates both an imperialist nostalgia for a past that is impossible to recreate, and the revival of nineteenth-century discourses on indigenous peoples.

---

which only two percent of all Type 2 diabetes patients suffer (see Velho & Froguel, "Type 2 Diabetes: Genetic Factors," 142, and Benyshek et al., "A Reconsideration of the Origins of the Type 2 Diabetes Epidemic among Native Americans and the Implications for Intervention Policy," 32).

[23] Kerin O'Dea, "Traditional Diet and Food Preferences of Australian Aboriginal Hunter–Gatherers," *Philosophical Transactions: Biological Sciences* 334/1270 (November 1991): 233–41, quoted in Zimmet et al., "Epidemiology, Evidence for Prevention: Type 2 Diabetes," in *The Epidemiology of Diabetes Mellitus: An International Perspective*, 45.

[24] O'Dea, quoted in "Epidemiology, Evidence for Prevention: Type 2 Diabetes," 45.

[25] O'Dea, quoted in "Epidemiology, Evidence for Prevention: Type 2 Diabetes," 45.

The Desert Walk

The Centre for Indigenous Peoples' Nutrition and Environment (CINE) is an independent research institute concerned with environmental and cultural changes affecting the traditional food systems and nutrition of Indigenous peoples worldwide.[26] One of their goals is to conserve or re-establish traditional agricultural techniques and cultivate historical crops and plants judged to have escaped modern scientific manipulation. Whereas this can be seen as a counter-movement to the contemporary food industry, it has also been inspired by biological discourses on the Indigenous body. CINE's research activities include identifying nutritional deficiencies that have resulted "from discontinued use of traditional food resources altered by degradation of the environment" and have increased "chronic diseases such as diabetes, cancer and heart disease when people move away from traditional diet and activity patterns."[27] This approach presupposes that there exists an original cuisine that needs to be rediscovered. It also assumes that food cultures are stable cultural phenomena resistant to changes: there is only a 'before' European contact and an 'after,' but nothing in between.

Beginning in the seventeenth century, a diet that included fried potatoes, butter, rice, and coffee with sugar was introduced to Native Americans. These 'new' staples, however, did not immediately replace the traditional menu; rather, they were added to the traditional diet of game, fish, fruit, and vegetables. After World War Two, the diet of all Americans changed dramatically and cases of obesity and food-related diseases increased in all populations. For Native Americans, this process was clearly forced by the Bureau of Indian Affairs' Urban Relocation Program of the 1950s, during which the US government's promotion of acculturation and assimilation used, among other methods, the encouragement of 'American' food habits, which consisted of a high-calorie and high-fat diet. These efforts, Brooke Olson speculates, combined with several other factors, including the stress the programme caused in most of its participants, were a major cause of the massive outbreak of diabetes in Native American populations during the 1970s.[28] These food patterns

---

[26] *Centre for Indigenous People's Nutrition and Environment*, http://www.mcgill.ca/cine/ (accessed 13 October 2010).

[27] "Research Activities and Publications," *Centre for Indigenous People's Nutrition and Environment*, http://www.mcgill.ca/cine/research/ (accessed 13 October 2010).

[28] Brooke Olson, "Meeting the Challenges of American Indian Diabetes: Anthropological Perspectives on Prevention and Treatment," in *Medicine Ways: Disease,*

established themselves especially on reservations where the local food store controlled the community's food supply. It has only been hesitantly that initiatives have been launched to break this vicious circle between unhealthy living habits and diabetes. One recently established programme to promote healthier living is the Tohono O'odham Community Action, which includes a campaign to return to the consumption of supposed-to-be traditional foods.

The Tohono O'odham Nation is based in Arizona in the heart of the Sonoran Desert. In 1963, a survey of rheumatoid arthritis was taken in the Gila River area, during which a high rate of diabetes was discovered. Since then, the community has made itself available for an intensive long-term study of diabetes – of which they have the highest reported incidence in the world – by NIDDK, the National Institute of Diabetes and Digestive and Kidney Disease.[29] More than half of all community members suffer from diabetes, among them children as young as seven, which is particularly unusual for the disease. Scientists have conducted a number of tests within the community in order to gain some understanding of the causes of diabetes and its consequences. A key segment of the study is identifying the genes responsible for the disease and specific aspects of its behaviour. The idea is to break the genetic codes and determine as early as possible those individuals at risk, intervening before the disease has established itself.[30] Researchers have noted the "uniqueness of the community":[31] intermarriage with other ethnicities is rare and the same families have lived in the Gila River community for generations, both situations providing ideal conditions for carrying out genetic studies. Despite this hospitable environment, progress in detecting the particular genes responsible for diabetes remains at a standstill.

The Tohono O'odham, however, have drawn their own conclusions from the study and are in the process of reviving traditional diet and food rituals. On 10 March 2000, a group of about forty Tohono O'odham, Comcaac, and Yoeme Indians set out to walk 240 miles across the desert, following in the footsteps of their ancestors. Their journey lasted twelve days, during which

---

Health, and Survival among Native Americans, ed. Clifford E. Trafzer & Diane Weiner (New York, Toronto & Plymouth: AltaMira, 2001): 163–84, here 166.

[29] K.M. Venkat Narayan et al., "Non-Caucasian North American Populations: Native Americans," in *The Epidemiology of Diabetes Mellitus: An International Perspective*, 184.

[30] DeMouy, "Pathfinders for Health."

[31] "Pathfinders for Health."

they ate only foods of the desert and used only medicines made from Native plants. The idea was to call attention to the high rate of diabetes in the Tohono O'odham population and to promote a return to traditional foods and ways of living. Daniel Lopez, an elder who participated in the walk, says (referring to the traditional diet):

> These foods are low in fat and sugar but high in the complex carbohydrates and soluble fibers that studies among desert dwellers, both in Arizona and Australia, have shown to lower blood glucose, insulin and cholesterol. But their connection to desert tribes is more complex than the dance of biochemicals.[32]

He further observes that

> We were once healthy people who ran, walked, and worked in the fields. We didn't have TV, and didn't sit a lot. Now we have to do drastic things, like walk 240 miles in the desert, to make our people realize that today's illnesses come from not eating healthy. Our people have to get back to eating foods from the desert.[33]

Both of the above statements allude to the notion of the Indigenous body as manipulated by white Europeans. Once healthy and vital, the Indigenous body has been severely compromised by foreign influences and, along with that body, the culture in which it was placed. And while scientists have been unable to work their way through the genetic maze of diabetes, they have managed to prove that a healthy diet minimizes the risk of diabetes, as Lopez, the elder, understood. The Desert Walk highlighted the history of food and food loss and one of its main consequences for these communities – diabetes. The walk was seen as a first step, both toward better health and toward cultural revival.

One of the actions taken to foster the shift in Tohono O'odham eating habits was the opening in March 2009 of the Desert Rain Café, created by the Tohono O'odham Community Action (TOCA), a nonprofit organization dedicated to "creating a healthy, culturally vital and sustainable Tohono O'odham community."[34] The re-introduction of traditional food and a healthy diet

---

[32] Daniel Lopez, quoted in "Desert Walk," http://www.ausbcomp.com/redman/desert_walk.htm (accessed 5 October 2010).

[33] Daniel Lopez, Museum Label, National Museum of the American Indian, 2000.

[34] *Tohono O'odham Community Action*, http://www.tocaonline.org/www.tocaonline.org/Home.html (accessed 19 September 2010).

became one of TOCA's top priorities. The café offers fresh, organic, regional food, low in fat and high in fibre and vitamins. At its busiest during lunch time, Desert Rain has become a place of social contact and exchange. The menu includes dishes such as prickly-pear chicken sandwiches and tepary bean, wild rice, and quinoa salad. The tepary bean is a traditional ingredient of Tohono O'odham meals and also plays a role in several traditional stories. The prickly pear, fruit of the cactus, is both a sweet addition to many meals and a popular ingredient in many drinks. Once comprising the basics of life, traditional ingredients began disappearing from the local food market at the start of the twentieth century. Initially, it was water-rights disputes, forced migrant labour, Indian boarding schools, and relocation programmes that made it impossible to continue to practise traditional agriculture and hunting. But, as mentioned earlier, Indigenous food cultures were also dealt a heavy blow by the US government's introduction of processed food, which became easily available and slowly replaced the traditional diet.

The Perfect Medicine

> Christine Johnson heard her son, Tony, sing the songs she had nearly forgotten, after a feast of rabbit, venison, beans, giant tortillas and mesquite cookies at an O'odham community center in the settlement of Little Tucson. […] [When she watched him leave for the Desert Walk] she ate a bowl of cholla cactus buds, which are full of the complex carbohydrates and soluble fibers that doctors now think can not only protect her from diabetes, but control the disease by regulating glucose and insulin levels. […] Native foods, in other words, may be her best medicine.[35]

It sounds simple – in order to be healthy, just consume Indigenous foods moderately and avoid westernized meals. But health and well-being are culturally defined, which in turn leads to different understandings of the disease, its causes, treatment, and its control.

> The Sandy Lake community [studied by Gittelsohn et al.] identified 'bad diet' and eating too much 'white man's' food as causes of diabetes but did not link lack of physical activity and obesity directly with diabetes.[36]

[35] Lopez, "Desert Walk."

[36] Zimmet et al., "Epidemiology, Evidence for Prevention: Type 2 Diabetes," 46.

Being overweight or obese is, in fact, considered 'normal' in many communities,[37] and weight loss may even be negatively perceived among Indigenous community members, seen as representing a communal history of poverty, hunger, and illness.[38] As in most cultures throughout the world, food plays a major role among Native Americans, hence it should come as no surprise that it is considered impolite to reject offers of food from friends and family.[39] This attitude is accompanied by the belief that the Indigenous body is genetically programmed to fall prey to diabetes anyway. As the Native American Diabetes Project explains,

> Diabetes means your body has trouble using food for energy. Diabetes is in the genes you were born with, just like the color of your hair and eyes! Do not blame yourself for having diabetes, it is not your fault. You did not get diabetes because you did something wrong.[40]

Based on this explanation, one might conclude that any attempts to prevent the disease are pointless. Genetics are often interpreted by Indigenous people as a complex entanglement of biological, social, and cultural identity.[41] Whereas "notions of genes are perceived and presented by scientists and health professionals as 'culture-free facts'" to Native Americans, they appear to be "social constructs."[42] Consequently, the notion of the Indigenous body as essentially different from the non-Native body goes beyond a biological perception. As one elder explains, diabetes runs in families:

> You inherited it from your mother. Well, of course, you watched your mother cook and all the things she gave you to eat, and now you serve yourself that and the kids that. [...] Yes, it's inherited that way.[43]

---

[37] Roubideaux & Acton, "Diabetes in American Indians," 205.

[38] Olson, "Meeting the Challenges of American Indian Diabetes. Anthropological Perspectives on Prevention and Treatment," 172.

[39] Diane Weiner, "Interpreting Ideas about Diabetes, Genetics, and Inheritance," in *Medicine Ways: Disease, Health, and Survival among Native Americans*, ed. Clifford E. Trafzer & Diane Weiner (New York, Toronto & Plymouth: AltaMira, 2001): 120.

[40] "Meeting 1: Exercise More," *Native American Diabetes Project*, http://www.laplaza.org/health/dwc/nadp/mtg1.htm#diabetes (accessed 14 March 2011).

[41] Weiner, "Interpreting Ideas about Diabetes, Genetics, and Inheritance," 109.

[42] "Interpreting Ideas about Diabetes, Genetics, and Inheritance," 109.

[43] "Interpreting Ideas about Diabetes, Genetics, and Inheritance," 119.

In the Indigenous world-view, the body is inseparable from culture, whereas Western science identifies the body as an independent entity. Both notions view the Indigenous body as exceptional, with diabetes seen by the former as the consequence of a loss of cultural heritage, and by the latter as a biochemical reaction to cultural change. Consequently, to cure diabetes requires returning to the original cultural habits and practices, which also presumes that an original, uninfluenced cuisine exists. A nostalgic past is perceived to be the perfect medicine to cure the Indigenous body of the illnesses apparently imported by Western civilization.

Exotic Tastes

The Indigenous body is not the sole target of culinary revivals. Venues that promote Indigenous food are gaining popularity on a much broader scale. Whereas the Tohono O'odham Desert Rain Café functions primarily as a destination for its community members, similarly themed enterprises have become major tourist attractions. The Brambuk Living Cultural Centre near Budjy Budjy (Halls' Gap) in the Gariwerd, or Grampian region of Victoria, Australia, which opened in December 1990, consists of a permanent exhibition, a gift shop, a café, and a restaurant, elements that are characteristic of more conventional museums. Nevertheless, Brambuk regards itself as a "living cultural centre" and is especially recognized for its unique cuisine. In the Brambuk Café, visitors can experience contemporary interpretations of native dishes that use traditional ingredients, such as crocodile and kangaroo meat, wattle seed, and bunya nuts, which are in turn transformed into selections that include crocodile and emu sausages, roo burgers, and, instead of cappuccino, a drink dubbed "wattlechino."[44] Three nights a week the café becomes the Gugidjela Restaurant and serves dinner. Its menu features courses like *cumbungi djarj gadjin cress* (hearts of reed and water cress), *gdjin yabidj* (yabbies – similar to crayfish – with garlic, bush tomato concassé, and wild rice), and *midjun quandong* (kangaroo fillets pan-fried with wild peach and bush chutney relish).[45] These evenings in the Gugidjela Restaurant have become the Brambuk Living Cultural Centre's main attraction.

Mitsitam, the café at the Smithsonian Institution's National Museum of the American Indian has also earned praise and has become one of the most

[44] Moira Simpson, *Making Representations: Museums in the Post-Colonial Era* (New York: Routledge, 2001): 128.

[45] Simpson, *Making Representations: Museums in the Post-Colonial Era*, 128.

successful sections of the museum. Meaning 'Let's eat' in the Piscataway and Delaware languages, Mitsitam offers seasonal food from the Western Hemisphere. Five serving stations arranged in a semicircle represent five different regions: South America, Northern Woodlands, Great Plains, Mesoamerica, and the Northwest Coast. The food is freshly prepared every day and the ingredients are mostly organically grown and provided by Indigenous communities. Wild salmon is flown into Washington, D.C. several times a week from tribes in the western states, and buffalo comes from animals bred by members of the Intertribal Bison Cooperative.[46] Visitors are willing to wait in line – which can last up to almost an hour on busy days – to try the food. The menu includes buffalo burgers and fry-bread with cinnamon and honey, stewed Anasazi beans, hominy, smoked ancho chilies, and prickly-pear ice cream.

Although Mitsitam is highly popular, food critics and scholars have questioned the authenticity of its recipes. Deborah Duchon, a nutritional anthropologist and *Food Network* personality, is sceptical about the café's food selection. The prickly-pear ice cream, for example, is not traditionally Native American, she argues. The prickly pear is indeed an Indigenous fruit, but ice cream is not something Native Americans ate, since their traditional diet did not include dairy products.[47] Other items criticized on the menu include the buffalo burger (Native Americans traditionally did not grind meat) and salmon with a wild berry sauce (Native Americans did not cook with fancy sauces). Richard Hetzler, the executive chef at the Mititsam, admits to working with traditional ingredients in order to make them more accessible to the café's patrons. Traditional corn bread, for example, is very dense, not sweetened, and thus would not appeal to a wide audience.[48] Whereas Duchon's position on the café's food is informed by her disciplinary location: i.e. history, the ethnobotanist Gary Nabhan considers the food to be a great representative of contemporary Indigenous cuisine and praises the café's efforts.[49]

Reproaches such as Duchon's concerning the authenticity of food reveal a desire for the presence of static monocultures that are incapable of adapting to

[46] Donna Boss, "Mitsitam Native Foods Café at the National Museum of the American Indian in Washington, D.C.," *Foodservice Equipment and Supplies* 60.5 (2007): 44.

[47] Gabriella Boston, "American Indian Variations at Café," *Washington Times* (30 July 2008): B1.

[48] Boston, "American Indian Variations at Café," B1.

[49] "American Indian Variations at Café," B1.

globalized modernity. The two cafés and one restaurant discussed here constitute a strong statement against any kind of assumed 'cultural authenticity'. At the same time, they are perceived as exotic by a growing crowd of culinary tourists. In the age of processed and flavoured food and the promotion of multiculturalism, Indigenous cuisines claim their place on the food-arts stage by employing the natural and the unusual. In an age of artificial flavours, the desire for the natural, or the *original*, is reflected in new recipes and eating facilities.

Conclusion: The Cultural Memory of Food

The desire for the culinary exotic by "the agents of colonialism" reflects a yearning "for the very forms of life they intentionally altered or destroyed."[50] This ongoing imperialist nostalgia is shaping not only taste in the arts but the taste of the palate as well. But beyond influencing these cultural shifts, this nostalgic turn has taken up residence in the sciences, where, at least in some quarters, the Indigenous body is perceived as different from the non-Indigenous one, a stance that perpetuates a discourse of oppositions. Diabetes, a product of Western civilization, therefore seems to confirm the notion that the Indigenous body is better off in the wilderness, thus moving Indigenous people off the stage of modernity. And regardless of the lack of scientific proof, diabetes is seen as a genetically determined disease mainly affecting Indigenous populations. Taking a broader view, diabetes also appears to be more prevalent in African Americans, Latinos, Asian Americans, Native Hawaiians and other Pacific Islanders, as well as the aged population. This information opens the door to the possibility that diabetes may after all be a "'political disease' whose roots lie deep within the structural inequalities engendered through conquest, colonization, and capitalist 'development'."[51] Once more the Indigenous body proves to be out of sync with civilization, rejecting the modern high-calorie/high-fat/low-fibre diet. Genetically, Native Americans once more become a monoculture whose death rates far exceed those of the rest of the American population in several areas – diabetes (249%), pneumonia and influenza (71%), tuberculosis (533%), and alcohol-

---

[50] Renato Rosaldo, "Imperialist Nostalgia," *Representations* 26 (Spring 1989): 107–108.

[51] Benyshek et al., "A Reconsideration of the Origins of the Type 2 Diabetes Epidemic among Native Americans and the Implications for Intervention Policy," 52.

ism (627%).[52] The notion of genetic determination greatly influences programmes aimed at the prevention and control of these serious health threats; the roots, however, might be located somewhere else.

During the sixth century, the Brahman period of Hindu medicine, diabetes was identified as "the disease of the rich and one that is brought about by the gluttonous overindulgence in oil, flour and sugar."[53] But whereas this is a description of aspects of a privileged life-style in sixth-century India, the trend has by now reversed, and those affected are predominantly from socially disadvantaged populations. Indigenous foods carry within them the cultural memory of their peoples' history, of destruction, starvation, and assimilation. They are fresh, organic, and rich in vitamins – a luxury in the age of artificiality, and a treat to any body.

WORKS CITED

Anon. "Confronting Diabetes with Tradition," *Indian Country Today* (20 December 2000), repr. in *America is Indian Country: Opinions and Perspectives from Indian Country Today*, ed. José Barreiro & Tim Johnson (Golden CO: Fulcrum, 2005): 236–38.

Benyshek, Daniel C., John F. Martin & Carol S. Johnston. "A Reconsideration of the Origins of the Type 2 Diabetes Epidemic among Native Americans and the Implications for Intervention Policy," *Medical Anthropology* 20.1 (2001): 25–64.

Boss, Donna. "Mitsitam Native Foods Café at the National Museum of the American Indian in Washington, D.C.," *Foodservice Equipment and Supplies* 60.5 (2007): 44.

Boston, Gabriella. "American Indian Variations at Café," *Washington Times* (30 July 2008): B1.

Codru, Neculai et al. "Diabetes in Relation to Serum Levels of Polychlorinated Biphenyls and Chlorinated Pesticides in Adult Native Americans," *Environmental Health Perspectives* 115.10 (October 2007): 1442–47.

---

[52] Senator Ben Nighthorse Campbell, "Charting a New Course in Indian Health Care," *Indian Country Today* (24 January 2003) in *America is Indian Country: Opinions and Perspectives from Indian Country Today*, 45.

[53] Quoted in Tim Mann & Monika Toeller, "Type 2 Diabetes: Aetiology and Environmental Factors," in *The Epidemiology of Diabetes Mellitus: An International Perspective*, 133.

DeMouy, Jane. "Genetic Research," *The Pima Indians: Pathfinders for Health*, http: //diabetes.niddk.nih.gov/dm/pubs/pima/genetic/genetic.htm (accessed 23 September 2010).

Flammang, Janet A. *The Taste of Civilization: Food, Politics, and Civil Society* (Chicago: U of Illinois P, 2009).

Foer, Jonathan Safran. *Eating Animals* (New York: Penguin, 2009).

Mann, Jim, & Monika Toeller. "Type 2 Diabetes: Aetiology and Environmental Factors," in *The Epidemiology of Diabetes Mellitus. An International Perspective*, ed. Jean–Marie Ekoé, Paul Zimmet & Rhys Williams (New York: John Wiley, 2001): 133–40.

Nakanishi, Shigetada et al. "A Comparison Between Japanese-Americans Living in Hawaii and Los Angeles and Native Japanese: The Impact of Lifestyle Westernization on Diabetes Mellitus," *Biomedicine and Pharmacotherapy* 58.10 (December 2004): 571–77.

Narayan, K.M Venka et al. "Non-Caucasion North American Populations: Native Americans," in *The Epidemiology of Diabetes Mellitus: An International Perspective*, ed. Jean–Marie Ekoé, Paul Zimmet & Rhys Williams (New York: John Wiley 2001): 184–91.

Neel, James. "Diabetes Mellitus: A 'Thrifty' Genotype Rendered Detrimental By 'Progress'," *American Journal of Human Genetics* 14 (1962): 353–62.

Nighthorse Campbell, Ben. "Charting a New Course in Indian Health Care," *Indian Country Today* (24 January 2003), repr. in *America is Indian Country. Opinions and Perspectives from Indian Country Today*, ed. José Barreiro & Tim Johnson (Golden CO: Fulcrum, 2005): 44–49.

O'Dea, Kerin. "Traditional Diet and Food Preferences of Australian Aboriginal Hunter–Gatherers," *Philosophical Transactions: Biological Sciences* 334/1270 (November 1991): 233–41.

Olson, Brooke. "Meeting the Challenges of American Indian Diabetes: Anthropological Perspectives on Prevention and Treatment," in *Medicine Ways: Disease, Health, and Survival among Native Americans*, ed. Clifford E. Trafzer & Diane Weiner (New York, Toronto & Plymouth: AltaMira, 2001): 163–84.

Pollan, Michael. *In Defense of Food: An Eater's Manifesto* (New York: Penguin, 2008).

Rosaldo, Renato. "Imperialist Nostalgia," *Representations* 26 (Spring 1989): 107–22.

Roubideaux, Yvette, & Kelly Acton. "Diabetes in American Indians," in *Promises to Keep*, ed. Mim Dixon & Yvette Roubideaux (Washington DC: American Public Health Association, 2001): 193–208.

Sahmoun, Abe E., Mary J. Markland & Steven D. Helgerson. "Mental Health Status and Diabetes among Whites and Native Americans: Is Race an Effect Modifier?" *Journal of Health Care for the Poor and Underserved* 18.3 (August 2007): 599–608.

Sietsema, Tom. "Mitsitam Café," *Washington Post Magazine* (16 October 2005), http://www.washingtonpost.com/gog/restaurants/mitsitam-cafe,1113026/critic-review.html (accessed 22 October 2010).

Simpson, Moira. *Making Representations: Museums in the Post-Colonial Era* (New York: Routledge, 2001).

Szathmáry, Emőke J.E. "Non-Insulin Dependent Diabetes Mellitus among Aboriginal North Americans," *Annual Review of Anthropology* 23 (1994): 457–82.

Velho, Gilberto, & Philippe Froguel. "Type 2 Diabetes: Genetic Factors," in *The Epidemiology of Diabetes Mellitus: An International Perspective*, ed. Jean–Marie Ekoé, Paul Zimmet & Rhys Williams (New York: John Wiley, 2001): 141–53.

Weiner, Diane. "Interpreting Ideas about Diabetes, Genetics, and Inheritance," in *Medicine Ways: Disease, Health, and Survival among Native Americans*, ed. Clifford E. Trafzer & Diane Weiner (New York, Toronto & Plymouth: AltaMira, 2001): 108–33.

Zimmet, Paul et al. "Epidemiology, Evidence for Prevention: Type 2 Diabetes," in *The Epidemiology of Diabetes Mellitus: An International Perspective*, ed. Jean–Marie Ekoé, Paul Zimmet & Rhys Williams (New York: John Wiley, 2001): 41–49.

Websites

The Centre for Indigenous Peoples' Nutrition and Environment (CINE): http://www.mcgill.ca/cine/ (accessed 13 October 2010).

The Desert Walk: http://www.ausbcomp.com/redman/desert_walk.htm (accessed 5 October 2010).

The Native American Diabetes Project: http://www.laplaza.org/health/dwc/nadp/mtg1.htm#diabetes (accessed 14 March 2011).

The Strong Heart Study: http://strongheart.ouhsc.edu/ (accessed 12 September 2010).

Tohono O'odham Community Action: http://www.tocaonline.org/www.tocaonline.org/Home.html (accessed 19 September 2010).

⌘

Part IV

Opposing and Reclaiming

Indigenous Opposition to Genetics Research

Views from Aboriginal Australia[1]

SHEILA COLLINGWOOD–WHITTICK

BEFORE BEGINNING THIS ESSAY, I WISH TO ACKNOWLEDGE the paradoxical nature of the position from which I speak – namely, that of a Western researcher reporting on the reasons for Indigenous opposition to Western research. Of equal importance, I think, is to recognize at the outset that the academy from which I draw status and authority has been (and too often still is) deeply implicated in the subjugation and the silencing of Indigenous peoples. While I cannot shed the cultural identity that gives me right of entry into that academy, I can and do unequivocally reject all areas of academic knowledge production in which colonial ideology continues to find comfort and support. Knowing the importance Indigenous peoples attach to speaking for themselves in their own voices, I have neither the wish nor the pretension to speak on their behalf here or anywhere else. But, as a white sexagenerian academic who has spent a lifetime studying the impact of colonization on the lives, cultures, and environments of colonized peoples, I am concerned by the rhetorical arsenal of denialism, rationalization, and trivialization that, as Ernest Hunter rightly points out, is currently being deployed in white-settler colonies like Australia to decredibilize attacks on their iniquitous colonial pasts.[2]

---

[1] Throughout this article, I use the terms 'Indigenous' and 'Aboriginal' to refer to both Aborigines and Torres Strait Islanders.

[2] Ernest Hunter, "'The Deep Sleep of Forgetfulness': Reflecting on Disremembering," Presentation to the Third World Conference for the International Society for Traumatic Stress Studies, Melbourne (17 March 2000): http://www.iap.org.au/ehunter .pdf (accessed 4 January 2011): 14.

"Unless trials become thinkable for us," the Australian philosopher Raymond Gaita has argued, "we cannot claim fully to understand the moral dimensions of our past."[3] My own conviction is that *until* trials become thinkable for us, we have to find other ways of testifying to the crimes against humanity that were committed worldwide during the colonial era. Aboriginal Australians have been doing that successfully in Australia for several decades now, but there are large sections of international opinion that Indigenous voices, regrettably, do not often reach. Academics in Europe, on the other hand, have access to forums from which they can inform a wider public of the criminal acts and lethal contemporary consequences of nineteenth-century colonialism. Not to make use of that privileged position to expose the inhumanity with which colonized peoples have been treated by their colonizers is, in my view, tantamount to "complicity in rationalizing and trivializing the harms done."[4]

Indigenous Australians and Genetic Research

Although there has been some acknowledgement in recent years that the prejudices and a-priori assumptions underpinning earlier 'scientific' theories on race were deeply implicated in the cataclysmic damage Indigenous societies sustained during the colonial era, there remains a persistent failure to imagine how that earlier experience of scientific racism might explain the view that Indigenous peoples have of science today. Thus, Indigenous opposition to research – generally represented by its advocates as having positive outcomes for the whole of humanity – frequently meets with incomprehension, not to say exasperation, in the Western world.

Mindful of Linda Tuhiwai Smith's assertion that "The ways in which scientific research is implicated in the worst excesses of colonialism remains a powerful remembered history for many of the world's colonized peoples,"[5] mindful also of the many examples of scientific abuse of Aboriginal peoples I

---

[3] Raymond Gaita, *A Common Humanity: Thinking about Love and Truth and Justice* (Melbourne: Text Publishing, 1999): 128, quoted in Hunter, "'The Deep Sleep of Forgetfulness'," 16.

[4] Hunter, "'The Deep Sleep of Forgetfulness'," 3.

[5] Linda Tuhiwai Smith, *Decolonizing Methodologies: Research and Indigenous Peoples* (London & New York, Zed, 1999): 1.

have encountered in the course of my own research,[6] I argue that Western incomprehension of Indigenous hostility to prestigious scientific undertakings like the Human Genome Diversity Project (HGDP) and the Genographic is, to say the least, disingenuous. Theorizing that Indigenous attitudes to science generally, and to biomapping in particular, are, for the most part, a direct result of the treatment to which those populations were subjected by men of science in the past, I wanted to test the validity of that hypothesis by talking with Indigenous peoples themselves. It was with the intention of gathering Aboriginal Australian opinion on the subject that I embarked on a study trip to Australia in 2009.

Although ideally I wanted to discuss the question of HGD research with as wide a range of Indigenous Australians as possible, I quickly realized that random wo/man-in-the-street interviews would probably not be the most suitable option. First, I feared that ordinary Aboriginal people might not know very much about biomapping projects (few Europeans, in my experience, have even heard of either the HGDP or the Genographic). Secondly, since, by its very nature, street interviewing allows little if any scope for the establishment of trust on the part of interviewees, I had reservations about the value of the results that kind of interview would yield. Finally, on a more personal level, I felt very ill at ease at the thought of asking strangers questions that could possibly re-activate painful, not to say traumatic, memories of the humiliation and suffering their people had endured as a result of Western scientific interest in them.

I therefore decided to contact a number of Indigenous intellectuals prior to my departure for Australia, reasoning that i) this type of interlocutor would undoubtedly be aware of HGD research and thus be better placed to foresee and comment on the ways it might be perceived by/influence Australia's Aboriginal populations; and ii), having been informed in advance of the subject of my research, anyone who agreed to speak with me would be prepared for the kind of issues I was likely to raise.

---

[6] See "Scientific Discourse and the West's Construction of its 'Others'," *Middle Ground: Journal of Literary and Cultural Encounters* 1 (2007): 44–54, and "Skeletons in the Cupboard: Imperial Science and the collection and museumization of Indigenous remains," in *Science and Empire in the Nineteenth Century: A Journey of Imperial Conquest and Scientific Progress*, ed. Catherine Delmas et al. (Newcastle upon Tyne: Cambridge Scholars, 2010): 65–82.

None of the people I subsequently emailed was known to me personally. If my choice of interlocutors was, in some respects, random, it was also, I acknowledge, guided by the opinion I had formed (mainly through reading written works they had published) that these were people whose analytical acuity and sensitivity towards matters of concern to the Aboriginal community made them potentially invaluable commentators on the subject I wanted to discuss. As a result of the relatively high level of positive response I received (roughly seventy-five percent of those contacted), I was able to carry out ten in-depth interviews with Aboriginal academics from a broad range of disciplines, including literature, history, archaeology, sociology, cultural studies, and population health.

The questionnaire I devised to elicit opinion on the various ethical, cultural, and political issues raised by HGD research consisted of a total of twelve questions, of which half were innocent of any theoretical orientation, the rest proceeding from my hypothesis that it was Aboriginal experience of the misuse of race-based science in the nineteenth century that was at the root of present Indigenous refusal to collaborate with Genographic researchers. However, fearing that a rigid question–answer format in which I would be the controlling agent might reduce the possibility of the spontaneous dialogic exchange I hoped for, I preferred, in the event, to assume a less directive role in what turned out to be wide-ranging discussions of at least an hour in length and in some cases considerably longer.

Once in Australia, I also decided (despite earlier misgivings) to carry out a similar number of unscheduled interviews with Indigenous Australians whom I encountered fortuitously in diverse settings and situations. Feeling that it would be unreasonable to detain for very long people who might not only never have heard of the Genographic project, but who might also feel wary of talking to an unknown white academic on any subject, I pared down the number of questions I asked during this type of interview to a minimum.

Apart from the fact that they were based on a more limited range of questions and were generally much shorter, the interviews conducted in public places also differed from the meetings that had been scheduled in advance, to the extent that they involved people of a much wider age-span and from a variety of backgrounds. As I had anticipated, this group of informants had little knowledge of HGD research, though all of them expressed (some of them vehemently) a negative view of the ways in which Aboriginal peoples have been treated by scientists.[7]

In this essay, I will both discuss some of the results of the two sets of interviews and present the preliminary findings of the research I have subsequently undertaken in response to information given to me by interlocutors. What I hope to demonstrate is that, notwithstanding the opinion often expressed within the scientific community, there is nothing remotely 'incomprehensible' or 'irrational' about the refusal of Indigenous Australians to collaborate with geneticists on studies of human diversity. On the contrary, I argue, Aboriginal attitudes towards HGD research are the result of justifiable scepticism, legitimate concerns, and a long-established, empirical understanding of the ways in which Western scientists tend to 'deal with' the non-Western objects of their study.

To judge from observations made by my interlocutors, Indigenous Australian objections to genetics research are shaped by three dominant perceptions:

— that Indigenous peoples have had more than enough of being the perennial objects of study by Western scientists;
— that research of the kind being carried out by population geneticists is either irrelevant or of peripheral concern to populations that are in urgent need of solutions to problems that threaten their very survival;
— that the irreparable damage Aboriginal peoples have sustained *and continue to suffer from* as a result of scientific interest in them makes it extremely difficult for them to respond positively to any research project that requires their collaboration.

Examining these three points in order, Linda Tuhiwai Smith's observation that "the term 'research,' is probably one of the dirtiest words in the indigenous world's vocabulary"[8] was very largely confirmed by the views expressed by those I consulted. In the opinion of one academic,

---

[7] Several people recounted personal stories that were deeply saddening. Two had been removed from their family as children; a third testified to multiple instances of discriminatory treatment that had scarred her life. Significantly, almost all of these interviewees identified scientific theories representing Aborigines as inferior as being the root cause of their suffering. That Australia remains a racist country was a statement I heard many times during both types of interviews.

[8] Tuhiwai Smith, *Decolonizing Methodologies*, 1.

> The conceit, arrogance, and sense of authority of scientists are unchanging whatever the enterprise on which they are engaged. All such projects are involved in the control of Indigenous people. Western pursuit of knowledge knows no bounds.[9]

A Bidjarra woman from south-east Queensland wearily pointed out that "Aboriginal people have been studied to death."[10] However, as Indigenous activists have not failed to remark, while non-Indigenous researchers often establish their academic reputations and earn comfortable livings from the studies they carry out, few advantages accrue to Indigenous peoples from investigations that habitually explore the most intimate expressions of their corporeal and spiritual existence. To quote the Aboriginal researcher Bronwyn Fredericks,

> Aboriginal people have been weighed, given blood, urine, faeces and hair samples, given their stories, explained their existence, been interviewed, questioned, observed, followed, interpreted, analysed and written about for years. From the data reports, books and theses have been generated. Papers have been delivered at conferences and journal articles published.[11]

Turning to the second point: namely, the questionable usefulness of research on human migation routes to peoples who are struggling merely to survive, sometimes in ethnocidal environments: one of the main grievances that Indigenous peoples express about research is its lack of positive impact on their lives.[12] As Justin Mohamed, chairperson of the Rumbalara Aboriginal Co-

---

[9] Since only some of my interlocutors indicated a willingness to be identified, I have decided to keep references to all of them anonymous.

[10] Australia's Indigenous inhabitants are generally considered to be "the most researched group in the world." See Bronwyn Fredericks, "Making an impact researching with Australian Aboriginal and Torres Strait Islander peoples," *Studies in Learning, Evaluation Innovation and Development* 5.1 (April 2008): 25.

[11] Fredericks, "Making an impact," 25.

[12] Speaking at a debate on the repatriation of Indigenous remains, Maurice Davies, deputy director of the Museums Associations, affirmed: "I have asked every scientist who has come to us whether there are any examples of medical research carried out on contested human remains in collections in this country that have led to medical benefits. And not one example has been found of that." See Jane Hubert et al., "Human remains: objects to study or ancestors to bury?" transcript of a debate held at the Royal

operative in Victoria, explains, there is an expectation in Aboriginal communities that "when you research a group of people in an area it really needs to benefit the people [...] where the research is taking place."[13]

Yet, the reality is that HGD projects have a poor track record when it comes to demonstrating concern for the lives of the human beings whose genetic samples they seek to collect. One has only to read the words of the North American Regional Committee of the Human Genome Diversity Project to understand that what *really* drives the scientists taking part in it is less an interest in improving the lives of Indigenous peoples than the excitement aroused by the 'increased' knowledge such research will afford them.

> The proposed HGDP is an exciting effort to increase our knowledge of the human family: its evolution, history, diversity, and essential unity. Its breadth may be unique among major scientific endeavours – it seeks to study our entire species in ways that will engage the attention of tens of millions of people. This scientific project intrigues people who are interested in origins, history, languages, cultures, medicine, and a host of other topics.[14]

Since it is the migratory routes of Aboriginal populations that Genographic scientists hope to trace through the former's DNA, their research is naturally not designed to address the critical problems with which many native peoples

---

College of Physicians (2 May 2003), http://www.instituteofideas.com/transcripts/human_remains.pdf (accessed 4 January 2011): 12.

[13] Justin Mohamed, "We Don't Like Research – But in Koori Hands It Could Make a Difference" (Melbourne: VicHealth Koori Health Research and Community Development Unit & University of Melbourne, 2000): 9.

[14] K.M. Weiss et al., "Proposed Model Ethical Protocol for Collecting DNA Samples," *Houston Law Review* 33 (1997) quoted by Matthew Rimmer in "The genographic project: traditional knowledge and population genetics," *Australian Indigenous Law Reporter* 11.2 (2007): 36. Participants in the Genographic Project place similar emphasis on the excitement of discovering answers to previously unanswered questions about the deep past of Indigenous peoples. Two good examples can be found in i) "Help unravel the mystery of our species' journey," *The Genographic Project*, IBM website and ii) "Field Research," *The Genographic Project*, National Geographic website. See also the comments of Brian Sykes – professor of human genetics at Oxford University – "I am driven by curiosity. Curiosity should drive all science. [...] There should be the desire to explore things not explored before." Martin Baker, "Curiosity drives the gene genie to a £1m turnover," *The Telegraph* (15 February 2008), online.

of former white-settler colonies are confronted today. Indeed, like a great deal of research on Indigenous people, the Genographic project is, to quote one interlocutor, "at the foundation rather than the application end of knowledge," greater scientific priority invariably being given to descriptive research than to studies that might lead to intervention in, for example, the life-threatening health problems from which many Aboriginal peoples now suffer.[15]

How to explain, then, that in diametrical opposition to Indigenous expectations that scientific research be of benefit to those it studies, it is precisely the *non-interventionist* nature of their research that the Genographic's promoters emphasize? What purpose is served by stressing the project's lack of interest in either searching for medical information or gathering clinically relevant data?

One possible explanation might be that Genographic scientists' overdetermined attempts to dispel Indigenous fears about medical research are simply a diversionary tactic – a means of shifting attention away from the fact that their project is simply irrelevant to the issues that blight the existence of Indigenous peoples today.[16]

For, from the point of view of the Aboriginal populations of former white-settler colonies who have yet to overcome the multiple transgenerational traumas resulting from their colonization, genetic research aimed at producing abstract knowledge of the migratory history of the human species ("the pursuit of useless knowledge," as one interlocutor described it) is wholly incommensurate with their most urgent concerns:

> All over the world we are being killed, we are being displaced. And while this is going on, the Genographic Project is spending millions of

---

[15] The accuracy of the above assessment is emphatically confirmed by a recent critical review of outputs in Indigenous Health Research in Australia, Canada, New Zealand, and the USA over the period 1987 to 2003. What that survey shows is that, though the proportion of "intervention publications" has gradually increased since 1987, there remains nonetheless "a consistent dominance of descriptive research in the Indigenous health field over time." In the case of Australia, the authors report, at least 78% of published research between 2001 and 2003 was descriptive. Robert W. Sanson–Fisher et al., "Indigenous health research: a critical review of outputs over time," *Medical Journal of Australia* 184.10 (2006), online: np.

[16] Such as the extremely high incidence of physical and mental ill-health experienced in many of their communities or the continuing war of attrition waged against their cultures by the Anglo-Celtic governments under whose rule they live.

> dollars on a study that hopes to show the patterns of population migrations. It's hard to see how this is a collaboration. Why don't they bring that money to us and ask us what we really need?[17]

Unless Aboriginal peoples are prepared to identify with the scientific perspective from which the pursuit of knowledge is held to be a self-evident good, there are, on the face of it, few compelling reasons why they *should* cooperate with geneticists on biomapping projects like the Genographic.[18] There are numerous reasons, on the other hand, why Indigenous populations might choose to withold their collaboration.

Not least of these is the entrenched conviction that, as one interlocutor put it, Western science is always "in the service of something – the handmaiden of capitalism in the West, in fact." While others expressed their views in less overtly political terms, all rejected Western culture's representation of science as value-neutral. The very idea, one person commented, was "laughable."

The ethnocentricity of Western modes of knowledge-production has, of course, been highlighted by both Indigenous and non-Indigenous critics for some time now, with the human sciences in particular being singled out for their susceptibility to ideological and political influences. Bearing these critiques in mind, and in the light of the fact that the Aboriginal Australians who spoke with me saw Genographic research as being completely disconnected from the main concerns of Indigenous peoples, it seems appropriate to consider the question posed by the Indigenous activists Debra Harry and Le`a Malia Kanehe: "If the Genographic Project will not benefit indigenous peoples, who will it benefit?"[19]

---

[17] The Indigenous activist Victoria Tauli–Corpuz, quoted in Charles Furniss, "Blood feud," *Geographical* 78 (September 2006), online: np.

[18] Two of the academics with whom I spoke felt nonetheless that Genographic research had, as one of them put it, "the potential to contribute to Indigenous peoples' knowledge of their deep past." The other suggested that such research might possibly be of interest to Tasmanian Aborigines "whose mitochondrial DNA could prove their disputed Aboriginality." While recognizing the usefulness of DNA tests when used by Aboriginal organizations such as Link-up which work to reunite members of the Stolen Generations with their Aboriginal families, a third interlocutor was implacably hostile to the Genographic.

[19] Debra Harry & Le`a Malia Kanehe, "Genetic Research: Collecting Blood to Preserve Culture?" *Cultural Survival Quarterly* 29.4 (Winter 2005), online: np.

A first major point to be noted is that, as the brainchild of a research partnership between the National Geographic and IBM, the Genographic Project is in the hands of two iconic American institutions, both of which may be justifiably regarded as having a vested interest in the continuing dominance of a Western scientific view. Furthermore, it is difficult to keep knowledge of the fact that IBM is providing "the core computational knowledge and infrastructure that manages the hundreds of thousands of genetic codes being analysed by the Genographic Project,"[20] in a mental compartment that is hermetically separated from awareness of the fact that the information technology of the American computing giant played a key role in facilitating the Holocaust.[21]

A further crucial point is the fact that the Genographic Project is *privately* funded, which, as Harry and Kanehe contend, means first and foremost that it "does not have to undergo the same depth of public scrutiny" as government-funded research. Consequently, Indigenous peoples who provide researchers with DNA samples will be left "with fewer mechanisms for accountability."[22]

Moreover, as the Secretariat to the United Nations Working Group on Indigenous Peoples warned in 1998, "the steady move from State-sponsored to privately funded research" tends to result in scientific studies falling increasingly under "the inevitable influence of the profit motive."[23] That there is

---

[20] "Frequently Asked Question," *The Genographic Project*, National Geographic website.

[21] See the extraordinarily well-documented and utterly damning case made by the investigative reporter Edwin Black in his book *IBM and the Holocaust: The Strategic Alliance Between Nazi Germany and America's Most Powerful Corporation* (New York: Crown, 2001).

[22] Harry, "Genetic Research: Collecting Blood to Preserve Culture?"

[23] *Working Group on Indigenous Populations, Standard-Setting Activities: Evolution of Standards Concerning the Rights of Indigenous Peoples: Human genome diversity research and indigenous peoples* Note by the Secretariat, UN Doc: E/CN.4 /Sub.2/AC.4/1998/4 (4 June 1998), quoted by Mick Dodson, "The protection of genetic information of Indigenous peoples," submission to the Australian Law Reform Commission inquiry into the protection of genetic information, *Australian Human Rights Commission* (May 13 2002), online: np. It is worth noting, I think, that when asked about the motivation for IBM's association with the Nazis, Edwin Black replied: "It was never about the Nazism. It was never about the anti-Semitism. It was only about the money. [...] It wasn't personal – it was just business." Reported by Paul Festa, "Probing IBM's Nazi Connection," CNET News.com (28 June 2001), online.

legitimate cause for such concern is suggested by the volume of articles published over the last fifteen years highlighting the immensely lucrative rewards to be reaped from genetics research. One example: in an article entitled "The Bioinformatics Gold Rush," Gene Myers, Jr., the vice-president of informatics research at Celera Genomics, is quoted as predicting that the information generated by genetic research "will be phenomenal, and everyone will be overwhelmed by it [...]. The race and competition will be who can mine it best. There will be such a wealth of riches."[24] Significantly, the analogy of the 'gold-rush' crops up with ominous frequency in reporting on genetics research.[25]

There can be little doubt that the first to benefit from genetics research are the bioinformatics and pharmaceutical companies that have invested in it. But there are other parties for whom the knowledge sought by HGD studies such as the Genographic could be both politically and economically advantageous.

For former white-settler colonies like Australia, the kind of data the Genographic project hopes to produce is a potential source of two important 'benefits'. First, if Genographic scientists are able to demonstrate – as they claim they can – that Indigenous peoples, like the rest of humankind, arrived in their current homelands after migrating from Africa, then the difference between Aboriginal occupation of Australia and that of Europeans could easily be reduced by those with a political interest in doing so to a matter of mere chronology. Indigenous people could then be construed, "in common with Australia's colonial settlers," as "invaders."[26]

Which is not to forget, of course, that most non-Indigenous Australians are prepared, nowadays, to accept the truth of what Indigenous groups have asserted and archaeologists have been confirming for some time – that Aboriginal Australians have inhabited the continent for at least 50,000 years.[27] Never-

---

[24] Ken Howard, "The Bioinformatics Gold Rush," *Scientific American* 283.1 (July 2000): 58.

[25] See, for instance, Lawrence M. Fisher, "Mining the Genome: Big Science as Big Business," Special report, *New York Times* (30 January 1994), online; Eliot Marshall, "Gene Prospecting in Remote Populations," *Science* 278/5338 (24 October 1997): 565; and the issue of *Business Week* (12 June 2000), online, in which numerous articles are devoted to the cover story "The Genome Gold Rush."

[26] Sarah Colley, *Uncovering Australia: Archaeology, Indigenous People and the Public* (Crows Nest, NSW: Allen & Unwin, 2002): 182.

[27] My thanks to Emma Kowal for her advice that I give due emphasis to this point.

theless, by gathering genetic evidence to prove that Aborigines *migrated* to Australia, HGD research may be interpreted as validating that current of minority opinion (always present in populations of settler descent) which "*wishes* to find scientific support to legitimize colonial dispossession of Aboriginal lands and to delegitimize Aboriginal claims for Native Title rights."[28] And, after all, as Harry and Kanehe remind us,

> Governments have a long history of trying to divest indigenous peoples of their land rights and undermine their cultural integrity, by any means necessary. Despite the speculative nature of genetic research on human histories, the findings of the Genographic Project will carry the weight of science, which could be used to trump indigenous peoples' unique political status and rights.[29]

Secondly, if IPCB predictions are right, and DNA tests replace traditional laws and customs as the normative means for determining Indigenous identity, then logically, as Mick Dodson argues, genetic evidence could become the grounds for deciding whether an individual is entitled to benefits, say, or native land rights.[30]

Referring to debates on identity-politics in Australia, Lucinda Aberdeen cites the precedent of "calls for the testing of Aboriginal blood as a mechanism to determine eligibility for government assistance."[31] If members of the minority of Aboriginal Australians still living a semi-traditional life in remote areas were to be judged on the evidence of a DNA test to be non-Indigenous,

---

[28] Ian J. McNiven & Lynette Russell, *Appropriated Pasts: Indigenous Peoples and the Colonial Culture of Archaeology* (New York, Toronto & Plymouth: AltaMira, 2005): 92. (My emphasis.)

[29] Harry & Kanehe, "Genetic Research: Collecting Blood to Preserve Culture?"

[30] Dodson, "The Protection of Genetic Information." Yet, Dodson points out, since *self-identification* constitutes "a fundamental criterion" of the provisions of the ILO Convention (No. 169) concerning Indigenous and Tribal Peoples in Independent Countries, then inevitably "a test of whether a person is indigenous based on biology is [...] contrary to international human rights principles."

[31] Lucinda Aberdeen, "Australian Scientific Research, 'Aboriginal Blood' and the Racial Imaginary," in *'A Race for a Place': Eugenics, Darwinism and Social Thought and Practice in Australia*, ed. Martin Crotty et al. (Callaghan, NSW: University of Newcastle, 2000): 108. One example quoted by Aberdeen is that of the national president of the Returned Servicemen's League, who proposed compulsory blood tests to "check the racial mix of Aboriginal people receiving social security pay" (fn 81).

they could, without the meagre benefits accorded them by the state, and having been deprived of virtually all their traditional means of economic subsistence, be forced by material need out of the communities in which they currently live. Economic pressure of that kind leaves people with no option but to pursue, in Patrick Dodson's words, "a new mobile Indigenous individualism [...] seek[ing] employment opportunities dictated by market forces at distant locations,"[32] Since it is the sense of *collective* identity that binds the members of Indigenous communities together, it is difficult to imagine how, in the long term, those few remaining communities in which Aboriginal languages and cultural practices still thrive could survive this kind of assault.[33]

While a small majority of the academics I spoke with did not believe that the Australian government would exploit the results of HGD research for the purpose discussed above,[34] several others thought it was a definite possibility. And the consensus was that, whatever they themselves might think, Aboriginal people in general were unlikely to be convinced by the reassurances of scientists – a view confirmed by opinions expressed in the second set of interviews. It would appear, then, that, despite its advocates' claims not only that it springs from benevolent intentions but that, ultimately, it will demolish the very foundations of racist ideologies, the Genographic project is perceived by many Indigenous Australians as potentially hostile to their best interests. If historical precedents are anything to go by, they have, as I will now discuss, good reason to be worried on that score.

---

[32] Patrick Dodson, "Whatever happened to reconciliation?" in *Coercive Reconciliation: Stabilise, Normalise, Exit Aboriginal Australia*, ed. Jon Altman & Melinda Hinkson (Melbourne: Arena, 2007): 22–23.

[33] The disintegration of 'traditional' Aboriginal culture may well, of course, be the ultimate goal of a State which has long advocated assimilation, the 'absorption' of Australia's Indigenous peoples into an Anglo-dominant society. See Dodson, "Whatever happened?" 25.

[34] Although, one interlocutor suggested, a certain section of non-Indigenous Australia would "jump for joy at the idea that Aborigines were also immigrants to Australia. *Quadrant*, for example, would jump at any story of this kind as it would constitute further ammunition, another string to their bow," such ideas, he believed, would not fundamentally change the status quo. Another was of the opinion that "only extremists and ratbags would do that and they don't have the clout to influence things in any important way." A third felt confident that "Australia has matured in the last three to four decades."

By privileging the construction of knowledge that could be used to contest the unique and privileged position of 'Indigenous peoples' that many colonized populations have only latterly, and after centuries of abuse, managed to establish, HGD research threatens to weaken the only safeguard that these peoples dispose of today to protect themselves from the rapacity of multinational corporations. Without their recently acknowledged special status, the world's Indigenous peoples would have even less chance of defending their land – with all that it represents in terms of spiritual and identitarian values– against the relentless predations of mining, agricultural, logging, and petrochemical industries.[35] In the specific case of Australia, much of the forty percent of the world's recoverable uranium that is known to exist on the continent is on Aboriginal territory. Since Russia, China, and India are all seeking to develop civilian nuclear energy, they are extremely eager to buy Australian uranium. In these times of economic crisis, the possibility of opening up mineral-rich land to which Aboriginal peoples claim native title to the multibillion-dollar mining and nuclear-power industries itching to get their hands on such resources is an obvious temptation to the government. In the Northern Territory alone, there are, John Pilger points out, "deals to be done. The Territory contains extraordinary mineral wealth, especially uranium. And Aboriginal land is wanted as a radioactive waste dump. This is very big business, and foreign companies want a piece of the action."[36]

The last of the three factors identified by my interlocutors as explaining Indigenous refusal to cooperate with projects like the Genographic is the deep-seated mistrust that Australia's Aboriginal peoples harbour towards Western science and scientists in general. In the view of one interlocutor, Indigenous Australians have been "burnt too often [by scientific research] in the past" to have confidence in *any* scientific project. As another commented, "there is no history of trust between scientists and Indigenous populations on which contemporary research can build. Generally speaking, Aboriginal people don't have a good relationship with any elite institution."

---

[35] As Mick Dodson puts it, "The most important factor which defines indigenous people is their cultural and spiritual relationship with their land." "The Protection of Genetic Information."

[36] "Breaking the Great Australian Silence," speech given at the Sydney Opera House on the occasion of Pilger being awarded the Sydney Peace Prize (5 November 2009), http://www.johnpilger.com/articles/breaking-the-great-australian-silence (accessed 4 January 2011).

Among the causes cited as accounting for this lack of trust were the following:

—hierarchizing Western taxonomies that, in positioning Aborigines on the bottom rung of the racial ladder as a sub-human life-form, constituted the primary cause of most of the suffering Aboriginal people have endured since colonization;
—social-Darwinist theories and the concomitant perception of Aborigines as a dying race, which had led to the outrage of Indigenous remains being illegally dug up and removed from burial sites for study by European scientists or for exhibition in anthropological museums;[37]
—eugenicist policies that began by attempts to 'biologically absorb' the Aboriginal population into the dominant white society through 'reproductive management', and ultimately resulted in the removal of their children for the purpose of cultural assimilation;
—the marginalization of traditional Indigenous knowledges by Western scientists and the imposition of Western understanding on Indigenous experience;
—nuclear tests carried out by scientists in the 1950s in areas inhabited by Aboriginal peoples.[38]

Since almost all of the above refer to established historical facts which, thanks to recent and ongoing historiography, are beginning to enter the domain of general public knowledge, I will not dwell on them here. A number of interlocutors spoke, however, of scientific practices such as medical experiments, the removal of tissue and body parts from Aboriginal corpses, and the

[37] Citing the William Lanney case, one interviewee spoke with incredulity of the "fragmentation of Aboriginal bodies," 'specimens' treated as "deader than dead" by the scientists who mishandled them. Another mentioned that in the late-1800s Bass Strait Islanders preferred not to have headstones on their graves for fear that their bodies might be dug up. The repatriation issue was generally considered to be of great importance, and several people expressed anger and incomprehension at the refusal of British museums to return Aboriginal remains.

[38] See Jim Green, "Radioactive racism in Australia," Friends of the Earth Australia website, for details of the nuclear-weapons tests carried out on Aboriginal territories between 1952 and 1963. The Friends of the Earth Australia website contains a large archive of articles dealing with the siting of radio-active waste on Aboriginal land.

coercive sterilization of women – all carried out without informed consent having been granted, let alone sought – to which Aboriginal Australians peoples have been subjected over the last century, and of these there is comparatively little awareness. Shocked to discover the existence of such practices, I have begun to delve further into these more recent examples of scientific abuse. It is the results of my preliminary investigations that I discuss in the second part of this essay.

Scientific and Medical Experiments on Aboriginal Adults and Children

In the last few years, there has been a consistent trickle of information about experiments conducted, mostly without consent, on Indigenous Australians of all ages. In 2002, the Vice-Chancellor of Adelaide University, Cliff Blake, acknowledged in a media release on the University website that "many of the tests and experiments carried out [by Adelaide University researchers] on Aboriginal people in South Australia in the name of science in the 1920s and 1930s were degrading and, in some cases, barbarous."[39]

Two years later, the Australian government published a report on Australians who had experienced institutional or out-of-home care as children. One of the findings of the enquiry that led to this report was that "children in orphanages and Homes have been used for medical experiments for many decades." Expressing their consternation about this discovery, the authors of the report wondered: "if these experiments were known, what other experiments may have occurred that were not officially reported?"[40]

The answer to that question was not long in coming. In a second government report published the following year, it was acknowledged that "further trials involving children in care have come to light." These included experiments carried out between 1959 and 1961 in which Salk vaccine known to be "contaminated with a monkey virus, SV40, which has been linked to cancer" had been tested on babies under twelve months old.[41]

---

[39] The University of Adelaide, "Apology for Past Experiments on Aboriginal People," *News and Events* (8 February 2002): http://www.adelaide.edu.au/news/news314.html (accessed 4 January 2011).

[40] Senate Community Affairs References Committee, *Forgotten Australians: A Report on Australians who experienced institutional or out-of-home care as children* (Canberra: Commonwealth of Australia 2004): 114.

[41] Senate Community Affairs References Committee, *Protecting Vulnerable Child-*

Summarizing the respective reports' findings on the numerous experiments that had been conducted on institutionalized children, John Murray stated in 2007 that

> A known list of the experimental agents run through the orphanages of Australia since the Nuremberg Trials include vaccines for diphtheria, whooping cough, herpes, polio, influenza, measles, rubella, quadruple antigen, and human pituitary hormones. It is also thought the testing of antipsychotic medications, anti-rejection medications (for use in organ transplants) and psychosurgical procedures were perfected in child welfare institutions before 'going public'.
>
> One of the experiments disclosed in 1997 was a 1950's trial of a vaccine for the sexually transmitted disease herpes. Eighty-three babies aged six to eight months old had been infected with the disease when the experimental agent was found to be worthless.[42]

Since neither of the government reports deals specifically with *Indigenous* children, it is essential to stress here that, due to the assimilation policy that Australia aggressively pursued with regard to its Aboriginal population for more than seventy years, a significant percentage of Aboriginal youngsters had to be among the "Australians who had experienced institutional or out-of-home care as children." As such, there can be little doubt that many of them were victims of the kind of experimentation described above.[43]

At a Senate inquiry into the Stolen Generations conducted in 2008, the Aboriginal elder and former Australian of the year 'Aunty' Kathleen Mills,

---

ren: A National Challenge, Second report on the inquiry into children in institutional or out-of-home care (Canberra: Commonwealth of Australia, 2005): 11–12. Significantly, the same Salk vaccine had been used in the early 1960s in experiments in which researchers at Cleveland Metropolitan General Hospital, USA, "inoculated newborns *from mostly lower-income black families* with doses ranging up to more than 100 times the dose recommended for adults." William Carlesen, "Rogue virus in the vaccine: Early polio vaccine harbored virus now feared to cause cancer in humans," *San Francisco Chronicle* (15 July 2001), online: np. (My emphasis.)

[42] John Murray, "From the Cradle to the Grave," *Origins Harp Healing and Recovery Project for Forgotten Australians* (14 August 2007): http://www.originsharp.com/experiment.html (accessed 4 January 2011).

[43] The notorious Child Removal Policy had resulted in tens of thousands of Indigenous children (many of them extremely young) being systematically removed from their parents and placed in Church- or State-run children's homes, orphanages, and educational institutions.

gave evidence which seemed to corroborate this supposition. It was, she testified, "a common experience and a common practice" at the Kahlin Compound in Darwin in the 1920s and 1930s for institutionalized Aboriginal children to be used as guinea-pigs for leprosy treatments.[44] The Health Minister Nicola Roxon responded to Ms Mills' claims by ordering an enquiry – a decision that was rubbished in a vitriolic article published in Australia's best-selling, tabloid newspaper.[45]

Once the results of the enquiry had been published, another conservative newspaper, *The Australian*, triumphantly announced: "No evidence stolen generation were used as guinea pigs for leprosy drug."[46] The truth is, however, rather different from that suggested by *The Australian*'s crowing headline. For, if the enquiry concluded that there was no evidence to support Ms Mills claims, it recognized with a certain embarrassment that there was derisorily little evidence to refute them. The cyclone that devastated the Darwin area in 1937, the bombing of the Northern Territory capitol in 1942 during World War Two, and the damage caused by Cyclone Tracey in 1974 have together resulted in most of the medical records for the period in question being irretrievably lost. The conclusions reached by the research team that investigated Kathleen Mills' allegations were, therefore, necessarily *inconclusive* because "based on the *limited evidence* [...] available."[47]

---

[44] AAP, "Stolen Generation Kids 'Used for Tests'," *Sydney Morning Herald* (15 April, 2008), online: np.

[45] In flagrant contradiction of the findings of the government reports of 2004 and 2005, the *Herald Sun* columnist Andrew Bolt (a Stolen Generations denialist) vehemently rejected any possibility of Ms Mills' testimony being true, fulminating "It is a sign of our sickness that claims we performed medical experiments on 'stolen' Aboriginal children are now taken seriously. [...] Only years of extreme propaganda could have produced this disposition to believe something so bizarre." "Forgotten leprosy facts," *Herald Sun* (18 April 2008), online: np.

[46] Sean Parnell, "No evidence stolen generation were used as guinea pigs for leprosy drug," *The Australian* (5 February 2010), online: np.

[47] Evolution Research, *Enquiry into claims regarding leprosy testing on Aboriginal children in the Northern Territory between 1920 and 1960* (Glenelg, SA: Australian Government, 2010): 4. (My emphasis.) The extremely slim report repeatedly stresses the difficulties encountered by the research team due to "the lack of primary data," the "scarce" availability of archival information, the "untraceability" of "the relevant records" etc. wherever they sought to carry out enquiries. See pages 4, 5, 8, 10, 18, 20, 21, 22, 23, 24 of the report.

In contrast to the equivocations, provisionality, and caution that hedge the findings of the enquiry into claims regarding the testing of leprosy treatments on Aboriginal children, Professor Ernest Hunter, Chair of Public Health at the North Queensland Clinical School of the University of Queensland, has no qualms in affirming that syphilis treatments were trialled on Indigenous Australians in Western Australia. In an article exploring the historical background to "the ambivalence with which medical professionals are viewed by Aborigines [...] in the Kimberley region," Dr Hunter mentions 1911 as the "year the doctor on the island started to experiment on the Aborigines with arsphenamine (salvarsan), with poor results."[48] The island Hunter mentions is one of the "isolated and desolated locations" to which Aborigines were exiled if they showed symptoms of the disease. Commenting on the inhumane methods by which Chief Medical Officer and Protector of the Aborigines, Dr Herbert Basedow, had suspected syphilis sufferers tracked down and herded together for transportation to the island, an observer of the period recounts that Aborigines were examined "by brute force," forcibly removed from their camps "chained together by their necks and [...] marched through the bush."[49]

As Hunter notes, it is hardly surprising that medical science has such a bad reputation among Indigenous Australians. In addition to the callous treatment they meted out to Aboriginal syphilitics, doctors in Western Australia were, he asserts, involved more generally in "reinforcing racist theory, defining policies (both of isolation and assimilation) informed by those theories, planning and participating in harsh processes that deprived Aborigines (but not non Aborigines) of their liberty, and *exposing Aborigines as uninformed and non-consenting subjects to experimentation*."[50]

Removal and Retention of Human Tissue and Body Parts from Aboriginal Corpses

The removal and retention of human tissue and body parts from Aboriginal corpses without the consent of their next of kin was indicated to me by an eminent Indigenous academic who had had direct experience of this practice in her own family. In carrying out my own research on the subject, I have

[48] Ernest Hunter, "Stains on the Mantle," *Medical Journal of Australia* 155 (2–16 December 1991): 780. Further page references are in the main text.

[49] E.L. Grant Watson, quoted in Hunter, "Stains on the mantle," 781.

[50] Hunter, "Stains on the Mantle," 782. (My emphasis.)

discovered an immense amount of published information, of which I can give only the broadest of outlines here.

Public anxiety about this question first surfaced in the late-1990s following the setting-up of enquiries in the UK into organ-stripping at the Bristol Royal Infirmary (1999–2001) and the Alder Hey Children's Hospital (1999–2001). In March 2001, sensational allegations made during an Australian TV programme about the grotesque treatment to which dead bodies at Sydney's Glebe Morgue were being subjected[51] triggered a raft of government enquiries. Two months later, the Australian Society for Medical Research openly admitted that organs removed for investigation at autopsy were likely *not* to be inside the body returned to the family for burial. The then President-elect also acknowledged that tissue samples obtained from such autopsies were routinely retained "for specific use in medical research or in the teaching of pathology."[52]

The report, *Organs Retained at Autopsy*, published by the Australian Health Ethics Committee in August of the same year, confirmed that "Next-of-kin consent has not been and is not required for the conduct of coronial autopsies."[53] It further stated that the laws, as they stood at that time,

> also allow for tissues and organs removed for the purposes of autopsy, whether coronial or non-coronial, to be retained for other therapeutic, medical or scientific purposes without any consent having been sought or given by the deceased or the next-of-kin for that subsequent retention and use.[54]

Although it was by no means just Indigenous bodies that were subjected to the post-mortem practices referred to in the report, there are nonetheless many factors pointing to a disproportionately high percentage of Aboriginal families having been affected by the phenomenon. To mention the most obvious, much of the human tissue used in research is harvested during autopsies that

---

[51] See Tanya Nolan, "Alledged [sic] Illegal Behaviour at Sydney Morgue,"*AM*, ABC Local Radio (19 March 2001), online: np.

[52] Peter R. Schofield, "Letter to Mr Bret Walker re: Inquiry into matters arising from the post mortem and anatomical examination practices of the Institute of Forensic Medicine," *Australian Society for Medical Research* (9 May 2001): http://www.asmr.org.au/IFM.pdf (accessed 4 January 2011).

[53] National Health and Medical Research Council (NHMRC), *Organs Retained at Autopsy* (Canberra: Commonwealth of Australia, 2001): here 7.

[54] NHMRC, *Organs Retained at Autopsy*, 7.

are demanded in the course of coronial investigations. And, as Carpenter and Tait point out with regard to the Coroners Act 2003 (Queeensland),

> The purpose of the coronial investigation is to determine a cause of death finding for any death that is considered suspicious, violent or unnatural. The coronial system is also called upon to determine cause of death when the death is neither suspicious, violent, nor unnatural, but where a medical officer cannot, or will not, write a cause of death certificate. More recently, the coronial system has been required to investigate deaths in institutions like prisons.[55]

To judge by the criteria that are used to determine the need for an autopsy, it is clear that (due to the epidemic levels of Indigenous suicide, the high incidence of violent deaths, homicides, drug overdose, and road-accident fatalities,[56] the low incidence of Aboriginal deaths in a medical setting, and the extraordinarily high percentage of Aboriginal people in prison), there is bound to be an over-representation of Aboriginal bodies among those subjected to organ and tissue removal. It is also highly probable that the bodies of the many Aboriginal youngsters who have died in such great numbers in the institutions in which the Australian state has confined them (orphanages and industrial schools in the past, detention centres and prisons today) have been autopsied. As Carpenter and Tait observe, statistically, the difference between the number of Indigenous bodies that are autopsied and the number of autopsies carried out on non-Indigenous cadavers is "highly significant" (34).

A further particularly 'significant' aspect of the situation is that, contrary to European cultures in which autopsy is an accepted practice, such procedures are regarded as abhorrent and therefore proscribed in Aboriginal cultures. As Aboriginal Australians see it, the physically invasive investigations that forensic scientists carry out on dead bodies have "implications for [their] cultural beliefs about the journey of the spirit after death" (36).[57]

---

[55] Belinda Carpenter & Gordon Tait, "Health, Death and Indigenous Australians in the Coronial System," *Australian Aboriginal Studies* 2009.1 (September 2009): 29. Further page references are in the main text.

[56] For Indigenous Australians, "death by suicide is between two and 2.6 times higher, assault mortality rates are between five and ten times higher, and deaths due to motor vehicle accidents are 3.4 times higher" than the figures for non-Indigenous Australians. "Health, Death and Indigenous Australians in the Coronial System," 31.

[57] For an idea of the distress that autopsies cause in Aboriginal communities, see the extract from affidavit material presented in an appeal against the coronial autopsy of a

The *Organs Retained at Autopsy* report reflects the ethical unease engendered by this clash of cultural perspectives in the stress it places on the need for further recommendations on "matters that are sensitive to the distinctive concerns and needs of Aboriginal and Torres Strait Islander people,"[58] and its (albeit elliptical) emphasis on the "the special history of Aboriginal and Torres Strait Islander people in Australia [that] warrants separate attention to their concerns."[59]

For populations that have yet to recover from the widespread desecration of their ancestors' remains by European men of science in the nineteenth and early-twentieth centuries, the recent revelations about organs of deceased Indigenous Australians having been removed and retained without their families' consent constitutes one demonstration too many of Western researchers' flagrant disrespect for Aboriginal peoples and their mortuary customs. Given the grief and bitterness that many families have experienced after learning belatedly that the body of a loved one they buried in the past was, as a result of scientific intervention, lacking certain body parts, it is hardly surprising that Indigenous Australians are reluctant to collaborate with the scientific community.

Sterilization of Aboriginal Women and Girls Without Consent

During the course of the discussions I had with Indigenous Australians, several individuals referred to the fact that Aboriginal women and girls either had been or were being sterilized without their consent. Two women attested to first-hand knowledge of members of their own family having been subjected to abusive sterilization treatment, a third person spoke of an academic paper he had heard at a conference in Western Australia which had presented detailed evidence on the subject, and several other interlocutors mentioned having been told stories about sterilization, adding that such experiences are frequently reported by Aboriginal people.

At this stage in my own research I have not been able to locate any published full-length study of the compulsory sterilization of Aboriginal women in Australia; nor have I discovered any hard documentary evidence to suggest

---

child in Western Australia in Tony Wilson, "Body rites – Aspects of possession and control of the dead human body," *Australian Property Law Bulletin* 13.5 (1998): 49, quoted in "Health, Death and Indigenous Australians in the Coronial System," 30.

[58] NHMRC, *Organs Retained at Autopsy*, 8.

[59] NHMRC, *Organs Retained at Autopsy*, 9.

that the forced sterilization of Indigenous women and girls is or has been practised on the kind of scale that has been recorded in North America, for example.[60] What I have found, on the other hand, are innumerable references to instances of Aboriginal women and girls having been made infertile without their consent either i) as a result of surgical procedures or ii) through the abusive prescription of contraceptive drugs (used for the purpose of 'population control') to teenagers who are given little, if any, information on the risks entailed by agreeing to the injections or implants they have been prescribed.

Two highly respected Indigenous Australian academics who have spoken about the compulsory sterilization of Aboriginal Australian women are the historian Jackie Huggins, who observes in *Sister Girl* that, in the past, "where white women's demands to control their fertility were related to contraception and abortion, Aboriginal women were subject to unwanted sterilisation,"[61] and Lynette Russell, a professor of Indigenous Studies, who testifies, apropos of her research in Australian libraries and archives, that she has "encountered records where the forced sterilization of Aboriginal women is described and documented."[62]

Quoting from documents in Queensland's State Archives, the historian Rosalind Kidd reminds us that, in a communication to Governor Leslie Wilson on the subject of 'half-castes' in 1934, the Under-Secretary for the Home Department, William Gall, proposed that "governments, sooner or later, will have seriously to consider the question of sterilization [because] inferior races will have to go."[63]

In 1987, a Queensland nurse, Ms Gwen Deemal–Hall, claimed to have hospital records which proved that Aboriginal women in that state "had been

---

60 Sheila Collingwood–Whittick, "Indigenous Peoples and Western Science," in this volume.

61 Jackie Huggins, *Sister Girl* (St Lucia: U of Queensland P, 1998): 27. Rebecca Marie Kluchin makes the same comparison with regard to the USA: namely, that "White women of different socioeconomic classes struggled to obtain contraceptive sterilization while poor women, predominantly women of color, struggled to resist coercive sterilization." "Introduction" to *Fit to be Tied: Sterilization and Reproductive Rights in America, 1950–1980* (New Brunswick NJ: Rutgers UP, 2009): 8.

62 Lynette Russell, "Indigenous knowledge and archives: accessing hidden history and understandings," *Australian Academic & Research Libraries* 36.2 (June 2005): 173.

63 QSA A/8724, 13/8/34 quoted in Rosalind Kidd, *The Way We Civilise* (St Lucia: U of Queensland P, 1997): 137.

given tubal ligations following caesarian sections" without their permission.[64] Deemal–Hall also asserted at a conference in Townsville that teenage Aboriginal girls, some as young as thirteen, were being forced to have injections of Depo Provera (a drug known to have serious side-effects) as "part of the Queensland government's 'subtle policy' of keeping down the numbers of Aborigines."[65]

An article published by the sociology professor Sharyn L. Roach two years later lends strong support to the credibility of Deemal–Hall's testimony. Asserting that governments and the medical profession often "define fertility among the poor and minority segments of industrialized societies [...] as deviant," Roach argues that the perceived necessity of family-planning programmes for population control

> frequently involves coerced or induced sterilization or the administration of often harmful contraceptives [...]. Despite serious questions of safety and bans or restrictions in the United States, the United Kingdom, and Australia, Depo Provera is being administered widely in poor, urban, minority segments of the population, for example among the Asian and West Indian women in the United Kingdom, Polynesian women in Zealand and Aboriginal women in Australia.[66]

Not that Queensland stands alone in being accused of coercive sterilization. Joy Williams, an Aboriginal woman whose mother was subjected to an involuntary hysterectomy at the age of eighteen, is quoted by the investigative journalist John Pilger as saying that Western Australia was "notorious" for

---

[64] Jennifer Conley, "'Depo Provera given to Queensland blacks' says nurse," *The Age* (27 August 1987): 3. As well as being a nurse, Ms Deemal–Hall was a postgraduate student at James Cook University in Townsville, where she had been carrying out investigations into medical practices in the state.

[65] "Pharmaceutical 'genocide'," *New Scientist* 1576 (3 September 1987): 23.

[66] Sharyn L. Roach, "New Reproductive Technologies and Legal Reform," *Reproductive and Genetic Engineering* 2.1 (1989): 25. Roach's claims are echoed by many other critics of Depo Provera use, including the Canadian researcher Laura Shea, who points out that the controversial drug is "disproportionately prescribed to society's most marginalized and disadvantaged groups. [...] The patterns are telling; in the United Kingdom Depo is used most often by Asian and West Indian women, in Australia by Aboriginal women, and in New Zealand by Maori and Pacific Island women." "Reflections on Depo Provera: Contributions to Improving Drug Regulation in Canada" (Toronto: Women and Health Protection, 2007): 10–11.

such practices.[67] The senior Indigenous researcher Mary Ann Bin-Sallik recalls, of her experience as a young nurse in Darwin, Northern Territory in the 1960s, that "Aboriginal women were sterilized without their permission. Very few of these women could speak English and therefore did not understand what was happening to them,"[68] while the non-Indigenous academic Gisela Kaplan recalls that in Canberra, during the "Women and Politics Conference" held in 1975, Aboriginal women demanded that the government "Stop forced sterilisation on our black women in Australia." Like Deemal–Hall and Roach, Kaplan points out that "family planning was at times seen as a deliberate attempt to limit the size of the Aboriginal population."[69]

More recently, in a parliamentary debate in 2001, the New South Wales MP The Hon. P.J. Breen drew his colleagues' attention to the fact that "research conducted in Victoria during the past few years, indicates that State wards or children in care were victims of [...] sterilisation procedures."[70] Questioned about allegations made in 2003 by Arnhem Land women who claimed they had been sterilized without their consent in the 1970s, the Northern Territory Health Minister Jane Aagaard's lame response was that her Department "ha[d] not received any information" on the matter.[71] Several articles published in the mainstream Australian press five years later reported claims by opposition MPs in Queensland that Aboriginal girls as young as twelve had been given the long-lasting contraceptive implant Implanon.[72]

As the genocide specialist Colin Tatz implied in 1999, the sterilization issue won't go away:

---

[67] John Pilger, *A Secret Country: The Hidden Australia* (London: Vintage, 1990): 68–69.

[68] MaryAnn Bin-Sallek, "Beyond Expectations: From Nursing to Academia," in *In Our Own Right: Black Australian Nurses Stories*, ed. Sally Goold & Kerrynne Liddle (Sydney: eContent Management, 2005): 31.

[69] Gisela Kaplan, *The Meagre Harvest: The Australian Women's Movement, 1950s–1990s* (St Leonards, NSW: Allen & Unwin, 1996): 146.

[70] P.J. Breen, "Human Tissue Retention," Hansard & Papers, Legislative Council, Parliament of New South Wales (27 March 2001), http://www.parliament.nsw.gov.au/Prod/Parlment/HansArt.nsf/V3Key/LC20010327049 (accessed 4 January 2011).

[71] "NT government denies sterilisation claims," *ABC News* (5 December 2003), online: np.

[72] See, for example, Tim Dick, "Girls aged 12 'temporarily' sterilised," *Sydney Morning Herald* (16 April 2008), online: np.

> From time to time allegations surface that state medical services engage in unilateral temporary or permanent sterilisation of women: in Western Australia, the use of Depo-Provera, producing three-to-six month infertility, and in Queensland, the "non-explained" tubal ligations. [...] Depo-Provera, by injection, has alarming side effects, dire warnings about contra-indications and the need for stringent physical examination before administration.[73]

More than a decade later, not only does Tatz's conclusion that "the whole matter of unilateral sterilisation needs careful research,"[74] remain equally valid, but the same could be said of the other scientific practices evoked above – namely, medical experiments carried out on Indigenous Australian adults and children, and the removal (in disproportionate numbers) of tissue and organs from Indigenous bodies in contempt of Aboriginal spiritual beliefs and in the absence of authorization from the dead person's kin.

One further concern voiced by many of those I spoke with was that DNA samples gathered by Genographic researchers might find their way onto DNA databases and thereafter be used by the police for racial profiling.[75] That there is real justification for concern about the amount of Aboriginal genetic material being collected by police for storage on population databases is amply confirmed by Greg Gardiner's critical analysis of "the new DNA forensic procedures regime [...] following the passage of the Victorian *Crimes (DNA Database) Act*, 2002 (VIC)."[76]

Just as in the UK (which has the world's largest national DNA database and where in 2009 the police were accused by the Human Genetics Commission of arresting people – particularly young black men – for the specific

---

[73] Colin Tatz, *Genocide in Australia* (Research Discussion Paper 8; Canberra: Australian Institute of Aboriginal and Torres Strait Islander Studies–AIATSIS, 1999): 1–50, fn 73.

[74] Tatz, "Genocide in Australia," fn 73.

[75] The assurances given by the organizers of the project – "All samples collected will be held under strict conditions maintaining confidentiality and may not be used for any purpose inconsistent with the strictly limited scientific objectives of the project" ("Ethical Framework," *The Genographic Project*, National Geographic Website) – were regarded with the utmost scepticism by most of my interlocutors.

[76] Greg Gardiner, "'Racial Profiling': DNA Forensic Procedures and Indigenous People in Victoria," *Current Issues in Criminal Justice* 17.1 (July 2005): 47. Further page references are in the main text.

purpose of expanding their DNA files),[77] young Indigenous Australians are, Gardiner points out, routinely subjected to DNA tests whenever they are arrested – as they are in hugely disproportionate numbers "in every jurisdiction in Australia" (47). One of the immediate problems arising from this situation is that,

> because the report of a match between a crime scene profile and a suspect or offender profile does not provide conclusive proof of identity, any individual whose profile exists on a database bears a greater risk of misidentification by chance than those not represented on the database. (51)

There is, however, a problem of a different order raised by the mass, coercive collection (and permanent retention) of Aboriginal DNA – an issue I shall return to in my concluding remarks.

Conclusion

Before the discussions I had with Indigenous Australians' in 2009, I had hypothesized that Aboriginal resistance to the Genographic project was probably attributable to collective memories of scientific racism – a phenomenon of which so many of Australia's Aboriginal peoples were victims in the nineteenth and early-twentieth centuries. While the discussions confirmed that hypothesis, they equally indicated that Aboriginal Australian attitudes to science have also been shaped by a much more recent experience of abusive scientific praxis.

One further surprising result that emerged as I followed up information I was given during discussions is that, owing to the control Western scientists have consistently exercised over Indigenous Australian bodies, there already exists a plethora of sources from which Aboriginal DNA could be extracted if scientists failed to obtain it in a more consensual manner. First, and notwithstanding Aboriginal demands for repatriation, there are thousands of 'osteological specimens' that still remain in the possession of Western museums and universities. Secondly, at least thirty-seven genetic research projects have already been carried out in Aboriginal communities since the 1960s.[78] Thirdly, while Australia does not as yet possess a comprehensive Human Genetic

---

[77] Alan Travis, "Police routinely arresting people to get DNA, inquiry claims," *The Guardian* (24 November 2009), online.

[78] This figure was given to me by Emma Kowal in a personal communication.

Research Database (HGRD), what it does have, according to David Weisbrot of the Australian Law Reform Commission, is "a large number of smaller or unsystematized HGRDs."[79] It is no doubt databanks such as these that provide the supply of human tissue products which, according to a draft document published in 2009 by the National Health and Medical Research Council, is "a well-established practice amongst profit and not for profit agencies in Australia."[80] There is also, Weisbrot reports,

> a major *inchoate* national HGRD in Australia, in the form of vast numbers of neonatal blood spot screening cards ('Guthrie cards'), stored in children's hospitals around the country (with varying degrees of care) – representing a complete, if unorganised, DNA collection of virtually everyone born in Australia in the past 50 years.[81]

Finally, although the concerns of the Indigenous Australians with whom I spoke were mainly focused on the possibility of DNA samples collected by genetic researchers being handed over to the police, what Gardiner points to is the likelihood of DNA samples collected in the course of a criminal inquiry transiting in the opposite direction: i.e. *from* databases consisting of 'forensic samples' *towards*, for example, genetics researchers.

Underlining the over-representation of Indigenous Australians in the criminal-justice system, the fact that an Indigenous person is many times more likely to be DNA-tested than a non-Indigenous person in Australia, and the resulting situation of "Indigenous over-representation on the state's DNA database system" (a situation that "will be replicated in the state's DNA sample storage facility"), Gardiner warns:

> The capacity of future governments to utilize, turn-over or 'on-sell' genetic material to researchers, corporations, institutions, and other government agencies, or to re-use such material for other purposes not previously stated, is only limited by their capacity to amend legisla-

---

[79] Maintained, he points out, by "universities, research centres and biotechnology companies; public and private hospitals (eg tissue banks, blood banks, pathology samples, paraffin blocks); pathology labs and familial cancer registers." "Strong Public Engagement and Public Policy: Critical Underpinnings for Human Genetic Research," in *Human Biology and Public Trust: Trends, Perception and Regulation*, ed. Mark Stranger (Hobart, Tasmania: Centre for Law and Genetics, 2007): 229.

[80] NHMRC, "Ethics and the Exchange, Sale of and Profit From Products Derived From Human Tissue: An Issues Paper" (April 2009): 20.

[81] Weisbrot, "Strong Public Engagement and Public Policy," 229.

> tion. [...] it's clear that the provisions that govern the collection and use of genetic material under the [Crimes] Act [1958 (Vic)] can be easily altered at a future date. (54)

Possibly hastening the realization of the scenario hypothesized by Gardiner are the concerns increasingly voiced by experts in the field about the number of false DNA database matches resulting from the cross-contamination of samples and other sundry laboratory errors.[82] In 2009, for instance, *Science* published a letter, signed by thirty-nine leading scientists, demanding access to CODIS, America's national DNA database because, they claimed,

> Analysis of NDIS [National DNA Index System] can [...] yield valuable insights into the frequency and circumstances under which certain typing errors may occur. A review of a government database from Victoria, Australia, containing 15,021 9-locus STR profiles shows how important such a review can be for "quality control purposes" [...]. The study found an error rate of about 1 in 300 for the typing of reference samples.[83]

Now, it is quite clear that more careful screening of DNA forensic procedures for the purpose of eliminating errors that "'lead to false identifications' of suspects,"[84] must be in the interests of Aboriginal Australians. The irony is, however, that the signatories to the letter also emphasize the following:

> Open access to data is a fundamental tenet of science. The need for openness was reinforced by the recent National Research Council report, which decried the insularity of forensic science and called for greater involvement of the academic community in assessment, validation, and improvement of forensic science methods [...]. Law enforcement should honor the norms of science and open the NDIS and other government DNA databases to independent scientific scrutiny.[85]

And this argument, if Australian state governments were to respond to it, could lead to Indigenous DNA being made freely available to *all* scientific

---

[82] See, for example, Maura Dolan & Jason Felch, "The Peril of DNA: It's not perfect," *L.A. Times* (26 December 2008), online: np.

[83] Dan E. Krane et al., "Time for DNA Disclosure," *Science* 326 (18 December 2009), Letters: 1631.

[84] Dolan & Felch, "The Peril of DNA."

[85] Krane et al., "Time for DNA Disclosure," 1632.

researchers – including those working on human genome diversity and the migratory routes of Indigenous populations.

Given the multiple opportunities, both real and potential, that geneticists have for procuring Aboriginal DNA without having to go to the trouble of obtaining consent, and bearing in mind the many precedents of scientific researchers' riding roughshod over ethical considerations once their 'sense of scientific curiosity' has been aroused,[86] perhaps the burning question as far as Indigenous Australians are concerned is not *why* they should agree to collaborate on the Genographic project but *what chance they actually have* of preventing biomapping geneticists from achieving their objectives.

WORKS CITED

Aberdeen, Lucinda. "Australian Scientific Research, 'Aboriginal Blood' and the Racial Imaginary," in *'A Race for a Place': Eugenics, Darwinism and Social Thought and Practice in Australia*, ed. Martin Crotty et al. (Callaghan, NSW: U of Newcastle P, 2000): 101–10.

AAP. "Stolen Generation Kids 'Used for Tests'," *Sydney Morning Herald* (15 April, 2008), http://www.smh.com.au/articles/2008/04/15/1208025169822.html (accessed 4 January 2011).

Anon. "The Genome Gold Rush," *Business Week* (12 June 2000), http://www .business week.com/datedtoc/2000/0024.htm (accessed 4 January 2011).

Anon. "NT government denies sterilisation claims," *ABC News* (5 December 2003), http://www.abc.net.au/news/stories/2003/12/05/1004386.htm (accessed 4 January 2011).

Anon. "Pharmaceutical 'genocide'," *New Scientist* 1576 (3 September 1987): 23.

Baker, Martin. "Curiosity drives the gene genie to a £1m turnover," *The Telegraph* (15 February 2008), http:www.telegraph.co.uk/finance/2784261/Curiosity-drives-the-gene-genie-to-a-1m-turnover.html (accessed 4 January 2011).

Bin-Sallek, MaryAnn. "Beyond Expectations: From Nursing to Academia," in *In Our Own Right: Black Australian Nurses Stories*, ed. Sally Goold & Kerrynne Liddle (Sydney: eContent Management, 2005): 23–35.

Black, Edwin. *IBM and the Holocaust: The Strategic Alliance Between Nazi Germany and America's Most Powerful Corporation* (New York: Crown, 2001).

Bolt, Andrew. "Forgotten leprosy facts," *Herald Sun* (18 April 2008), http://www .heraldsun.com.au/opinion-old/forgotten-leprosy-facts/story-e6frfifx-111116093117 (accessed 4 January 2011).

---

[86] Sheila Collingwood–Whittick, "Indigenous Peoples and Western Science," in this volume.

Breen, P.J. "Human Tissues Act," *Legislative Council: Parliament of New South Wales*, Hansard Transcript (27 March 2001), http://www.parliament.nsw.gov.au/Prod/Parlment/HansArt.nsf/V3Key/LC20010327049 (accessed 4 January 2011).

Carlesen, William. "Rogue virus in the vaccine: Early polio vaccine harbored virus now feared to cause cancer in humans," *San Francisco Chronicle* (15 July 2001), http://articles.sfgate.com/2001-07-15/news/17607195_1_sv40-human-tumors-polio-vaccine/2 (accessed 4 January 2011).

Carpenter, Belinda, & Gordon Tait. "Health, Death and Indigenous Australians in the Coronial System," *Australian Aboriginal Studies* 2009.1 (September 2009): 29–41.

Colley, Sarah. *Uncovering Australia: Archaeology, Indigenous People and the Public* (Crows Nest, NSW: Allen & Unwin, 2002).

Collingwood–Whittick, Sheila. "Scientific Discourse and the West's Construction of its 'Others'," *Middle Ground: Journal of Literary and Cultural Encounters* 1 (2007): 44–54.

——. "Skeletons in the Cupboard: Imperial Science and the collection and museum-ization of Indigenous remains," in *Science and Empire in the Nineteenth Century: A Journey of Imperial Conquest and Scientific Progress* ed. Catherine Delmas et al. (Newcastle upon Tyne: Cambridge Scholars, 2010): 65–82.

Conley, Jennifer. "'Depo Provera given to Queensland blacks' says nurse," *The Age* (27 August 1987): 3.

Dick, Tim. "Girls aged 12 'temporarily' sterilised,'" *Sydney Morning Herald* (April 16 2008), http://smh.com.au/news/national/girls-aged-12-temporarily-sterilised/2008/04/15/1208025189880.html (accessed 4 January 2011).

Dodson, Michael. "The protection of genetic information of Indigenous peoples: Submission to the Australian Law Reform Commission inquiry into the protection of genetic information," *Australian Human Rights Commission* (13 May 2002), http://www.hreoc.gov.au/legal/submissions/genetic_information.html (accessed 4 January 2011).

Dodson, Patrick. "Whatever happened to reconciliation?" in *Coercive Reconciliation: Stabilise, Normalise, Exit Aboriginal Australia*, ed. Jon Altman & Melinda Hinkson (Melbourne: Arena, 2007): 21–29.

Dolan, Maura, & Jason Felch. "The Peril of DNA: It's not perfect," *L.A. Times* (26 December 2008): http://articles.latimes.com/print/2008/dec/26/local/me-dna26 (accessed 4 January 2011).

Evolution Research, *Enquiry into claims regarding leprosy testing on Aboriginal children in the Northern Territory between 1920 and 1960* (Glenelg, SA, 2010): 1–25, http://www.health.gov.au/internet/main/publishing.nsf/Content/D19F17E110 F01 074CA2576BF0013 E6DE/$file/leprosy_testing.pdf (accessed 4 January 2011).

Festa, Paul. "Probing IBM's Nazi Connection," *CNET News.com* (28 June 2001), http://news.cnet.com/2009-1082-269157.html (accessed 4 January 2011).

Fisher, Lawrence M. "Mining the Genome: Big Science as Big Business," *New York Times* (30 January 1994), http://www.nytimes.com/1994/01/30/us/mining-genome-big-science-big-business-special-report-profits-ethics-clash.html?pagewanted=all (accessed 4 January 2011).

Fredericks, Bronwyn. "Making an impact researching with Australian Aboriginal and Torres Strait Islander peoples," *Studies in Learning, Evaluation Innovation and Development* 5.1 (April 2008): 24–33, http://sleid.cqu.edu.au (accessed 4 January 2011).

Furniss, Charles. "Blood feud," *Geographical* 78 (September 2006), http://www.geographical.co.uk/Magazine/Dossiers/The_Genographic_Project_-_September_2006.html (accessed 4 January 2011): 43–53.

Gardiner, Greg. "'Racial Profiling': DNA Forensic Procedures and Indigenous People in Victoria," *Current Issues in Criminal Justice* 17.1 (July 2005): 47–67.

Green, Jim. "Radioactive racism in Australia," *Friends of the Earth Australia*, http://www.foe.org.au/resources/chain-reaction/editions/96/radracism (accessed 4 January 2011).

Harry, Debra, & Le`a Malia Kanehe. "Genetic Research: Collecting Blood to Preserve Culture?" *Cultural Survival Quarterly* 29.4 (Winter 2005), http://www.culturalsurvival.org/publications/cultural-survival-quarterly/none//article/genetic-research-collecting-blood-preserve-culture (accessed 4 January 2011).

Howard, Ken. "The Bioinformatics Gold Rush," *Scientific American* 283.1 (July 2000): 58–63.

Hubert, Jane et al. *Human remains: objects to study or ancestors to bury?* Transcript of a debate held at the Royal College of Physicians (2 May 2003): 1–19, http: //www.instituteofideas.com/transcripts/human_remains.pdf (accessed 4 January 2011).

Huggins, Jackie. *Sister Girl* (St Lucia: U of Queensland P, 1998).

Hunter, Ernest. "'The Deep Sleep of Forgetfulness': Reflecting on Disremembering," Presentation to the Third World Conference for the International Society for Traumatic Stress Studies," Melbourne (17 March 2000): 1–21, http://www.iap.org.au/ehunter.pdf (accessed 4 January 2011).

——. "Stains on the Mantle: Doctors in Aboriginal Australia have a history," *The Medical Journal of Australia* 155 (2–16 December 1991): 779–83.

International Business Machines (IBM). *The Genographic Project*, http://www.ibm.com/solutions/genographic/us/en (accessed 4 January 2011).

Kaplan, Gisela. *The Meagre Harvest: The Australian Women's Movement, 1950s–1990s* (St Leonards, NSW: Allen & Unwin, 1996).

Kidd, Rosalind. *The Way We Civilise* (St Lucia: U of Queensland P, 1997).

Kluchin, Rebecca Marie. *Fit to be Tied: Sterilization and Reproductive Rights in America, 1950–1980* (New Brunswick NJ: Rutgers UP, 2009).

Krane, Dan E. et al. "Time for DNA Disclosure," *Science* 326 (18 December 2009): 1631–32, www.sciencemag.org (accessed 4 January 2011).

McNiven, Ian J., & Lynette Russell. *Appropriated Pasts: Indigenous Peoples and the Colonial Culture of Archaeology* (New York, Toronto & Plymouth: AltaMira, 2005).

Marshall, Eliot. "Gene Prospecting in Remote Populations," *Science* 278/5338 (24 October 1997): 565.

Mohamed, Justin. "We Don't Like Research – But in Koori Hands It Could Make a Difference" (Melbourne: VicHealth Koori Health Research and Community Development Unit & University of Melbourne, 2000): 1–32, http://www.limenetwork.net.au/files/lime/CR1-WeDontLike%20Research.pdf (accessed 4 January 2011).

Murray, John. "From the Cradle to the Grave," *Origins Harp Healing and Recovery Project for Forgotten Australians* (14 August 2007), http://www.originsharp.com/experiment.html (accessed 4 January 2011).

The National Geographic. *The Genographic Project*, https://genographic.nationalgeographic.com/genographic/lan/en/index.html (accessed 4 January 2011).

National Health and Medical Research Council. *Organs Retained at Autopsy* (Canberra: Commonwealth of Australia, 2001): 1–31, http://www.nhmrc.gov.au/_files_nhmrc/file/publications/synopses/e41.pdf (accessed 4 January 2011).

National Health and Medical Research Council. "Ethics and the exchange, sale of and profit from products derived from human tissue: An Issues Paper" (April 2009): 1–83, http://www.nhmrc.gov.au/_files_nhmrc/file/health_ethics/ahec/public_consultation_draft.pdf (accessed 4 January 2011).

Nolan, Tanya. "Alledged [sic] Illegal Behaviour at Sydney Morgue," *AM*, ABC Local Radio (19 March 2001), http://www.abc.net.au/am/stories/s262339.htm (accessed 4 January 2011).

Parnell, Sean. "No evidence stolen generation were used as guinea pigs for leprosy drug," *The Australian* (5 February 2010), http://www.theaustralian.com.au/news/nation/no-evidence-stolen-generation-were-used-as-guinea-pigs-for-leprosy-drug/story-e6frg6nf-1225827046813 (accessed 4 January 2011).

Pilger, John. "Breaking the great Australian silence," City of Sydney peace prize lecture (5 November 2009), http://www.johnpilger.com/articles/breaking-the-great-australian-silence (accessed 4 January 2011).

——. *A Secret Country: the Hidden Australia* (London: Vintage, 1990).

Rimmer, Matthew. "The genographic project: traditional knowledge and population genetics," *Australian Indigenous Law Reporter* 11.2 (2007): 33–54.

Roach, Sharyn L. "New Reproductive Technologies and Legal Reform," *Reproductive and Genetic Engineering* 2.1 (1989): 11–27.

Russell, Lynette. "Indigenous knowledge and archives: accessing hidden history and understandings," *Australian Academic & Research Libraries* 36.2 (June 2005): 169–80.

Sanson–Fisher Robert W. et al. "Indigenous health research: a critical review of outputs over time," *Medical Journal of Australia* 184.10 (2006): 502–505. http:

//www.mja.com.au/public/issues/184_10_150506/san10764_fm.html (accessed 4 January 2011).

Schofield, Peter R. "Letter to Mr Bret Walker re: Inquiry into matters arising from the post mortem and anatomical examination practices of the Institute of Forensic Medicine," *Australian Society for Medical Research* (9 May 2001), http://www.asmr.org.au/IFM.pdf (accessed 4 January 2011).

Senate Community Affairs References Committee. *Forgotten Australians: A Report on Australians who experienced institutional or out-of-home care as children* (Canberra: Commonwealth of Australia, 2004): 1–410, http://www.cyf.vic.gov.au/__data/assets/pdf_file/0006/16728/forgotten_australians_report.pdf (accessed 4 January 2011).

Senate Community Affairs References Committee. *Protecting Vulnerable Children: A National Challenge, Second report on the inquiry into children in institutional or out-of-home care* (Canberra: Commonwealth of Australia 2005): 1–221, hhttp://www.aph.gov.au/senate/committee/clac_ctte/completed_inquiries/2004-07/inst_care/report2/c01.pdf (accessed 4 January 2011).

Shea, Laura. "Reflections on Depo Provera: Contributions to Improving Drug Regulation in Canada" (Toronto: Women and Health Protection, October 2007), 1-22, http://www.whp-apsf.ca/pdf/reflectionsOnDepoProvera.pdf (accessed 4 January 2011).

Tatz, Colin. "Genocide in Australia" (Research Discussion Paper 8; Canberra: Australian Institute of Aboriginal and Torres Strait Islander Studies, AIATSIS, 1999): 1–50, http://www.aiatsis.gov.au/research/docs/dp/DP08.pdf (accessed 4 January 2011).

Travis, Alan. "Police routinely arresting people to get DNA, inquiry claims," *The Guardian* (24 November, 2009): http://www.guardian.co.uk/politics/2009/nov/24/dna-database-inquiry (accessed 4 January 2011).

Tuhiwai Smith, Linda. *Decolonizing Methodologies: Research and Indigenous Peoples* (London & New York: Zed, 1999).

University of Adelaide. "Apology for Past Experiments on Aboriginal People," *News and Events* (8 February 2002), http://www.adelaide.edu.au/news/news314.html (accessed 4 January 2011).

Weisbrot, David. "Strong Public Engagement and Public Policy: Critical Underpinnings for Human Genetic Research," in *Human Biology and Public Trust: Trends, Perception and Regulation*, ed. Mark Stranger (Hobart: Centre for Law and Genetics, 2007): 219–36.

⌘

Disturbing Pasts and Promising Futures

The Politics of Indigenous Genetic Research in Australia[1]

EMMA KOWAL

IN A WORLD WHERE KNOWLEDGE ABOUT HUMAN GENETIC VARIATION is increasingly circulated, Indigenous Australians are prominent dissenters. Since refusing to participate in the Human Genome Diversity Project in the 1990s, Indigenous Australians have largely continued to eschew genetic research and are rarely included in surveys of 'worldwide' genetic variation.[2] Within the scientific literature, only thirty-seven genetic studies of any kind in Indigenous communities have been published since 1979.[3] As a comparison, a search of the US National Library of Medicine (PubMed)

---

[1] Research for this article was supported by a National Health and Medical Research Council Aboriginal and Torres Strait Islander Training Fellowship (Grant no. 454813).

[2] See, for example, Jun Z. Li, Devin M. Absher, Hua Tang et al., "Worldwide Human Relationships Inferred from Genome-Wide Patterns of Variation," *Science* 319.5866 (2008): 1100–104, and Mattias Jakobsson, Sonja W. Scholz, Paul Scheet et al., "Genotype, Haplotype and Copy-Number Variation in Worldwide Human Populations," *Nature* 451/7181 (2008): 998–1003. Note that 'Oceania' is often included in such articles, but they will almost always consist of samples from the Pacific (and predominantly Melanesia) which is seen as a far more 'hospitable' site for collecting DNA. See Aroha Te Pareake Mead & Steven Ratuva, *Pacific Genes and Life Patents: Pacific Experiences and Analysis of the Commodification and Ownership of Life* (Tokyo: Call of the Earth / Llamado de la Tierra, United Nations University–Institute of Advanced Studies, 2007).

[3] Emma Kowal, Glenn Pearson, Lobna Rouhani et al., "Genetic Research and Aboriginal and Torres Strait Islander Australians" (MS under review).

database for the keywords 'Native Americans' and 'genetics' yields 1,642 studies published since 1963. Given this discrepency, it is not surprising that a recent review of public-health genomics in Australia concluded that "very little is known about the specific genetic issues relevant to Indigenous Australians."[4]

Why have Indigenous Australians resisted genetic research, and resisted it so successfully? Some of the reasons relate to risks that apply to all participants in genetic research (such as fears of implications for insurance). Other factors apply to Indigenous people the world over, such as the potential for population genetics to contradict cultural beliefs about human origins. Still other factors I will explore in this essay are specific to the Australian context.

In attempting to explain why Indigenous Australians are largely absent from the global stock of genetic knowledge, I will provide an overview of the politics of Indigenous genetic research in Australia, a context about which very little has been written. I will draw on published literature and media coverage of Indigenous genetics in Australia and beyond, as well as my ongoing ethnographic research with geneticists who work in Indigenous communities.[5] To begin with, however, I briefly outline some characteristics of the Indigenous population of Australia, the health status of Indigenous Australians, and the efforts that have been made to improve Indigenous health through effective and culturally appropriate research.

I Indigenous Australians and Indigenous Health Research

Indigenous Australians are those who are descended from the precolonial inhabitants of the Australian continent and associated islands. At the most recent census in 2006, Indigenous Australians numbered 517,000 people, amounting to approximately 2.5% of the total population.[6] It is important to appreciate the diversity of the Indigenous population of Australia. Before

---

[4] Sylvia A. Metcalfe, Alan H. Bittles, Peter C. O'Leary et al., "Australia: Public Health Genomics," *Public Health Genomics* 12.2 (October 2009): 121–28.

[5] Note that the question of whether this rejection of genetic research by Indigenous Australians is beneficial or detrimental is beyond the scope of this essay.

[6] Indigenous Australians are also known as Aboriginal and Torres Strait Islander peoples. 'Aboriginal' refers to all mainland and Tasmanian Indigenous people, with Torres Strait Islanders comprising a separate group. In 2006, 90% of Indigenous Australians identified as Aboriginal only, 6% as Torres Strait Islander only, and 4% as both Aboriginal and Torres Strait Islander.

European colonization, there were between 350 and 750 distinct social groupings in Australia, each with a distinct language.[7] Today, only 12% of Indigenous people speak an Indigenous language at home, and only twenty Indigenous languages are widely spoken (including many creoles). The remainder of the Indigenous population speaks Standard and/or Aboriginal English as their first language.[8] The contemporary Indigenous population is also diverse geographically, with 32% living in major cities, 43% living in regional centres, and 26% living in remote areas. While the definition of indigeneity in Australia does not refer to blood quantums, sustained high rates of intermarriage, as well as sexual violence perpetrated by white men on Aboriginal women, mean that most Indigenous Australians have both Indigenous and non-Indigenous ancestors.

Overall, the Indigenous population is younger, less employed, less educated, and has higher mortality rates than the general population. An excess of chronic diseases (such as heart and kidney disease and diabetes) and injuries means that there is a substantial life-expectancy gap.[9] At birth, an Indigenous male baby can expect to live 11.5 years fewer than a non-Indigenous child, and the figure is 9.7 years for females.[10] The reasons behind the poor health of Indigenous Australians are complex and multifactorial. The proximal reasons include poor housing, poor nutrition, high rates of smoking and substance use, and discrimination especially within health care. The more 'upstream' causes include the social determinants of health such as poverty, poor education, and high rates of unemployment. Colonization is seen by many as the ultimate cause of poor Indigenous health status.[11]

---

[7] Michael Walsh, "Overview of Indigenous Languages of Australia," in *Language in Australia*, ed. Suzanne Romaine (Cambridge: Cambridge UP, 1991): 27–48.

[8] Australian Institute of Health and Welfare and Australian Bureau of Statistics, *The Health and Welfare of Australia's Aboriginal and Torres Strait Islander Peoples, 2008* (Canberra: Australian Bureau of Statistics, 2008).

[9] Australian Institute of Health and Welfare and Australian Bureau of Statistics, *The Health and Welfare of Australia's Aboriginal and Torres Strait Islander Peoples, 2008*.

[10] Australian Bureau of Statistics, *Experimental Life Tables for Aboriginal and Torres Strait Islander Australians, 2005–2007* (Canberra: Australian Bureau of Statistics, 2009).

[11] Ian Ring & Ngaire Brown, "The Health Status of Indigenous Peoples and Others: The Gap Is Narrowing in the United States, Canada, and New Zealand, but a Lot More Is Needed," *British Medical Journal* 327/7412 (August 2003): 404–405; Chris

One way in which Australian governments have attempted to address Indigenous health disparities is by funding health research. Funds for research into Indigenous health problems increased throughout the 1990s. In 2009, the major funder of health research in Australia, the National Health and Medical Research Council (NHMRC), committed to spending at least 5% of its research funds on Indigenous health research projects.[12]

Indigenous people have long stressed that research in their communities can only be beneficial to them if it is conducted in a culturally appropriate way and with full Indigenous participation.[13] One way in which the NHMRC has responded to these concerns is by establishing Indigenous ethics guidelines. In 1991, the first guidelines for the conduct of research in Indigenous communities was published.[14] This was followed in 2003 with a revision entitled *Values and Ethics*.[15] The latter manual is structured around six core Aboriginal and Torres Strait Islander values relevant to health-research ethics: reciprocity, respect, equality, survival and protection, responsibility, and spirit and integrity (see Figure 1).

Cunningham & Fiona Stanley, "Indigenous by Definition, Experience, or World View," *British Medical Journal* 327/7412 (August 2003): 403–404.

[12] NHMRC, *Road Map II: A Strategic Framework for Improving the Health of Aboriginal and Torres Strait Islander People through Research* (Canberra: National Health and Medical Research Council, 2010).

[13] Justin Mohamed, "We Don't Like Research – But in Koori Hands It Could Make a Difference" (Melbourne: VicHealth Koori Health Research and Community Development Unit & University of Melbourne, 2000); Kim Humphery, "Dirty Questions: Indigenous Health and 'Western Research'," *Australian and New Zealand Journal of Public Health* 25.3 (June 2001): 197–202; John Henry, Terry Dunbar, Allan Arnott et al., *Indigenous Research Reform Agenda: Rethinking Research Methodologies* (Darwin, NT: Cooperative Research Centre for Aboriginal and Tropical Health, 2002).

[14] Kim Humphery, "The Development of the National Health and Medical Research Council Guidelines on Ethical Matters in Aboriginal and Torres Strait Islander Health Research: A Brief Documentary and Oral History," Discussion Paper Number 8 (Melbourne: VicHealth Koori Health Research and Community Development Unit, 2002); NHMRC, *Guidelines on Ethical Matters in Aboriginal and Torres Strait Islander Health Research* (Canberra: National Health and Medical Research Council, 1991).

[15] NHMRC, *Values and Ethics: Guidelines for Conduct of Aboriginal and Torres Strait Islander Health Research* (Canberra: National Health and Medical Research Council, 2003).

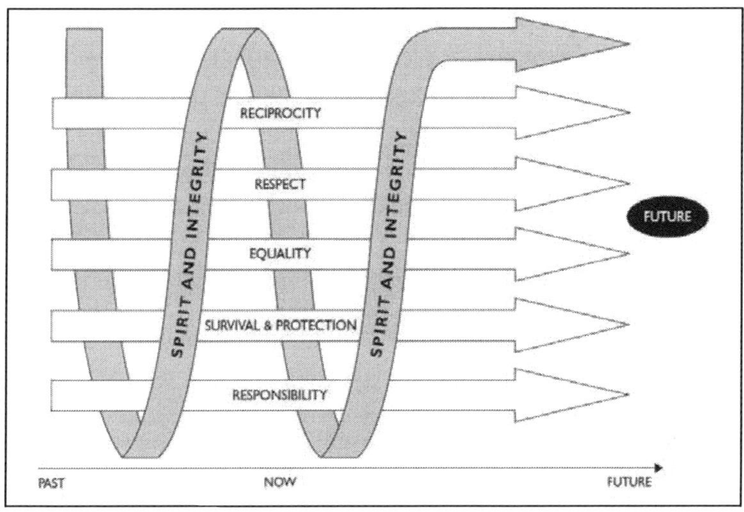

FIGURE 1. Aboriginal and Torres Strait Islander values relevant to health research ethics[16]

A powerful voice in the Indigenous health sector is that of the health centres based in Aboriginal communities and governed by local Aboriginal boards. Their peak body, the National Aboriginal Community Controlled Health Organisation (NACCHO), represents around 150 Aboriginal community-controlled health services across the country.[17] Another set of stakeholders in Indigenous health research are the Aboriginal ethical committees. In most states in Australia, in addition to ethics committees within universities, hospitals and other institutions there are special ethics committees which are made up solely of Indigenous members. Health researchers working in Indigenous communities are not obliged to submit their ethics applications to

[16] Source: NHMRC, *Values and Ethics: Guidelines for Conduct of Aboriginal and Torres Strait Islander Health Research* (Canberra: National Health and Medical Research Council, 1991): 9.

[17] See http://www.naccho.org.au. With regard to the above discussion, note that NACCHO does not endorse the Values and Ethics guidelines NHMRC, *Values and Ethics: Guidelines for Conduct of Aboriginal and Torres Strait Islander Health Research*, considering that "they are incomplete and fail to provide sufficient guidance to researchers on how ethical research should be conducted" (http://www.naccho.org.au/activities/ethical/html)

these committees, but, increasingly, 'mainstream' ethics committees are referring Indigenous health-research ethics applications to the Aboriginal ethics sector. In general, researchers consider that Aboriginal ethics committees require higher ethical standards than mainstream committees, particularly in terms of demonstrating direct research benefits to Indigenous communities. Some Aboriginal committees require that all publications from a research project are reviewed by the committee, a practice which is not standard in health-research ethics.[18]

II Explaining the Dearth of Genetic Research in Australia

The reasons why so little genetic research has occurred in Indigenous Australia include: ethical issues; cultural issues; legal issues; issues related to the technical requirements of genetics; and issues related to the public dissemination of results. I will consider these issues in turn, drawing on the Australian experience and also on Indigenous genetic-research 'scandals' that have occurred in other Indigenous populations.

The one factor that has had the most direct effect on Indigenous genetic research in the past few decades is the Human Genome Diversity Project (HGDP). This project was envisaged as the companion to the Human Genome Project which carried out its sequencing in 2000. The HGDP sought to collect DNA from Indigenous groups in order to understand the diversity of human species. It created much controversy both in Australia and internationally by calling Indigenous groups "Isolates of Historical Interest" that had to be sampled before they "vanished."[19] Indigenous advocates have stressed the links between researchers' desire for Indigenous DNA and the rush to

[18] For example, the ethics committee of the Aboriginal Health and Medical Research Council requires that "A final draft report must be provided to the AH&MRC Ethics Committee to be reviewed for compliance with ethical and cultural criteria." See http://www.ahmrc.org.au/Ethics%20and%20Research.htm

[19] Michael Dodson & Robert Williamson, "Indigenous Peoples and the Morality of the Human Genome Diversity Project," *Journal of Medical Ethics* 25.2 (April 1999): 204–208; Jenny Reardon, *Race to the Finish: Identity and Governance in an Age of Genomics* (Princeton NJ: Princeton UP, 2005); Jonathan M. Marks, "Human Genome Diversity Project (HGDP): Impact on Indigenous Communities," in *Living with the Genome: Ethical and Social Aspects of Human Genetics*, ed. Angus Clarke & Flo Ticehurst (London: Palgrave Macmillan, 2006): 49–55.

exploit indigenous lands in the colonial era.[20] The current Genographic project has also been resisted by Indigenous groups.[21]

The Australian reaction to the HGDP provides an excellent summary of Indigenous objections to genetic research. In 1994, the Indigenous director of Central Australian Aboriginal Congress, one of the oldest and most powerful Aboriginal community-controlled health services in Australia, spoke out against the HGDP. As quoted in the media, he explained:

> The Human Genome Diversity (or 'Vampire') Project wants to visit 600 indigenous communities around the world and take blood and hair samples from the people. These samples would then be flown back to the United States and frozen in a 'genetic bank'. They would then be used by scientists to do research on the genetic make-up of indigenous people. Congress is very concerned about the Vampire Project. First, there has been no consultation, as far as we know, with our people or their organisations. Second, our people would have no control over how these samples are to be used and what research is to be performed on them. The most worrying aspect is the possibility of patenting of Aboriginal genetic samples and any drugs developed from them. This means that some company or government could legally own the genetic information present in our people's bodies.[22]

From this example, many of the ethical issues related to genetic research are clear. These include: the tension between individual informed consent and community consent and consultation; control over the storage and secondary use of samples; ownership of samples; and arrangements to share any benefits

---

[20] Aroha Te Pareake Mead, "Genealogy, Sacredness, and the Commodities Market," *Cultural Survival Quarterly* 20.2 (Summer 1996): 46–52; Jonathan M. Marks, "Your Body, My Property: The Problem of Colonial Genetics in a Post-Colonial World," in *Embedding Ethics*, ed. Lynn Meskel & Peter Pels (Oxford: Berg, 2005): 29–45; Indigenous Peoples' Council on Biocolonialism, *Indigenous Peoples, Genes and Genetics: What Indigenous Peoples Should Know About Biocolonialism* 2000, Indigenous Peoples Council on Biocolonialism, http://www.ipcb.org/publications /primers/index .html (accessed 15 August 2010).

[21] Kimberly TallBear, "Narratives of Race and Indigeneity in the Genographic Project," *Journal of Law, Medicine and Ethics* 35.3 (Fall 2007): 412–24.

[22] Anon., "Concerns at 'Vampire' Project," *Green Left Weekly* (Wednesday 2 February 1994): http://www.greenleft.org.au/node/7328 (accessed 15 August 2010).

that may arise from genetic research.[23] Most of these issues are 'generic', in that they may apply to all groups participating in genetic research. However, special care must be taken when negotiating these issues with Indigenous communities, given the relative vulnerability of these communities and their historically low levels of trust in researchers and research.[24]

All of the ethical issues mentioned contribute to the reluctance of Indigenous people to participate in genetic research. An important factor reinforcing this reluctance is the relative political strength of Indigenous Australian leaders. By the 1990s, the Indigenous community-controlled health sector was quite strong and well integrated into the Indigenous political community. As a result, opposition to the HGDP was widespread and uniform. This was illustrated in the 1996 report of the Aboriginal and Torres Strait Islander Social Justice Commissioner, a national report to the Attorney-General on "the exercise and enjoyment of human rights by Australia's Indigenous peoples."[25]

The 1996 report included a section entitled the 'Vampire project' that strongly criticized the project and linked it to attempts to patent human genetic material from "a Papuan and a Solomon Islander woman, without their informed consent."[26] It also emphasized the importance of Indigenous ethics committees in protecting Indigenous communities from harmful research.

---

[23] For examples of literature dealing with these issues, see Bartha Maria Knoppers & Ruth Chadwick, "Human Genetic Research: Emerging Trends in Ethics," *Nature Reviews: Genetics* 6.1 (January 2005): 75–79; Doris Schroeder & Carolina Lasén–Diaz, "Sharing the Benefits of Genetic Resources: From Biodiversity to Human Genetics," *Developing World Bioethics* 6.3 (December 2006): 135–43; United Nations Educational Scientific and Cultural Organization, *Bioethics and Human Population Genetics Research* (Paris: UNESCO International Bioethics Committee, 1995); Rebecca Tsosie & Joan L. McGregor, "Genome Justice: Genetics and Group Rights," *Journal of Law, Medicine and Ethics* 35.3 (Fall 2007): 352–55.

[24] Justin Mohamed, "We Don't Like Research – But in Koori Hands It Could Make a Difference" (Melbourne: VicHealth Koori Health Research and Community Development Unit & University of Melbourne, 2000); Linda T. Smith, *Decolonizing Methodologies, Research and Indigenous Peoples* (London & New York: Zed, 1999).

[25] http://www.hreoc.gov.au/social_justice/sj_report/index.html. In 1992, the federal government created the position of the Aboriginal and Torres Strait Islander Social Justice Commissioner (within the Human Rights and Equal Opportunity Commission).

[26] See Hilary Cunningham, "Colonial Encounters in Postcolonial Contexts: Patenting Indigenous DNA and the Human Genome Diversity Project," *Critique of Anthropology* 18.2 (June 1998): 205–33.

Most striking, however, was where the report repeated the strongest claims against the HGDP made by the prominent international NGO, the Indigenous Peoples' Council on Biocolonialism (IPCB):[27]

> The Vampire Project not only jeopardises the rights and safety of the peoples targeted, but could also lead to the cultural, political and social complexity of Indigenous identity and Aboriginal rights being reduced to an arbitrary genetic test. It is not unreasonable to speculate that such research could be used for biological warfare in some form, given exchanges have already taken place between medical researchers and the United States military establishment.[28]

The claim that human genetic researchers are in league with the government in a plot to develop a biological weapon genetically designed to target Indigenous people is by far the most radical assertion made by the IPCB. That this claim was included in such a high-level national report reflects the depth of suspicion about genetics among the Australian Indigenous community.

In addition to the ethical issues outlined above, there are cultural matters that must be heeded in accounting for the lack of genetic research in Indigenous communities. In Indigenous communities the world over, DNA is considered to be collective cultural property. For example, the Indigenous leader Professor Mick Dodson explains that blood and tissue samples are "treated with cultural respect as part of the inheritance of the group."[29] The late Frank Dukepoo, a Native American geneticist, expressed a similar sentiment:

> To us, any part of ourselves is sacred. Scientists say it's just DNA. For an Indian, it's not just DNA, it's part of a person, it is sacred, with deep religious significance. It is part of the essence of a person.[30]

The Canadian guidelines for research in Aboriginal communities explains Canadian Aboriginal philosophies regarding "full embodiment," in which it is

---

[27] See Indigenous Peoples' Council on Biocolonialism, *Indigenous Peoples, Genes and Genetics: What Indigenous Peoples Should Know About Biocolonialism.*

[28] Human Rights and Equal Opportunity Commission, *Aboriginal and Torres Strait Islander Social Justice Commissioner, Fourth Report* (Sydney: Human Rights and Equal Opportunity Commission, 1996): 109.

[29] Dodson & Williamson, "Indigenous Peoples and the Morality of the Human Genome Diversity Project," 205.

[30] Frank C. Dukepoo, "Genetic Services in the New Era: Native American Perspectives," *Community Genetics* 1.3 (1998): 130–33.

held that every part and product of the body is sacred, and constitutes an essential part of the person.[31]

Another set of obstacles to the conduct of genetic research in Indigenous communities arises from the nature of genetic research itself. As inherited biological substance, genes tell us where we are from in terms of our direct relatives and descendants, and our more distant ancestors. In the case of Indigenous peoples, genetic research can contradict cultural (or religious) beliefs. In the USA, for example, the views of some Native Americans that the Bering Strait hypothesis conflicts with their cosmological beliefs has generated much debate.[32] In addition, some Indigenous people believe genetic research will have political consequences such as the removal or lack of recognition of customary rights to land or 'native title'. This latter issue was prominent in a recent genetic-research scandal in the USA.

The Havasupai case involved a small tribe who live on the floor of the Grand Canyon and who provided samples for genetic study on diabetes in the early 1990s. Over the following decade, their samples were used for a number of studies without their knowledge, including population genetics studies which they felt threatened their sovereignty rights.[33] In the media coverage of the case, Havasupai elders emphasized the gravity of this threat:

> Though some Havasupai knew already that their ancestors most likely came from Asia, "when people tell us, 'No, this is not where you are from,' and your own blood says so – it is confusing to us," Rex Tilousi said. "It hurts the elders who have been telling these stories to our grandchildren." Others questioned whether they could have unwittingly contributed to research that could threaten the tribe's rights to its land. "Our coming from the canyon, that is the basis of our sovereign rights," said Edmond Tilousi, the tribe's vice chairman.[34]

---

[31] CIHR, "CIHR Guidelines for Health Research Involving Aboriginal People," *Canadian Institutes of Health Research* (2007): 1–40.

[32] See Adam Kuper, "The Return of the Native," *Current Anthropology* 44.3 (June 2003): 389–402.

[33] A similar scandal concerned the Canadian First Nations Nuuchahnulth community who initially participated in a project on the genetics of arthritis, and the researcher subsequently used their samples for population genetics without their specific consent. David Wiwchar, "Nuu-Chah-Nulth Blood Returns to West Coast," *Ha-Shilth-Sa Newsletter* 31.25 (2004): 1–4.

[34] Amy Harmon, "Indian Tribe Wins Fight to Limit Research of Its DNA," *New York Times* (21 April 2010): A1.

The case was widely publicized in 2010 when the tribes settled with Arizona State University, which held the samples, and tribal members reclaimed the remaining samples from the freezer in a moving ceremony. In the wake of the case, scholars have closely considered the informed consent process and the Indigenous governance of the project.[35]

One set of issues that has not been directly considered in the literature on Indigenous genetic research relates to the 'technical' requirements of genetics, including 'population structure' and 'admixture' (which can involve researchers reporting on the percentages of 'Indigenous' and 'Caucasian' ancestry in a sample), and 'low heterozygosity' (whereby researchers report whether there is high level of 'inbreeding' within a population). These issues can be the source of much concern for Indigenous communities participating in genetic research. They were recently discussed in an Australian context at a meeting of genetic researchers, Indigenous leaders, and other key stakeholders (see Kowal and Anderson, this volume).

The final source of concern surrounding Indigenous genetic research arises from the publication of research findings in academic journals, particularly the potential harm produced when and if these findings are reported in the media. Concerns of this nature are well illustrated by another recent Indigenous genetic-research scandal, this time in Aotearoa New Zealand. In fact, the case of the Rakaipaaka Health and Ancestry Study illustrates the potential for genomic research to both benefit and harm Indigenous communities.

Māori and non-Māori health researchers at New Zealand's Environmental Science Research Institute have collaborated with Te Iwi o Rakaipaaka (the organization representing members of the Rakaipaaka community) on the Rakaipaaka Health and Ancestry Study, based predominantly in Nuhaka (Hawke's Bay). It aimed to recruit 3,000 Māori to participate in a longitudinal "envirogenomics" project that will investigate common diseases affecting families, such as diabetes, gout, heart disease, and cancer. The research team was composed of Māori and Pākehā (white) researchers, led by Rod Lea and Marino Lea, the latter of whom is a member of the Raikapaaka *iwi* or extended kinship group.

The research team took into account many of the ethical issues raised above. An incorporated community organization had control of the project,

[35] Michelle M. Mello & Leslie Wolf, "The Havasupai Indian Tribe Case – Lessons for Research Involving Stored Biologic Samples," *New England Journal of Medicine* 363.3 (July 2010): 204–207.

and retained ownership of the genetic information. They also formed a "Māori *kaitiaki* [guardianship] group" to oversee research that uses Māori genetic information and develop policy regarding secondary use of samples.[36] The research team also expressed concern about possible stereotyping in the media if their study findings are interpreted as proof of 'Māori genetic susceptibility' to certain diseases. They proposed to manage this by taking care not to extrapolate their findings beyond the community they studied.[37]

However, the lead researcher Rod Lea generated worldwide controversy in 2006 when he presented findings from a separate research project that found that Māori were twice as likely as non-Māori to carry a gene associated with alcohol and tobacco use.[38] The particular polymorphism of the gene that encodes the enzyme monoamine oxidase (MOA) has also been associated with risk-taking and aggression, and is consequently known as the "warrior gene."[39] This episode was widely reported in the international media as proving that Māori were genetically predetermined to be violent, a depiction that the researchers argued was a misrepresentation of their research.[40] A recent critique by a Māori academic argued that linking the MOA allele with high levels of violence among Māori is scientifically unsound, effectively makes being Māori a 'disease', and may lead to genetic and racial discrimination by insurance companies. Further, "contributions to racial stereotyping by trained scientists are unethical and scandalous."[41]

---

[36] One example of such an arrangement for culturally-appropriate procedures for the storage of blood and other samples comes from the New Zealand cancer tissue bank, which since 2004 has offered all donors the option of having their sample disposed of with a Maori blessing or *karakia*. Helen Morrin, Sarah Gunningham, Margaret Currie et al., "The Christchurch Tissue Bank to Support Cancer Research," *Journal of the New Zealand Medical Association* 118/1225 (November 2005): 1–12.

[37] Maui L. Hudson, Annabel L.M. Ahuriri–Driscoll, Marino G. Lea et al., "Whakapapa – a Foundation for Genetic Research?" *Journal of Bioethical Inquiry* 4.1 (2007): 43–49.

[38] Rod Lea & Geoffrey Chambers, "Monoamine oxidase, addiction, and the 'warrior' gene hypothesis," *Journal of the New Zealand Medical Association* 120/1250 (2007): 1–6.

[39] Jon Stokes, "Family Tragedy Behind Gene Work," *New Zealand Herald* (12 August 2006).

[40] Lea & Chambers, "Monoamine oxidase, addiction."

[41] Gary Raumati Hook, "'Warrior Genes' and the Disease of Being Maori," *MAI Review* 2 (2009): 6.

This leaves us with an apparent dilemma: are we to perceive Rod Lea as the culturally appropriate lead co-researcher of the Rakaipaaka Health and Ancestry Study or as the proponent of the potentially damaging 'warrior gene' hypothesis? Rather than indicating a split personality, this apparent contradiction may reflect the dangers of conducting Indigenous genetic research (particularly some areas such as behavioural traits) in an environment where both long-held racial stereotypes and health inequalities are widespread.

III Conclusion

The combination of all the factors I have discussed above results in genetics being seen as 'too hard' to do in Indigenous communities. Whether this is true or not, it quickly becomes a self-fulfilling prophecy. If health researchers, grant reviewers, ethics committees, and Indigenous leaders think it is 'too hard' to do Indigenous genetic research, any attempt to do it will be quashed. My ethnographic research has recorded many instances of this.

Perhaps a more troubling consequence of this self-fulfilling prophecy is that there has been no discussion of the implications of genetic research or genetic technology for Indigenous Australians. This situation stands in sharp contrast to that obtaining in New Zealand, Canada, and the USA, where trans-disciplinary discussions have produced a growing literature on this issue.[42] In

[42] Ramari V. Port, John Arnold, Dale Kerr et al., "Cultural Enhancement of a Clinical Service to Meet the Needs of Indigenous People; Genetic Service Development in Response to Issues for New Zealand Maori," *Clinical Genetics* 73.2 (February 2008); 132–38; Linda Burhansstipanov, Lynne T. Bemis & Mark B. Dignan, "Native American Recommendations for Genetic Research to be Culturally Respectful," *Jurimetrics* 42.2 (2002): 149–57; Richard R. Sharp & Morris W. Foster, "Community Involvement in the Ethical Review of Genetic Research: Lessons from American Indian and Alaska Native Populations," *Environmental Health Perspectives* 110, Supplement 2 (April 2002): 145–48; Laura Arbour & Doris Cook, "DNA on Loan: Issues to Consider When Carrying Out Genetic Research with Aboriginal Families and Communities," *Community Genetics* 9.3 (2006): 153–60; Malcolm B. Bowekaty & Dena S. Davis, "Cultural Issues in Genetic Research with American Indian and Alaskan Native People," *IRB: Ethics and Human Research* 25.4 (July–August 2003): 12–15; Hudson, Ahuriri–Driscoll, Lea et al., "Whakapapa: A Foundation for Genetic Research?"; Mervyn Tano, "Interrelationships Among Native Peoples Genetic Research, and the Landscape: Need for Further Research Into Ethical, Legal, and Social Issues: (DNA Fingerprinting and Civil Liberties)," *Journal of Law, Medicine, and*

Canada, for instance, discussions on the use of genetics in Aboriginal health research began in the 1990s and continued through the first decade of the twenty-first century, while in Australia no such discussions took place.[43] The result of this discrepancy is clear: while the guidelines for Aboriginal health research in Canada clearly outline how genetic (and other biological) research should be conducted, issues relating to the ownership, governance and use of samples are not mentioned in the Australian guidelines.[44]

Given the significant risks involved in Indigenous genetic research, it is up to geneticists to strongly make the case for the potential benefits of genetic research to Indigenous people. Geneticists argue that there is untapped potential for genetics to contribute to solving health inequalities through improving our understanding of disease pathogenesis, using genetics to probe environmental risks, predicting disease risk, developing novel diagnostics and drug targets, or applying pharmacogenomics (tailoring drugs to individuals).[45] In-

---

Ethics 34.2 (Summer 2006): 301–309; LorrieAnn Santos, "Genetic Research in Native Communities," *Progress in Community Health Partnerships: Research, Education, and Action* 2.4 (Winter 2008): 321–27.

[43] National Council of Ethics in Human Research et al., 2001; Canadian Institutes of Health Research and Institute of Aboriginal Peoples' Health, 2001; Canadian Institutes of Health Research and Institute of Aboriginal Peoples' Health, 2002. An important opportunity for discussing these issues was missed in the early 2000s in the lead up to the Australian Law Reform Commission's extensive report on the protection of human genetic information. Australian Law Reform Commission, *Essentially Yours:The Protection of Human Genetic Information in Australia* (Canberra: Australian Government Publishing Service, 2003). The commission consulted with many Indigenous people and received many submissions. However, the sole focus of the chapter on Indigenous issues was to argue that indigeneity could not be defined genetically, as Indigenous notions of kinship are social and not biological. Australian Law Reform Commission, *Essentially Yours: The Protection of Human Genetic Information in Australia - Section 36: Kinship and Identity* (Canberra: Australian Goverment Publishing Service, 2003). While this point is important, the report's exclusive focus on it and the neglect of any other issues were disappointing, and reflect the lack of dialogue between Indigenous community and those concerned with the contribution genetics can make to human health.

[44] CIHR, "CIHR Guidelines for Health Research Involving Aboriginal People"; NHMRC, *Values and Ethics: Guidelines for Conduct of Aboriginal and Torres Strait Islander Health Research*.

[45] Edward Ramos & Charles Rotimi, "The A's, G's, C's, and T's of Health Disparities," *BMC Medical Genomics* 2.1 (July 2009): 29.

digenous advocates, on the other hand, discuss the risks of racialization in reinforcing racism, the diversion of resources and attention from the social determinants of health, and the likelihood that disadvantaged groups will miss out on any health benefits produced through genomic research.[46] Finding common ground is not an easy task.[47]

Whether or not one believes that genetics offer any tangible benefits to Indigenous people, and if one does, whether or not one believes these potential benefits outweigh the potential risks, it is clear that the lack of formal discussion of these issues has left Australia far behind comparable nations. Genetic researchers and funding bodies must do more to address issues of concern to Indigenous people and find ways to minimize the many risks outlined in this essay. In turn, Indigenous communities must keep up with the pace of change in genomic technologies so that they can assess the potential benefits and risks of participation from a position of knowledge. Only then can we ensure that Indigenous Australians can benefit from the promised genomic future without being subjected to the mistakes made in the past.

WORKS CITED

Anon. "Concerns at 'Vampire' Project," *Green Left Weekly* (Wednesday 2 February 1994): http://www.greenleft.org.au/node/7328 (accessed 15 August 2010).

Arbour, Laura, & Doris Cook. "DNA on Loan: Issues to Consider when Carrying Out Genetic Research with Aboriginal Families and Communities," *Community Genetics* 9.3 (2006): 153–60.

Australian Bureau of Statistics. *Experimental life tables for Aboriginal and Torres Strait Islander Australians, 2005–2007* (Canberra: Australian Bureau of Statistics, 2009).

Australian Institute of Health and Welfare, & Australian Bureau of Statistics. *The Health and Welfare of Australia's Aboriginal and Torres Strait Islander Peoples, 2008* (Canberra: Australian Bureau of Statistics, 2008).

---

[46] Immaculada Melo–Martin, "Genetic Research and Reduction of Health Disparities," *New Genetics and Society* 27.1 (March 2008): 57–68.

[47] Morris W. Foster, "Looking for Race in All the Wrong Places: Analyzing the Lack of Productivity in the Ongoing Debate About Race and Genetics," *Human Genetics* 126.3 (September 2009): 355–62, Michael Fortun, "For an Ethics of Promising, Or: A Few Kind Words About James Watson," *New Genetics and Society* 24.2 (August 2005): 157–73.

Australian Law Reform Commission. *Essentially Yours: The Protection of Human Genetic Information in Australia* (Canberra: Australian Government Publishing Service, 2003).

——. *Essentially Yours: The Protection of Human Genetic Information in Australia – Section 36: Kinship and Identity* (Canberra: Australian Goverment Publishing Service, 2003).

Bowekaty, Malcolm B., & Dena S. Davis. "Cultural Issues in Genetic Research with American Indian and Alaskan Native People," *IRB: Ethics and Human Research* 25.4 (July–August 2003): 12–15.

Burhansstipanov, Linda, Lynne T. Bemis & Mark B. Dignan. "Native American Recommendations for Genetic Research to be Culturally Respectful," *Jurimetrics* 42.2 (Winter 2002): 149–57.

CIHR. "CIHR Guidelines for Health Research Involving Aboriginal People," *Canadian Institutes of Health Research* (2007): 1–40.

Cunningham, Chris, & Fiona Stanley. "Indigenous by definition, experience, or world view," *British Medical Journal* 327/7412 (August 2003): 403–404.

Cunningham, Hilary. "Colonial Encounters in Postcolonial Contexts: Patenting Indigenous DNA and the Human Genome Diversity Project," *Critique of Anthropology* 18.2 (June 1998): 205–33.

Dodson, Michael, & Robert Williamson. "Indigenous peoples and the morality of the Human Genome Diversity Project," *Journal of Medical Ethics* 25.2 (April 1999): 204–208.

Dukepoo, Frank C. "Genetic Services in the New Era: Native American Perspectives," *Community Genetics* 1.3 (1998): 130–33.

Fortun, Michael. "For an Ethics of Promising, Or: A Few Kind Words About James Watson," *New Genetics and Society* 24.2 (August 2005): 157–73.

Foster, Morris W. "Looking for Race in All the Wrong Places: Analyzing the Lack of Productivity in the Ongoing Debate About Race and Genetics," *Human Genetics* 126.3 (September 2009): 355–62.

Harmon, Amy. "Indian Tribe Wins Fight to Limit Research of Its DNA," *New York Times* (21 April 2010): A1.

Henry, John, Terry Dunbar, Allan Arnott et al. *Indigenous Research Reform Agenda: Rethinking Research Methodologies* (Darwin, NT: Cooperative Research Centre for Aboriginal and Tropical Health, 2002).

Hook, Gary Raumati. "'Warrior Genes' and the Disease of Being Maori," *MAI Review* 2 (2009): 1–11.

Hudson, Maui L., Annabel L.M. Ahuriri–Driscoll, Marino G. Lea et al. "Whakapapa: A Foundation for Genetic Research?" *Journal of Bioethical Inquiry* 4.1 (2007): 43–49.

Human Rights and Equal Opportunity Commission. *Aboriginal and Torres Strait Islander Social Justice Commissioner, Fourth Report* (Sydney: Human Rights and Equal Opportunity Commission, 1996).

Humphery, Kim. "The Development of the National Health and Medical Research Council Guidelines on Ethical Matters in Aboriginal and Torres Strait Islander Health Research: A Brief Documentary and Oral History" (Discussion Paper 8; Melbourne: VicHealth Koori Health Research and Community Development Unit, 2002).

——. "Dirty Questions: Indigenous health and 'Western research'," *Australian and New Zealand Journal of Public Health* 25.3 (June 2001): 197–202.

Indigenous Peoples' Council on Biocolonialism. *Indigenous Peoples, Genes and Genetics: What I Indigenous Peoples Should Know About Biocolonialism* (2000), Indigenous Peoples' Council on Biocolonialism, http://www.ipcb.org/publications/primers/index.html (accessed 15 August 2010).

Jakobsson, Mattias, Sonja W. Scholz, Paul Scheet et al. "Genotype, haplotype and copy-number variation in worldwide human populations," *Nature* 451/7181 (2008): 998–1003.

Knoppers, Bartha Maria, & Ruth Chadwick. "Human Genetic Research: Emerging Trends in Ethics," *Nature Reviews: Genetics* 6.1 (January 2005): 75–79.

Kowal, Emma, Glenn Pearson, Lobna Rouhani et al. "Genetic Research and Aboriginal and Torres Strait Islander Australians" (MS under review).

Kuper, Adam. "The Return of the Native," *Current Anthropology* 44.3 (June 2003): 389–402.

Lea, Rod, & Geoffrey Chambers. "Monoamine oxidase, addiction, and the 'warrior' gene hypothesis," *Journal of the New Zealand Medical Association* 120/1250 (2007): 1–6.

Li, Jun Z., Devin M. Absher, Hua Tang et al. "Worldwide Human Relationships Inferred from Genome-Wide Patterns of Variation," *Science* 319/5866 (2008): 1100–104.

Marks, Jonathan M. "Human Genome Diversity Project (HGDP): Impact on Indigenous Communities," in *Living with the Genome: Ethical and Social Aspects of Human Genetics*, ed. Angus Clarke & Flo Ticehurst (London: Palgrave Macmillan, 2006): 49–55.

——. "Your body, my property: The problem of colonial genetics in a post-colonial world," in *Embedding Ethics*, ed. Lynn Meskel & Peter Pels (Oxford: Berg, 2005): 29–45.

Mello, Michelle M., & Leslie Wolf. "The Havasupai Indian Tribe Case – Lessons for Research Involving Stored Biologic Samples," *New England Journal of Medicine* 363.3 (July 2010): 204–207.

Melo–Martin, Immaculada. "Genetic Research and Reduction of Health Disparities," *New Genetics and Society* 27.1 (March 2008): 57–68.

Metcalfe, Sylvia A., Alan H. Bittles, Peter C. O'Leary et al. "Australia: Public Health Genomics," *Public Health Genomics* 12.2 (October 2009): 121–28.

Mohamed, Justin. "We Don't Like Research – But in Koori Hands It Could Make a Difference" (Melbourne: VicHealth Koori Health Research and Community Development Unit & University of Melbourne, 2000).

Morrin, Helen, Sarah Gunningham, Margaret Currie et al. "The Christchurch Tissue Bank to support cancer research," *Journal of the New Zealand Medical Association* 118/1225 (November 2005): 1–12.

NHMRC. *Guidelines on Ethical Matters in Aboriginal and Torres Strait Islander Health Research* (Canberra: National Health and Medical Research Council, 1991).

——. *Road Map II: A Strategic Framework for Improving the Health of Aboriginal and Torres Strait Islander People Through Research* (Canberra: National Health and Medical Research Council, 2010).

——. *Values and Ethics: Guidelines for Conduct of Aboriginal and Torres Strait Islander Health Research* (Canberra: National Health and Medical Research Council, 2003).

Port, Ramari V., John Arnold, Dale Kerr et al. "Cultural Enhancement of a Clinical Service to Meet the Needs of Indigenous People: Genetic Service Development in Response to Issues for New Zealand Maori," *Clinical Genetics* 73.2 (February 2008): 132–38.

Ramos, Edward, & Charles Rotimi. "The A's, G's, C's, and T's of Health Disparities," *BMC Medical Genomics* 2.1 (July 2009): 29.

Reardon, Jenny. *Race to the Finish: Identity and Governance in an Age of Genomics* (Princeton NJ: Princeton UP, 2005).

Ring, Ian, & Ngaire Brown. "The health status of indigenous peoples and others: The gap is narrowing in the United States, Canada, and New Zealand, but a lot more is needed," *British Medical Journal* 327/7412 (August 2003): 404–405.

Santos, LorrieAnn. "Genetic Research in Native Communities," *Progress in Community Health Partnerships: Research, Education, and Action* 2.4 (Winter 2008): 321–27.

Schroeder, Doris, & Carolina Lasén–Diaz. "Sharing the Benefits of Genetic Resources: From Biodiversity to Human Genetics," *Developing World Bioethics* 6.3 (December 2006): 135–43.

Sharp, Richard R., & Morris W. Foster. "Community Involvement in the Ethical Review of Genetic Research: Lessons from American Indian and Alaska Native Populations," *Environmental Health Perspectives* 110, Supplement 2 (April 2002): 145–48.

Smith, Linda T. *Decolonizing Methodologies, Research and Indigenous Peoples* (London & New York: Zed, 1999).

Stokes, Jon. "Family Tragedy Behind Gene Work," *New Zealand Herald* (12 August 2006).

TallBear, Kimberly. "Narratives of Race and Indigeneity in the Genographic Project," *Journal of Law, Medicine and Ethics* 35.3 (Fall 2007): 412–24.

Tano, Mervyn. "Interrelationships Among Native Peoples, Genetic Research, and the Landscape: Need for Further Research Into Ethical, Legal, and Social Issues: (DNA Fingerprinting and Civil Liberties)," *Journal of Law, Medicine, and Ethics* 34.2 (Summer 2006): 301–309.

Te Pareake Mead, Aroha. "Genealogy, Sacredness, and the Commodities Market," *Cultural Survival Quarterly* 20.2 (Summer 1996): 46–52.

——, & Steven Ratuva, ed. *Pacific Genes and Life Patents: Pacific Experiences and Analysis of the Commodification and Ownership of Life* (Tokyo: Call of the Earth / Llamado de la Tierra, United Nations University – Institute of Advanced Studies, 2007).

Tsosie, Rebecca, & Joan L. McGregor. "Genome Justice: Genetics and Group Rights," *Journal of Law, Medicine and Ethics* 35.3 (Fall 2007): 352–55.

United Nations Educational Scientific and Cultural Organization. *Bioethics and Human Population Genetics Research* (Paris: UNESCO International Bioethics Committee, 1995).

Walsh, Michael. "Overview of indigenous languages of Australia," in *Language in Australia*, ed. Suzanne Romaine (Cambridge: Cambridge UP, 1991): 27–48.

Wiwchar, David. "Nuu-Chah-Nulth Blood Returns to West Coast," *Ha-Shilth-Sa Newsletter* 31.25 (2004): 1–4.

⌘

Difficult Conversations

Talking About Indigenous Genetic Health Research in Australia

EMMA KOWAL AND IAN ANDERSON

GENETICS IS AT THE FOREFRONT OF MEDICAL RESEARCH, but it is rarely used in Indigenous health-research projects. In the past, proposals to conduct genetic studies in Indigenous communities in Australia have been highly criticized and rarely funded. However, genetic researchers worldwide argue that genetics has the potential to reduce health disparities (including Indigenous health disparities) in multiple ways: through understanding disease pathogenesis; using genetics to probe environmental risk; predicting disease risk; finding novel diagnostics and drug targets; and pharmacogenomics.[1]

Understandably, many Indigenous people interpret genetic research in the context of their experiences of colonization. Fears of genetic theft or 'biopiracy', fears that genetics will be used to determine Aboriginality and may fuel racism, poor access to the potential health care innovations that genetics may bring, bad experiences of the Human Genome Diversity Project (known by some as the 'Vampire' project) and struggles over access to DNA extracted from human remains – all constitute barriers to effective research partnerships between Indigenous communities and genetic researchers.[2]

---

[1] Edward Ramos & Charles Rotimi, "The A's, G's, C's, and T's of Health Disparities," *BMC Medical Genomics* 2.1 (July 2009): 29; Tara Acharya et al., "Strengthening the Role of Genomics in Global Health," *PLoS Medicine* 1.3 (December 2004): e40; Sandra Soo–Jin Lee, "Pharmacogenomics and the Challenge of Health Disparities," *Public Health Genomics* 12.3 (February 2009): 170–79.

[2] Indigenous Peoples Council on Biocolonialism, "Indigenous Peoples, Genes and

Clearly, there are serious risks of genetic research for minorities as well as potential benefits. The only way in which these risks and benefits can be managed effectively is through open discussion among all of those parties with an interest in the conduct of genetic research in Indigenous communities.[3]

On 2 July 2010, a roundtable on Indigenous genetic research in Australia was organized by the authors on behalf of the Lowitja Institute, Australia's national institute for Aboriginal and Torres Strait Islander health research. The Lowitja Institute is an innovative research body that brings together Aboriginal organizations, academic institutions, and government agencies to facilitate collaborative, evidence-based research into Aboriginal and Torres Strait Islander health.[4] It is controlled by an Indigenous-majority board. The roundtable was organized under the auspices of Professor Ian Anderson, the Director of Research and Innovation.

The roundtable was attended by twenty-four Indigenous and non-Indigenous attendees including experts in genetics, Indigenous health research, Indigenous research ethics, genetic ethics, and genetic literacy. This debate was especially significant, as it was the first discussion of genetics and Indigenous health to have ever taken place in Australia.[5] While comparable nations such

---

Genetics: What Indigenous Peoples Should Know About Biocolonialism," *Indigenous Peoples Council on Biocolonialism* (June 2000), http://www.ipcb.org/publications/primers/index.html (accessed 15 August 2010); Jenny Reardon, *Race to the Finish: Identity and Governance in an Age of Genomics* (Princeton NJ: Princeton UP, 2005); Kimberly TallBear, "DNA, Blood, and Racializing the Tribe," *Wicazo SA Review* 18.1 (Spring 2003): 81–107; Rebecca Tsosie & Joan L. McGregor, "Genome Justice: Genetics and Group Rights," *Journal of Law, Medicine and Ethics* 35.3 (Fall 2007): 352–55; Morris W. Foster & Richard R. Sharp, "Genetic Research and Culturally Specific Risks: One Size Does Not Fit All," *Trends in Genetics* 16.2 (February 2000): 93–95.

[3] Sarah Knerr, Edward Ramos, Juleigh Nowinski et al., "Human Difference in the Genomic Era: Facilitating a Socially Responsible Dialogue," *BMC Medical Genomics* 3.1 (May 2010): 20; Morris W. Foster, "Looking for Race in All the Wrong Places: Analyzing the Lack of Productivity in the Ongoing Debate About Race and Genetics," *Human Genetics* 126.3 (September 2009): 355–62.

[4] See The Lowitja Institute, "About the Lowitja Institute," *The Lowitja Institute* (29 November 2010), http://www.lowitja.org.au/ (accessed 28 February 2011).

[5] The only other meeting pertaining to Indigenous genetics in Australia was a gathering at the Australian Institute of Aboriginal and Torres Strait Islander Studies in the 1990s to discuss the Human Genome Diversity Project and Indigenous population

as Canada have a long history of discussing these issues,[6] in Australia these conversations have not yet taken place (see Kowal, this volume). The aims of this conversation were to create a community of interest around Indigenous genetic research; to foster collaborative genetic research in Indigenous communities that is ethical, productive, and of high quality; and to begin to consider what tools and models are needed to make genetic research a culturally safe option for Indigenous people.

The Lowitja Institute was well aware of the potential difficulty of this conversation. As Ian Anderson commented,

> We have hosted conversations on lateral violence in Indigenous communities and other sensitive issues. These are topics that are in the 'too-hard' basket that can end up in unproductive conversations. But we know that it is important to have those conversations.[7]

Indeed, discussions were honest and constructive and considered many difficult issues. These included the sensitivities inherent in the reporting requirements of sample quality control (degree of heterozygosity or 'inbreeding' in the population, and corrections for population structure and diversity of ancestry within Indigenous populations); the risks involved when research findings are generalized to the entire Indigenous population; the existence of highly publicized examples of 'bad' Indigenous genetic research (including the Havasupai case in the USA and the 'Warrior gene' controversy in New Zealand); and the need to provide short-term benefits to participating Indigenous communities, given the long-term nature of the potential benefits of genetic research.

---

genetic research in general. I have heard verbal reports of this meeting but not found any documentation relating to it.

[6] National Council of Ethics in Human Research, Canadian Commission for UNESCO & Health Canada, *Continuing the Dialogue: Genetic Research with Aboriginal Individuals and Communities* (Vancouver: NCEHR, 2001); Canadian Institutes of Health Research & Institute of Aboriginal Peoples' Health, *An Exploratory Workshop on a Tribal Controlled DNA Bank: Part 1* (Vancouver: CIHR, 2001); Canadian Institutes of Health Research & Institute of Aboriginal Peoples' Health, *An Exploratory Workshop on a Tribal Controlled DNA Bank: Part 2* (Tofino, British Columbia: CIHR, 2002).

[7] Emma Kowal, Lobna Rouhani & Ian Anderson, *Genetic Research in Aboriginal and Torres Strait Islander Communities: Beginning the Conversation* (Melbourne: Lowitja Institute, 2011): 51. Further page references are in the main text.

This essay provides some insight into the conversations that took place at the roundtable. It draws on a record of the roundtable discussion published by the Lowitja Institute as part of a longer discussion paper.[8] Roundtable participants were offered the opportunity to amend or comment on the record of conversations that we draw on. As well as synthesizing the main themes of the discussion, at various points we offer our own analysis of the workshop and the broader politics of difficult conversations in Indigenous research. This analysis illustrates both the potential misunderstandings that can occur across very different world-views and the potential for finding common ground.

The Politics of Genetic Difference: Heterozygosity and Principal Components Analysis

The first theme we explore is taken from a presentation given by a senior genetic researcher who works in one remote Indigenous community on a genetic-research project relating to ear infections and cardiovascular disease, diseases that significantly impact on the well-being of children and adults respectively.

Having worked in Indigenous health for a few years, this researcher was well aware of the sensitivity of genetics in Indigenous contexts. For instance, she took time to stress that genetic research did *not* imply that a disease was inherited and therefore not a product of social disadvantage. Rather, genes and environment are intimately connected and can't be separated from each other. "You sample your environment through your genes. Your genes determine your response to your environment (genetics); your environment influences expression of your genes (epigenetics)" (52).

She discussed the tool that is currently the basis of most genetic research, the Genome Wide Association Study (GWAS). A Genome Wide Association Study typically examines over 500,000 single nucleotide polymorphisms (SNPs) using a 'SNP chip'. Either a case-control design or a family-based study is usually used, with ideally a large number of participants (at least 2000). Since the year 2007, known by some as 'the year of GWAS', many have been published and, as of March 2010, 779 highly significant associations had been published on 148 traits (53).

She then considered the various technical procedures that geneticists must perform on their data and their samples to ensure that they are of high quality

[8] Kowal, Rouhani & Anderson, *Genetic Research in Aboriginal and Torres Strait Islander Communities: Beginning the Conversation.*

and that their study findings are true. While the concepts discussed below are seen by geneticists as purely 'technical', they raise challenges for the conduct of ethical research and the reporting of findings in the media. At the core of this challenge is the possible risk that 'technical' constructs considered by scientists to be 'value-free' and objective can be deployed in public representations in ways that Indigenous people feel are damaging. Furthermore, this raises the challenge of how these representational risks might be communicated within the relationships between Indigenous peoples and researchers and ethically managed through research processes.

The first issue discussed was heterozygosity. At each locus on the genome, we carry two copies of a gene (alleles), one from our biological mother and one from our biological father. Heterozygosity is the proportion of points on the genome where people have two different alleles, as opposed to the same one. If there is a high degree of intermarriage within a group, over a long time people will have lower levels of heterozygosity because two people who have children together may have both inherited the same copies of genes from their shared ancestors. Geneticists need to know about this to make sure that the SNP chips will be effective in this population and to interpret their data correctly. Low or high levels of heterozygosity can occur for many reasons, including if there has been sample contamination in the lab. She explains that low heterozygosity "can be explained by a founder effect, or it could be that we are using the wrong chips, or because of consanguinity (partnerships between people who are related in some way), or 'inbreeding', which is a nasty word that comes up" (53).

It was apparent to us that, while the geneticist in question would prefer not to use the word herself, she is clearly aware that the word 'inbreeding' could be used to denigrate Indigenous communities that participate in genetic research because of its negative moral connotations. Research mentioning 'inbreeding' was, in fact, one of the three major objections the Havasupai tribe had to the research carried out on samples from tribal members held by Arizona State University.[9]

Another issue raised by this researcher at the roundtable relates to 'population structure'. When a genetic researcher compares two different groups, such as one group that has a particular disease, and another group that does

---

[9] Michelle Mello & Leslie Wolf, "The Havasupai Indian Tribe Case – Lessons for Research Involving Stored Biologic Samples," *New England Journal of Medicine* 363.3 (July 2010): 204–207.

not, it is important from a scientific perspective to check that the two groups are comparable in terms of their ancestry. Following this example, geneticists will look for genes that are found in different proportions in the two groups to explain why one group has the disease and the other group seems resistant to it. However, if the two groups are significantly different in terms of their ancestry, then any genetic differences between the two groups may be related to their different ancestry, and have nothing to do with the disease. For this reason, geneticists argue they must 'control' for the different ancestries the two groups may have, otherwise known as 'population structure'. If they do not control for population structure, they may think that a gene is related to a disease, when it is actually common to people with a particular ancestry and is unrelated to the disease.

Another reason for working out population structure is to determine the 'origin' of the genetic component of the disease. In discussing this reason for population structure, she gives the example of the genetic neurological disease Machado–Joseph that is found in Arnhem Land and where genetic research has established that the disease gene originated in Portugal and reached Arnhem Land via Macassan (now Indonesian) fisherman.[10] It was also mentioned by roundtable participants that genetic research could show that other diseases have been 'introduced' by genes from non-Indigenous ancestors, countering any perception that 'Indigenous' genes were to blame for the poor health status of the Indigenous population.[11]

The current method used by geneticists to discern population structure is Principal Component Analysis.[12] This method produces a schematic diagram comparing the genetic ancestry of different groups. Drawing on some of her work from Pakistan, the researcher showed how Principal Component Analysis found that Hindu and Muslim people from one village she worked in were actually genetically distinct. If she had found a gene that was present in the Muslim population and not the Hindu population, she might have mistakenly thought it was a disease gene when it really reflected a different population history.

---

[10] Tim Burt, Bart Currie, Charles Kilburn et al., "Machado–Joseph Disease in East Arnhem Land, Australia: Chromosome 14q32.1 Expanded Repeat Confirmed in Four Families," *Neurology* 46.4 (April 1996): 1118–22.

[11] Personal notes, Emma Kowal, 2 July 2010.

[12] David J. Balding, Martin Bishop & Chris Cannings, *Handbook of Statistical Genetics* (Chichester: Wiley, 2007).

She then showed a graph presenting the results of Principal Component Analysis for a number of different populations. The graph is an upside-down 'V' with Asian populations on the bottom left-hand side, African populations on the right-hand side, and Caucasian populations at the top, with South Asian and Middle Eastern populations in between (see Figure 1 below). She then indicated where the Indigenous people involved in this research were on the graph. The researcher pointed out a subset of the data representing Indigenous people with European 'admixture' (see section below) that was represented as dots "moving up" from the cluster of 'Indigenous' results towards the top of the inverted 'V'.

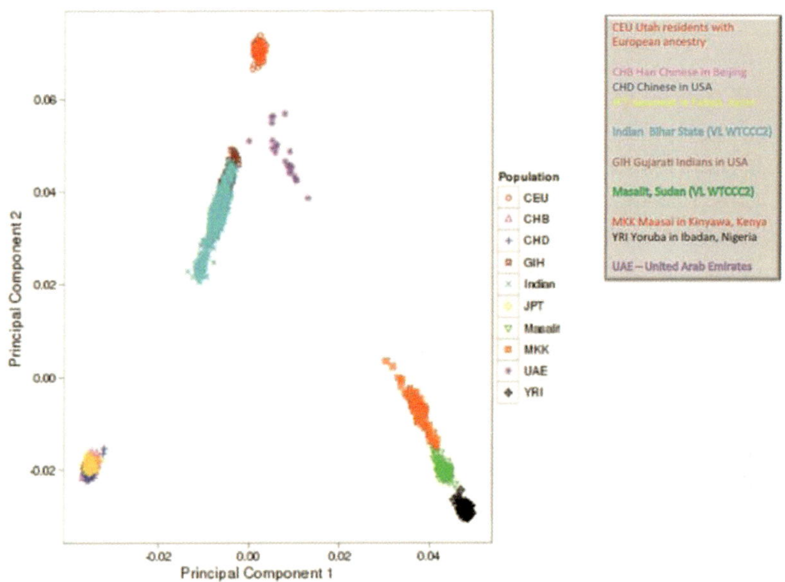

FIGURE 1. Principal Components 1 and 2 for a range of ethnic populations

She then addressed the issue of whether disease genes that might be found in Indigenous Australian populations are likely to be different from disease genes found in any other populations: "Will we find anything novel by studying Indigenous Australians?" In answering this question, she argues that out of fourteen diabetes genes found in Caucasian populations, only four have been found in Asian populations and only one has been found in the United Arab Emirates. The novel diabetes gene found in the United Arab Emirates is a gene that plays an important role in insulin secretion and may be a potential drug target. Based on experiences such as these, she argues that "there is a fair chance we would find novel genes in the Indigenous population" (55).

An Indigenous participant was invited to respond to the presentation. He highlighted the challenges that genetic research presents to Indigenous people, particularly Indigenous representatives who must make decisions about participation on behalf of their community. He raised concerns about research more generally, saying that "one of things I've noticed about research is that it can disconnect us as human beings, especially data collection" (57). He raised the question of how Indigenous people can build up their own capacity to understand these concepts when the technical language presents a serious obstacle to effective communication. The researchers can be thought of as "the great white centre" (57), and the roundtable goes some way towards countering this by providing a space for Indigenous people to talk about these issues.

He argued that power and control are central to research efforts, and that it is critical to ask the researcher why they want to do their research – to what end. These questions must be asked of researchers because of the historically negative relationship that indigenous people in Australia and elsewhere have had with both research and genetics.[13] "There are bells ringing for me when I hear this researcher speak. It speaks to this idea that we've broken and we are busted and we need to disappear and vanish. And it connects with our lived experience around not trusting whitefellas" (57), he said (also see Collingwood–Whittick, this volume).

He then refers directly to the Principal Component Analysis graph, an inverted 'V' that had Caucasian populations represented on the very top of the graph (see Figure 1 above). "When I see the graph and see how we are moving up towards the Caucasians again, I think, why can't we move down towards the Caucasians?" (57) he asked.

---

[13] Patrick Brantlinger, *Dark Vanishings: Discourse on the Extinction of Primitive Races, 1800–1930* (Ithaca NY & London: Cornell UP, 2003), Kim Humphery, "Dirty Questions: Indigenous Health and 'Western Research'," *Australian and New Zealand Journal of Public Health* 25.3 (June 2001): 197–202; Linda Tuhiwai Smith, *Decolonizing Methodologies, Research and Indigenous Peoples* (London & New York: Zed, 1999); Indigenous Peoples Council on Biocolonialism, "Indigenous Peoples, Genes and Genetics: What Indigenous Peoples Should Know About Biocolonialism"; Justin Mohamed, "We Don't Like Research – But in Koori Hands It Could Make a Difference" (Melbourne: VicHealth Koori Health Research and Community Development Unit & University of Melbourne, 2000).

From the discussion, it was apparent that the genetic researcher presenting this material meant no harm or disrespect through her use of the Principal Components Analysis diagram. Her use of the graph merely followed scientific convention and she did not anticipate the criticism of the Indigenous researcher who saw the graph as hierarchical. This indicates the difference in world-view between the researchers and the researched, and the importance of discussions like these as well as other forms of collaboration. While another genetic researcher in the audience suggested that the problem could be fixed by "putting [the graph] upside down" (57) so that the Caucasians are at the bottom, this incident provided a 'teachable moment' for the genetic researchers present, who had an opportunity to appreciate an Indigenous perspective.

"A Gold-Plated Christmas Gift": The Problem of 'Admixture'

The issues raised by Principal Component Analysis closely relate to the equally thorny issue of 'admixture'. This issue came up at a number of points during the roundtable. Most prominently, they arose when another researcher presented her research with a rural Indigenous community in New South Wales.

In the course of her research, which spans many years, she has had to justify her research against claims that genetics threatens Indigenous identity and land rights. In response, she argues that genetics does not threaten Aboriginality or native title, and that having mixed Indigenous and non-Indigenous parentage does not deny Aboriginality. "The criteria by which people identify as Aboriginal does not require genetic evidence" (58)[14] She explained that she takes care to make her position on this clear to her research participants in both her written and her verbal communication.

One Indigenous roundtable participant commented that while he finds the research to be very interesting, he was "concerned about the impact of this research" (58). Another colleague from an Aboriginal ethics committee commented that, of 160 projects that the committee had been asked to review in recent years, only three had required serious negotiation between the committee and researchers (58). All three of these projects concerned genetic re-

[14] See Australian Law Reform Commission, *Essentially Yours: The Protection of Human Genetic Information in Australia – Section 36: Kinship and Identity* (Canberra: Australian Government Publishing Service, 2003).

search. In discussing in general terms the concerns that the committee had had with the genetic research projects, he focused on the issue of 'admixture'.[15]

Some background analysis may be useful here. Australian Indigenous communities are diverse, and many Aboriginal and Torres Strait Islander people acknowledge ancestry from a variety of non-Indigenous sources in addition to their Indigenous ancestors. However, Indigenous identity is a very sensitive topic. The notion of 'authenticity' that imagines 'pure' and 'inauthentic' Indigenous people is a widespread source of interpersonal and systemic racism, as well as discrimination within Indigenous communities.[16] The governmental control of Indigenous Australians on the basis of their supposed blood quantum in the nineteenth and twentieth centuries is also the source of contemporary aversion to genetics among Indigenous people.[17]

'Admixture' is a common concept in genetic studies and there are many methods for measuring and analysing it.[18] While the concept of 'race' has been widely criticized for being a social and not biological construct, the concept of 'geographical ancestry' is emerging effectively in its place.[19] As

---

[15] See Duana Fullwiley, "The Biologistical Construction of Race: 'Admixture' Technology and the New Genetic Medicine," *Social Studies of Science* 38.5 (October 2008): 695–735; Yin C. Paradies, Michael J. Montoya & Stephanie M. Fullerton, "Racialized Genetics and the Study of Complex Diseases: The Thrifty Genotype Revisited," *Perspectives in Biology and Medicine* 50.2 (Spring 2007): 203–27; Kimberly TallBear, "Narratives of Race and Indigeneity in the Genographic Project," *Journal of Law, Medicine and Ethics* 35.3 (Fall 2007): 412–24.

[16] Yin C. Paradies, "Beyond Black and White: Essentialism, Hybridity and Indigeneity," *Journal of Sociology* 42.4 (December 2006): 355–67; Ian Anderson, "Introduction: The Aboriginal Critique of Colonial Knowing," in *Blacklines: Contemporary Critical Writing by Indigenous Australians*, ed. Michèle Grossman (Melbourne: Melbourne UP, 2003); Lynette Russell, *Savage Imaginings: Historical and Contemporary Constructions of Australian Aboriginalities* (Melbourne: Australian Scholarly, 2001); Lynette Russell, "Introduction" to *Boundary Writing: An Exploration of Race, Culture, and Gender Binaries in Contemporary Australia*, ed. Lynette Russell (Honolulu: U of Hawai'i P, 2006).

[17] *Political Theory and the Rights of Indigenous Peoples*, ed. Duncan Ivison, Paul Patton & Will Sanders (Cambridge: Cambridge UP, 2000).

[18] Balding, Bishop & Cannings, *Handbook of Statistical Genetics*.

[19] Timothy Caulfield, Stephanie M. Fullerton, Sarah E. Ali–Khan et al., "Race and Ancestry in Biomedical Research: Exploring the Challenges," *Genome Medicine* 1.8 (2009): 81–88; Charmaine D. Royal, John Novembre, Stephanie M. Fullerton et al.,

Royal et al. put it, "each segment of the genome has its own ancestral history, and various segments of an individual's genome may have ancestral histories that trace to different populations."[20] Although there are many ways to conceptualize ancestry, the most common way is geographically by way of continental ancestral labels such as Australian, Native American, European, Asian or African. If our recent ancestors came from different places on the globe, then geographical ancestry is a way of distinguishing between different parts of our genome that, via recombination, were inherited from different ancestors. Admixture is the common method of measuring the likely overall contribution to our genomes of ancestors with varying origins.

Some of the genetic rationales for discerning admixture were mentioned above in relation to population structure: to make sure that any disease associations are related to the disease, and not to ancestral differences between the two groups that are compared; and to find out which ancestral group a disease may have been inherited from. Admixture is commonly presented as percentages, with the proportion of each contributing ancestry in an individual or group of samples reported. A sample can, for example, be reported as 67% Indigenous and 33% Caucasian.

One representative of an Aboriginal ethics committee commented that a major danger of discussing 'admixture' in terms of percentages is that it "plays into the hands" of conservative commentators. "It's like offering Andrew Bolt [a conservative Australian columnist] a gold-plated Christmas present. He will say that benefits should only be given to those above a certain percentage" (59). While some geneticists at the roundtable believed that this issue can be dealt with by writing scientific papers without using percentages to discuss admixture, others disagreed, believing that reporting percentages was imperative for scientific validity (59). Whether genetic researchers report admixture percentages in their publications or whether they refrain from doing so, the challenge remains to disentangle genetic ancestry from social identity. Perhaps we are compelled to forever repeat the mantra that Indigenous identity is a social phenomenon not linked in any way to a particular

---

"Inferring Genetic Ancestry: Opportunities, Challenges, and Implications," *American Journal of Human Genetics* 86.5 (May 2010): 661–73.

[20] Royal, Novembre, Fullerton et al., "Inferring Genetic Ancestry: Opportunities, Challenges, and Implications," 661.

'blood quantum' of Indigenous ancestry, or necessarily to Indigenous ancestry per se.[21]

The Aboriginal ethics committee representative quoted above suggested that it's going to be very hard to entirely deal with these risks: "you can manage risks and explain them, but at the end of the day people will still be worried about the risks" (59). He argued that it is up to researchers to convince Indigenous people of the benefits, because the risks are high. In his reading of the literature, genetic studies that are about Indigenous people invariably generalize their findings to cover the entire Indigenous population. This perspective is confirmed by scandals such as the 'Warrior gene' controversy in Aotearoa New Zealand, where a study of a neurotransmitter gene in one group of Māori was reported in the international media in 2006 as proving that Māori were genetically susceptible to violence.[22]

The Politics of Representation: Who Decides?

Given the range of sensitive issues raised by genetics, it is not surprising that the question of who has control over the conduct of genetic research in Indigenous communities provoked much discussion. While concerns about the governance of Indigenous research are not unique to genetics, they are all the more pertinent in this area of research, given the sensitivities surrounding the use, storage, and ownership of DNA.[23]

---

[21] Australian Law Reform Commission, *Essentially Yours: The Protection of Human Genetic Information in Australia – Section 36: Kinship and Identity*.

[22] Gary Raumati Hook, "'Warrior Genes' and the Disease of Being Maori," *MAI Review* 2 (2009): 1–11.

[23] Ramari V. Port, John Arnold, Dale Kerr et al., "Cultural Enhancement of a Clinical Service to Meet the Needs of Indigenous People; Genetic Service Development in Response to Issues for New Zealand Maori," *Clinical Genetics* 73.2 (February 2008): 132–38; Linda Burhansstipanov, Lynne T. Bemis & Mark B. Dignan, "Native American Recommendations for Genetic Research to Be Culturally Respectful," *Jurimetrics* 42.2 (Winter 2002): 149–57; Richard R. Sharp & Morris W. Foster, "Community Involvement in the Ethical Review of Genetic Research: Lessons from American Indian and Alaska Native Populations," *Environmental Health Perspectives* 110, Supplement 2 (April 2002): 145–48; Laura Arbour & Doris Cook, "DNA on Loan: Issues to Consider When Carrying Out Genetic Research with Aboriginal Families and Communities," *Community Genetics* 9.3 (2006): 153–60; Malcolm B. Bowekaty & Dena S. Davis, "Cultural Issues in Genetic Research with American Indian and Alaskan Native People," *IRB: Ethics and Human Research* 25.4 (July–August 2003): 12–

Two areas of tension emerged in the roundtable discussions in relation to informed consent to participate in genetic research. One area of tension related to whether only Indigenous community organizations or individual Indigenous people and families had the right to consent to participate in genetic research. The guidelines for the conduct of health research in Indigenous communities advise that owing to the "the distinctive cultures and community life of Aboriginal and Torres Strait Islander Peoples [... it is] essential that researchers engage with Aboriginal and Torres Strait Islander communities collectively, not just with individuals."[24]

However, for one genetic researcher at the roundtable, working directly with families has been a cornerstone of her long-standing genetic research. She argued that community organizations had a right to agree to the research, but that it was ultimately up to individuals to make their own decisions about their own participation. She reported that some research participants felt insulted that ethics committees (including Aboriginal ethics committees) based in places geographically distant from their communities could override their decision to participate, or could impose conditions that imply that local participants and families are not capable of making decisions for themselves (58).

In response, a member of an Aboriginal ethics committee raised concerns about directly recruiting Indigenous families rather than collaborating with representative bodies, and suggested that working directly with families can look like avoiding conflict: "the idea of [working with] small groups can be seen as cherry picking people who will agree with you" (59).

The second area of tension related to whether any Human Research Ethics Committee[25] (HREC) could approve an Indigenous genetics project, or whether this should be done by an Aboriginal ethics committee (where there was one available). In Australia, all research on humans needs approval from

---

15; Maui Hudson, Annabel L.M. Ahuriri–Driscoll, Marino G. Lea et al., "Whakapapa: A Foundation for Genetic Research?" *Journal of Bioethical Inquiry* 4.1 (2007): 43–49; Mervyn Tano, "Interrelationships among Native Peoples Genetic Research, and the Landscape: Need for Further Research into Ethical, Legal, and Social Issues," *Journal of Law, Medicine, and Ethics* 34.2 (Summer 2006): 301–309.

[24] National Health & Medical Research Council, *Values and Ethics: Guidelines for Conduct of Aboriginal and Torres Strait Islander Health Research* (Canberra: NHMRC, 2003): 18.

[25] These are known by other names in other jurisdictions; for instance, in the USA they are known as Institutional Review Boards.

an HREC in order to proceed. A survey of Australian HRECs found that only 2% were Aboriginal HRECs (where all members of the committee are Aboriginal) and 12% had some Aboriginal representation on the committee. A further 7% of committees referred Indigenous project applications to an Aboriginal HREC or subcommittee, while 45% reported using ad-hoc methods such as co-opting an Indigenous person onto the committee or sending the proposal to an Indigenous person for external review.[26] The authors of the article argued that more "formal mechanisms" for having Indigenous representation on the HREC or referring proposals to an Aboriginal HREC "are more likely to support processes of ethical review that are more transparent and accountable both to researchers and to Indigenous communities."[27]

While most of the genetic researchers at the roundtable had ethical approval from an Aboriginal ethics committee (alone, or in addition to approval from their University HREC), some did not. One geneticist working in a remote community had ethics approval from their University HREC but not from an Aboriginal ethics committee. He recounted how the community he worked in did not want the researchers to seek approval from the Aboriginal committee in their jurisdiction:

> We got ethics [approval] from [the University ethics committee] in 1986 and that [approval] has continued on. In 2007 we thought we better go and see about ethics and we went to [the community] and asked them whether we should go to the [Aboriginal ethics] committee and they said "No we're not interested in them don't go [to them]." (64)

The researchers have followed the wishes of the community leaders and have continued working with the University ethics committee and not the relevant Aboriginal ethics committee. Another participant expressed concern that the Indigenous community leaders in that case did not want the researchers to have their project approved through an Aboriginal ethics committee in addition to the University ethics committee: "We need indigenous people to be speaking with one voice" (65).

---

[26] Paul Stewart, Sanchia Shibasaki, Ian Anderson et al., "Aboriginal and Torres Strait Islander Participation in Ethical Review of Health Research," *Australian New Zealand Journal of Public Health* 30.3 (June 2006): 291–92.

[27] Stewart, Shibasaki, Anderson et al., "Aboriginal and Torres Strait Islander Participation in Ethical Review of Health Research," 292.

However, it would seem to us that while the issue of the form and composition of the ethics committee is significant and worthy of debate, all genetic-research proposals will be considered by a HREC in some form, and ethics committees would benefit from the development of specific guidelines or resources to support their decision making.

Conclusion: More Guidelines or More Trust?

The roundtable ended on the question of what was needed in the Australian context to make genetic research a culturally safe option for Indigenous people. A variety of new strategies were discussed to address these issues, including the development of guidelines or a 'guiding note' for Indigenous genetic research to complement the *National Statement* and *Values & Ethics* documents;[28] genetic literacy tools for researchers working in Indigenous contexts; genetic literacy workshops for Indigenous health workers and Indigenous leaders offered before or after relevant conferences; media guidelines for reporting Indigenous genetic-research findings; and cultural-competence training for genetic researchers. Levels of support for each of these strategies differed among roundtable participants.

This was the case, for instance, regarding the need for guidelines for genetic research in Indigenous communities. There is a clear lack of guidelines specifically addressing this issue in Australia. This contrasts with other comparable countries such as Canada, the USA, and Aoteoroa New Zealand where both academic literature and official guidelines (in the case of Canada) provide much guidance for genetic researchers seeking to work in Indigenous communities.[29] However, one Indigenous roundtable participant questioned

---

[28] National Health & Medical Research Council, *Values and Ethics: Guidelines for Conduct of Aboriginal and Torres Strait Islander Health Research*; National Health & Medical Research Council, Australian Research Council & Australian Vice-Chancellors' Committee, *National Statement on Ethical Conduct in Human Research* (Canberra: Australian Government, 2007).

[29] Port, Arnold, Kerr et al., "Cultural Enhancement of a Clinical Service to Meet the Needs of Indigenous People"; Canadian Institutes of Health Research, "CIHR Guidelines for Health Research Involving Aboriginal People," *Canadian Institutes of Health Research* (May 2007), http://www.cihr-irsc.gc.ca/e/29134.html (accessed 28 February 2011); Burhansstipanov, Bemis & Dignan, "Native American Recommendations for Genetic Research to Be Culturally Respectful"; Sharp, "Community Involvement in the Ethical Review of Genetic Research: Lessons from American Indian

whether guidelines were what were needed: "is it a problem of guidelines, or a problem of trust?"[30]

This question is particularly pertinent, considering that many of the ethical challenges of Indigenous genetic research relate to the communication of potential risks and benefits to research participants and the ongoing management of what was referred to above as 'representational' risks. Ethical review, while important, is arguably insufficient in addressing the problems inherent in the development of the relationships and communicative practices that might be needed. It might be that a broader approach is required to developing strategies and tools that build a shared understanding of the relevant issues. Certainly, building relationships of trust between genetic researchers and Aboriginal and Torres Strait Islander communities is needed above and beyond the question of the adequacy of current guidelines.

Ultimately, it may be that these two processes – writing guidelines and building trust – are complementary. The process of writing guidelines would bring together the many disparate stakeholders in Indigenous genetic research, from genetic researchers through funding bodies to Aboriginal ethics committees and Indigenous communities. Deciding what issues needed to be addressed and debating an appropriate response to those issues would test the relationships between these stakeholders, and hopefully strengthen them. Only by facing these issues head-on through more 'difficult conversations' can genetic researchers build their capacity to conduct culturally appropriate research, and, at the same time, Indigenous communities be empowered to make informed decisions about participating in genetic research.

⌘

---

and Alaska Native Populations"; Arbour & Cook, "DNA on Loan: Issues to Consider When Carrying Out Genetic Research with Aboriginal Families and Communities"; Bowekaty & Davis, "Cultural Issues in Genetic Research with American Indian and Alaskan Native People"; Hudson, Ahuriri–Driscoll, Lea et al., "Whakapapa: A Foundation for Genetic Research?"; Tano, "Interrelationships Among Native Peoples Genetic Research, and the Landscape: Need for Further Research into Ethical, Legal, and Social Issues."

[30] Personal notes, Emma Kowal, 2 July 2010.

Works Cited

Acharya, Tara, et al. "Strengthening the Role of Genomics in Global Health," *PLoS Medicine* 1.3 (December 2004): e40.

Anderson, Ian. "Introduction: The Aboriginal Critique of Colonial Knowing," in *Blacklines: Contemporary Critical Writing by Indigenous Australians*, ed. Michèle Grossman (Melbourne: Melbourne UP, 2003): 17–24.

Arbour, Laura, & Doris Cook. "DNA on Loan: Issues to Consider When Carrying Out Genetic Research with Aboriginal Families and Communities," *Community Genetics* 9.3 (2006): 153–60.

Australian Law Reform Commission. *Essentially Yours: The Protection of Human Genetic Information in Australia – Section 36: Kinship and Identity*. Canberra: Australian Government Publishing Service, 2003.

Balding, David J., Martin Bishop & Chris Cannings. *Handbook of Statistical Genetics* (Chichester: Wiley, 2007).

Bowekaty, Malcolm B., & Dena S. Davis. "Cultural Issues in Genetic Research with American Indian and Alaskan Native People," *IRB: Ethics and Human Research* 25.4 (July–August 2003): 12–15.

Brantlinger, Patrick. *Dark Vanishings: Discourse on the Extinction of Primitive Races, 1800–1930* (Ithaca NY & London: Cornell UP, 2003).

Burhansstipanov, Linda, Lynne T. Bemis & Mark B. Dignan. "Native American Recommendations for Genetic Research to Be Culturally Respectful," *Jurimetrics* 42.2 (Winter 2002): 149–57.

Burt, Tim, Bart Currie, Charles Kilburn et al. "Machado–Joseph Disease in East Arnhem Land, Australia: Chromosome 14q32.1 Expanded Repeat Confirmed in Four Families," *Neurology* 46.4 (April 1996): 1118–22.

Canadian Institutes of Health Research & Institute of Aboriginal Peoples' Health. *An exploratory workshop on a tribal controlled DNA bank: Part 1* (Vancouver: CIHR, 2001).

——. *An exploratory workshop on a tribal controlled DNA bank: Part 2* (Tofino, British Columbia: CIHR, 2002).

——. "CIHR Guidelines for Health Research Involving Aboriginal People," *Canadian Institutes of Health Research* (May 2007), http://www.cihr-irsc.gc.ca/e/29134.html (accessed 28 February 2011).

Caulfield, Timothy, Stephanie M. Fullerton, Sarah E. Ali–Khan et al. "Race and Ancestry in Biomedical Research: Exploring the Challenges," *Genome Medicine* 1.8 (2009): 81–88.

Foster, Morris W. "Looking for Race in All the Wrong Places: Analyzing the Lack of Productivity in the Ongoing Debate About Race and Genetics," *Human Genetics* 126.3 (September 2009): 255–362.

——, & Richard R. Sharp. "Genetic Research and Culturally Specific Risks: One Size Does Not Fit All," *Trends in Genetics* 16.2 (February 2000): 93–95.

Fullwiley, Duana. “The Biologistical Construction of Race: ‘Admixture’ Technology and the New Genetic Medicine,” *Social Studies of Science* 38.5 (October 2008): 695–735.

Hudson, Maui, Annabel L.M. Ahuriri–Driscoll, Marino G. Lea et al. “Whakapapa: A Foundation for Genetic Research?” *Journal of Bioethical Inquiry* 4.1 (2007): 43–49.

Humphery, Kim. “Dirty Questions: Indigenous Health and ‘Western Research’,” *Australian and New Zealand Journal of Public Health* 25.3 (June 2001): 197–202.

Indigenous Peoples Council on Biocolonialism. “Indigenous Peoples, Genes and Genetics: What Indigenous Peoples Should Know About Biocolonialism,” *Indigenous Peoples Council on Biocolonialism* (June 2000), http://www.ipcb.org/publications/primers/index.html (accessed 15 August 2010).

Ivison, Duncan, Paul Patton & Will Sanders, ed. *Political Theory and the Rights of Indigenous Peoples* (Cambridge: Cambridge UP, 2000).

Knerr, Sarah, Edward Ramos, Juleigh Nowinski et al. “Human Difference in the Genomic Era: Facilitating a Socially Responsible Dialogue,” *BMC Medical Genomics* 3.1 (May 2010): 20.

Kowal, Emma, Ian Anderson, & Lobna Rouhani. *Indigenous Australians and Genetic Research: Beginning the Conversation* (Melbourne: Lowitja Institute, 2011).

Lee, Sandra Soo–Jin. “Pharmacogenomics and the Challenge of Health Disparities,” *Public Health Genomics* 12.3 (February 2009): 170–79.

Mello, Michelle, & Leslie Wolf. “The Havasupai Indian Tribe Case – Lessons for Research Involving Stored Biologic Samples,” *New England Journal of Medicine* 363.3 (July 2010): 204–207.

Mohamed, Justin. “We Don’t Like Research – But in Koori Hands It Could Make a Difference” (Melbourne: VicHealth Koori Health Research and Community Development Unit & University of Melbourne, 2000)

National Council of Ethics in Human Research, Canadian Commission for UNESCO, & Health Canada. *Continuing the Dialogue: Genetic Research with Aboriginal Individuals and Communities* (Vancouver: NCEHR, 2001).

National Health & Medical Research Council. *Values and Ethics: Guidelines for Conduct of Aboriginal and Torres Strait Islander Health Research* (Canberra: NHMRC, 2003).

——, Australian Research Council, & Australian Vice-Chancellors’ Committee. *National Statement on Ethical Conduct in Human Research* (Canberra: Australian Government, 2007).

Paradies, Yin C. “Beyond Black and White: Essentialism, Hybridity and Indigeneity,” *Journal of Sociology* 42.4 (December 2006): 355–67.

——, Michael J. Montoya & Stephanie M. Fullerton. “Racialized Genetics and the Study of Complex Diseases: The Thrifty Genotype Revisited,” *Perspectives in Biology and Medicine* 50.2 (Spring 2007): 203–27.

Port, Ramari V., John Arnold, Dale Kerr et al. "Cultural Enhancement of a Clinical Service to Meet the Needs of Indigenous People; Genetic Service Development in Response to Issues for New Zealand Maori," *Clinical Genetics* 73.2 (February 2008): 132–38.

Ramos, Edward, & Charles Rotimi. "The A's, G's, C's, and T's of Health Disparities," *BMC Medical Genomics* 2.1 (July 2009): 29.

Raumati Hook, Gary. "'Warrior Genes' and the Disease of Being Maori," *MAI Review* 2 (2009): 1–11.

Reardon, Jenny. *Race to the Finish: Identity and Governance in an Age of Genomics* (Princeton NJ: Princeton UP, 2005).

Royal, Charmaine D., John Novembre, Stephanie M. Fullerton et al. "Inferring Genetic Ancestry: Opportunities, Challenges, and Implications," *American Journal of Human Genetics* 86.5 (May 2010): 661–73.

Russell, Lynette. "Introduction" to *Boundary Writing: An Exploration of Race, Culture, and Gender Binaries in Contemporary Australia*, ed. Lynette Russell (Honolulu: U of Hawai'i P, 2006).

——. *Savage Imaginings: Historical and Contemporary Constructions of Australian Aboriginalities* (Melbourne: Australian Scholarly, 2001).

Sharp, Richard, & Morris W. Foster. "Community Involvement in the Ethical Review of Genetic Research: Lessons from American Indian and Alaska Native Populations," *Environmental Health Perspectives* 110, Supplement 2 (April 2002): 145–48.

Stewart, Paul, Sanchia Shibasaki, Ian Anderson, et al. "Aboriginal and Torres Strait Islander Participation in Ethical Review of Health Research," *Australian and New Zealand Journal of Public Health* 30.3 (June 2006): 291–92.

TallBear, Kimberly. "DNA, Blood, and Racializing the Tribe," *Wicazo SA Review* 18.1 (Spring 2003): 81–107.

——. "Narratives of Race and Indigeneity in the Genographic Project," *Journal of Law, Medicine and Ethics* 35.3 (Fall 2007): 412–24.

Tano, Mervyn. "Interrelationships among Native Peoples, Genetic Research, and the Landscape: Need for Further Research into Ethical, Legal, and Social Issues," *Journal of Law, Medicine, and Ethics* 34.2 (Summer 2006): 301–309.

Tsosie, Rebecca, & Joan L. McGregor. "Genome Justice: Genetics and Group Rights," *Journal of Law, Medicine and Ethics* 35.3 (Fall 2007): 352–55.

Tuhiwai Smith, Linda. *Decolonizing Methodologies, Research and Indigenous Peoples* (London & New York: Zed, 1999).

⌘

Travelling Bones

The Repatriation of Indigenous Ancestral Remains

MATTHEW RIMMER

IN THE ALBUM *JOURNEY*, ARCHIE ROACH – THE AUSTRALIAN INDIGENOUS singer–songwriter hailing from Mooroopna in Victoria – has a melancholy song called "Travellin' Bones."[1] It is about the repatriation of Indigenous ancestral remains to their rightful home. This essay considers the legal, ethical, and cultural conflicts over Australian Indigenous remains being held in museums, in Australia, the UK, the European Union, and the USA.[2] James Nason comments:

> The explosion of legal and extra legal attention on issues of cultural property and heritage was born of the frustration and anger of indigenous peoples whose rights and perspectives about cultural property and heritage issues had been largely absent and essentially unwanted by the museum of community.[3]

Part I focuses on disputes in Australia involving the repatriation of Indigenous Australian remains. In *Bropho v. HREOC*, there was controversy over a

---

[1] Archie Roach, "Travellin' Bones," on *Journey* (Liberation Music LBCD92595, 2007).

[2] There have been parallel issues involved with the law and ethics of genetic testing with respect to deceased public figures. See Jordan Paradise & Lori Andrews, "Tales from the Crypt: Scientific, Ethical, and Legal Considerations for Biohistorical Analysis of Deceased Historical Figures," *Temple Journal of Science, Technology and Environmental Law* 26.2 (Fall 2007): 223–99.

[3] James Nason, "Beyond Repatriation: Cultural Policy and Practice for the Twenty-First Century," in *Borrowed Power: Essays on Cultural Appropriation*, ed. Bruce Ziff & Pratima Rao (New Brunswick NJ: Rutgers UP, 1997): 309.

cartoon mocking the repatriation of the remains of Yagan, an Indigenous warrior, to Western Australia.[4] There was a discussion about the operation of the Racial Discrimination Act 1975 (Cth), and the exemptions available from the operation of the regime. Part II considers the efforts by Te Papa Tongarewa – the Museum of New Zealand – to repatriate Māori and Moriori ancestral remains to New Zealand, and to *iwi* communities of origin. The conclusion considers the relevance of the United Nations Declaration on the Rights of Indigenous Peoples 2007, and issues raised by ventures such as the Genographic Project.

I Australia

There have been a number of instances in which museums have yielded up the remains of Indigenous Australians to their original communities. The National Museum of Australia, based in the national capital, Canberra, has played a key role in the repatriation of Indigenous ancestral remains:

> The National Museum of Australia's repatriation team works to return ancestral human remains and secret and sacred objects to Aboriginal and Torres Strait Islander people. The Museum has been returning remains and objects since its inception in 1980 and is recognised nationally and internationally for its repatriation work. More than 1000 individuals and over 360 secret and sacred objects have been unconditionally returned to Indigenous communities. Museum staff continue to work closely with Indigenous communities to return remains and artefacts to their ancestral custodians.[5]

The National Museum of Australia's Aboriginal and Torres Strait Islander collection included human remains and sacred objects, largely derived from the collections originally held by the former Australian Institute of Anatomy. It has documented its efforts in repatriation of such ancestral remains and sacred objects. The National Museum of Australia has become the temporary repository and repatriation point for a range of collections to be returned overseas.

There have also been efforts to reclaim Indigenous human remains in overseas institutions – particularly in the UK in May 2007. Aboriginal remains

---

[4] *Bropho* v. *HREOC* FCAFC 16 (2004).

[5] National Museum of Australia, "Repatriation," *National Museum of Australia*, http://www.nma.gov.au/collections/repatriation/ (accessed 20 February 2011).

held in the Natural History Museum in London were repatriated to their original community in Tasmania, after much disputation and the threat of legal conflict.[6] The Tasmanian Aboriginal Centre had sought their return for more than twenty years. On the return of the Tasmanian remains, Carolyn Smallwood commented:

> This is both a joyful and deeply distressing occasion for Aborigines. These remains, together with those of 13 more of our people, were removed without any Aboriginal consent from Tasmania during the 1880s and we have been fighting since the 1980s for their return. Greg and I are both proud and honoured to be able to take them home to lay their tormented spirits to rest. [...] However, we are grieving also, because the only reason the Museum has allowed these four remains to be returned at this point is because scientists have finished their tests on them. We have been telling the Museum for over thirty years that physical interference of any kind with our dead is an absolute violation of Aboriginal spiritual beliefs and we are completely opposed to any form of it. The Museum has shown and continues to show no respect for our wishes or our religion.[7]

In 2008, the National Museum of Natural History at the Smithsonian Institute in Washington, D.C., returned the remains of thirty-three Indigenous people from the Gunbalanya and Groote Eylandt in Arnhemland, in the Northern Territory, after sixty years. There have been ongoing efforts to repatriate further ancestral remains.[8]

There have also been larger questions addressing the reportage of such controversies over the repatriation of Indigenous remains. In Western Austra-

---

[6] Kathleen Falko, *Grave Concerns: The Repatriation of Tasmanian Indigenous Remains and the Nature of Cultural Property Rights* (Canberra: ANU College of Law, Second Semester, 2007); Chris Davies & Kate Galloway, "The Story of Seventeen Tasmanians: the Tasmanian Aboriginal Centre and Repatriation from the Natural History Museum," *Newcastle Law Review* 11.1 (2008–2009): 11.

[7] Tasmanian Aboriginal Centre, "Aborigines are able to collect human remains now scientists at Natural History Museum have finished testing them" (27 April 2007), http://www.eniar.org/pdf/PR_%20TAC.pdf (accessed 20 February 2011).

[8] Anon., "Aboriginal Remains Repatriation," *Creative Spirits*, http://www.creativespirits.info/aboriginalculture/people/aboriginal-remains.html (accessed 20 February 2011).

lia, there was much debate about the repatriation of the remains of Yagan, a Nyungar warrior. Hannah McGlade provides a snapshot of the return:

> Old Nyungar men are singing and the clapping sticks can be heard throughout Perth's international airport late in the night. There are up to three hundred Nyungars who have come to meet the Aboriginal delegation due to arrive on the 11 pm flight from London. The delegation are bringing home the head of Yagan, the Ngunyar warrior. [...]
>
> Yagan was murdered in 1833 by two young brothers whom he had befriended and with whom he had shared a meal. His head was cut off in accordance with the barbaric English colonial practice of the time and then sent to England where it was displayed in various fairs and sideshows, in the Insect Room of the Royal Institution of Liverpool, and the Liverpool City Museum. In 1964 Yagan's head, along with two Maori heads and a Peruvian mummy, was buried in a local cemetery.[9]

She comments: "The story of Yagan is an extremely powerful story: it is a story of invasion, of early contact between two cultures, of colonialism and its racist, bloody nature."[10]

In the case of *Bropho v. HREOC*, there was controversy over a cartoon mocking the repatriation of the remains of the aforementioned Yagan to Western Australia.[11] The matter arose after a complaint to the Human Rights and Equal Opportunity Commission that its distribution was reasonably likely to offend, insult, humiliate or intimidate Nyoongar people and was done on account of their race. The complaint alleged that the conduct was unlawful by virtue of section 18C of the Racial Discrimination Act 1975 (Cth).

A Commissioner of the Human Rights Commission made the following findings about the cartoon:

> I am satisfied that, based on the reasonable victim test, a reasonable Nyungar or Aboriginal person would have found the contents of the cartoon offensive, insulting, humiliating or intimidating. [...] Firstly, the cartoon presents a demeaning portrayal of Yagan, an ancestor of the complainants – particularly the reference to a warm beer and a quiet pommie pub in the context of the widespread community view regarding the relationship between alcohol and Aboriginal people.

---

9 Hannah McGlade, "The Repatriation of Yagan: A Story of Manufacturing Dissent," *Law Text Culture* 4.1 (1998): 244–45.

10 McGlade, "The Repatriation of Yagan: A Story of Manufacturing Dissent," 252.

11 *Bropho* v. *HREOC* FCAFC 16 (2004).

> Secondly, the cartoon contains derogatory and demeaning references to the Waugyl, a religious figure. Thirdly, it treats the issue of death in a manner which causes offence to Aboriginal people. Fourthly, it provides intimate details of the ancestry of individuals in circumstances where the intercourse was not a matter of choice for the Aboriginal women concerned, and suggests a diminishing of the race by the resultant racial mix. Fifthly, it reinforces a misinformed and stereotypical view of Aboriginal people taking advantage of government grants.[12]

However, the Commissioner found that the newspaper could raise a defence under section 18D of the Racial Discrimination Act 1975 (Cth), because it acted "reasonably and in good faith," commenting: "The cartoon concerned was published after a series of articles and editorial comments dealing with this issue" and "it was an issue of importance for the West Australian community in general, as well as to the Aboriginal community, and was treated as such by the newspaper."[13]

A challenge to his decision by way of judicial review was dismissed by a judge of the Federal Court.[14] The Full Federal Court of Australia considered whether a cartoon about the repatriation of the Aboriginal leader Yagan's head amounted to racial vilification. Justice Robert French (who has since become the chief justice of the High Court of Australia) reflected upon the history of the tale:

> The story of the Western Australian Aboriginal leader, Yagan, and his death at the hands of two young settlers in 1833 is a tale of colonial tragedy. It has sadly familiar overtones of mutual incomprehension, fear, prejudice and retribution. The sequel to his death, the severing and smoking of his head and its removal to England for display in a museum, demonstrated a contempt for his humanity which is striking even at this historical remove. The recovery of Yagan's head from a graveyard in Liverpool and its return to Australia with a group of Aboriginal elders in 1997 was accompanied by a degree of sometimes undignified acrimony over who had the appropriate cultural claims, by descent, to bring the remains back.[15]

---

[12] *Bropho* v. *HREOC* FCA 1510 (2002).

[13] *Bropho* v. *HREOC* FCA 1510 (2002).

[14] As reported in *Bropho* v. *HREOC* FCA 1510 (2002).

[15] *Bropho* v. *HREOC* FCAFC 16 (2004);

Justice French noted that "the conduct of those involved in the controversy was lampooned in a cartoon published in the Western Australian newspaper in September 1997; in particular, "the cartoon reflected upon the mixed ancestry of some of the Aboriginal people involved."[16]

The Human Rights and Equal Opportunities Commission had found the cartoon was exempt under s. 18D of the Racial Discrimination Act 1975 (Cth), because it was done reasonably and in good faith. This finding was ultimately upheld by the majority of the Full Federal Court. With the support of Justice Christopher Carr, Justice French observed:

> An act will be done reasonably in the performance, exhibition or distribution of an artistic work if it is done for the purpose and in a manner calculated to advance the purpose of the artistic expression in question. An act is done reasonably in relation to statements, publications, discussions or debates for genuine academic, artistic or scientific purposes, if it bears a rational relationship to those purposes. The publication of a genuine scientific paper on the topic of genetic differences between particular human populations might, for one reason or another, be insulting or offensive to a group of people. Its discussion at a scientific conference would no doubt be reasonable.[17]

Nonetheless, there remain concerns about the interpretation of this defence or exception to claims of racial vilification, and what constitutes 'reasonable conduct'.[18]

Justice Carr concurred that the Commission had followed the correct procedure in its reasoning:

> First, the Commission expressed its satisfaction that the second respondent acted reasonably and in good faith. Then it placed the cartoon into context. It assessed the cartoon in the context of having been published after a series of articles and editorial comments dealing with the issue of the return of Yagan's head and the disunity it had generated. Next the Commission observed that the issue was treated as one which was important for the West Australian community in general, as well as to the Aboriginal community, and had been treated as such by the second respondent. There was then a reference to Commissioner Johnston's 'margin of tolerance'.[19]

---

[16] *Bropho* v. *HREOC* FCAFC 16 (2004);

[17] *Bropho* v. *HREOC* FCAFC 16 (2004).

[18] *Bropho* v. *HREOC* FCAFC 16 (2004).

[19] *Bropho* v. *HREOC* FCAFC 16 (2004).

Carr noted further:

> The Commission went on to say that its 'view in this area' (which I take to be its view that the second respondent acted reasonably and in good faith) was strengthened by reading other material published on this issue in 'The West Australian' which provided a balanced report of what took place, and an opinion which, in the main, encouraged unity in, and support of, the Aboriginal community.[20]

In a principled and persuasive dissent, Justice Lee observed that the Commission had failed to ask the correct question in the case:

> Contemporaneous, or prior, publication of anodyne material would not, in itself, make an act of publication done because of race and involving racially offensive material, an act done reasonably and in good faith. A publisher of a catholic range of opinions could not rely upon past publication of diverse material to show that it acted reasonably and in good faith by publishing, because of race, a work or material that is offensive, insulting, humiliating or intimidating to persons of that race, if it acts without regard to whether the act of publication would cause the harm the Act seeks to prevent, and does not attempt to show how the risk of harm from the otherwise prohibited act, was counterbalanced, or outweighed, by matters showing the act to have been done reasonably and in good faith.[21]

The judge concluded:

> The evidence of the editor, 'that he had made a judgment call, knowing that the cartoon would receive some opposition' did not show that the racially offensive consequences of the act of publication, an act done because of race, had been duly considered and on its own that evidence was incapable of satisfying the onus on WA Newspapers to show that the act of publication had been done reasonably and in good faith.[22]

There was a further attempt to get the High Court of Australia to consider a further appeal in the matter.[23] Chief Justice Anthony Gleeson and Justice Kenneth Madison Hayne rejected the application:

---

[20] *Bropho* v. *HREOC* FCAFC 16 (2004).

[21] *Bropho* v. *HREOC* FCAFC 16 (2004).

[22] *Bropho* v. *HREOC* FCAFC 16 (2004).

[23] *Bropho* v. *HREOC & Anor* HCATrans 9 (9 February 2005).

> Having regard to the reasoning of the Commission and to the particulars of the proposed ground of appeal, a majority of this Court is of the view that there are insufficient prospects of success of an appeal to warrant a grant of special leave and further that the case is not a suitable vehicle for the agitation of the questions sought to be raised by the applicant.[24]

The progressive Justice Michael Kirby[25] dissented that he would have granted special leave:

> I would do so principally having regard to: first, the possible significance of the Constitution to the issues in the case, which appears not to have been given weight in the courts below; secondly, the dissenting views expressed by Justice Lee in the Full Court of the Federal Court; thirdly, the importance of the issue and the interpretation and application of the *Racial Discrimination Act* 1975 (Cth) ss. 18C and 18D; and, fourthly, the significance of the issues for racial, ethnic and religious minorities in Australia, including for the indigenous people.[26]

There has been much discussion about the verdict and its reasoning, particularly among human-rights lawyers.[27] The case is also interesting in that it anticipates the later controversy over the so-called 'Danish cartoons',[28] which raised questions of religious and racial vilification, and countervailing concerns about freedom of speech.

---

[24] *Bropho* v. *HREOC & Anor* HCATrans 9 (4 February 2005).

[25] See *Appealing to the Future: Michael Kirby and his Legacy*, ed. Hugh Selby & Ian Freckleton (Sydney: Thomson Reuters, 2009).

[26] *Bropho* v. *HREOC & Anor* HCATrans 9 (4 February 2005).

[27] Hannah McGlade, "Race Vilification Before the Human Rights and Equal Opportunity Commission," *Indigenous Law Bulletin* 5.7 (2001): 8; Hannah McGlade, "Race Discrimination in Australia: A Challenge for Treaty Settlement," in *Honour Among Nations?: Treaties and Agreements with Indigenous People*, ed. Marcia Langton et al. (Melbourne: Melbourne UP, 2004): 273–87; Anna Chapman, "Australian Racial Hatred Law: Some Comments on Reasonableness and Adjudicative Method in Complaints Brought by Indigenous People," *Monash University Law Review* 30.1 (2004): 27–48.

[28] See, for instance, *Danish Muslim Organisations* v. *Jyllands-Posten* ('The Danish Cartoons case') (Denmark: City Court of Aarhus, 2006); *Muslim World League* v. *Charlie-Hebdo and Philippe Val* (France, 2006); and *Jamiat-ul-ulama of Transvaal* v. *Johncom Media Investment Ltd* 1127/06 (Unreported, High Court of South Africa, Witwatersrand Local Division, 2006).

II New Zealand

In 1993, nine Māori tribes of Mataatua hosted the First International Conference on the Cultural and Intellectual Property Rights of Indigenous Peoples. The delegates produced the Mataatua Declaration on Cultural and Intellectual Property Rights of Indigenous Peoples 1993, which, among other things, stressed:

> All human remains and burial objects of indigenous peoples held by museums and other institutions must be returned to their traditional areas in a culturally appropriate manner. Museums and other institutions must provide, to the country and indigenous peoples concerned, an inventory of any indigenous cultural objects still held in their possession. Indigenous cultural objects held in museums and other institutions must be offered back to their traditional owners.[29]

The Mataatua Declaration on Cultural and Intellectual Property Rights of Indigenous Peoples 1993 called upon the United Nations to "monitor and take action against any States whose persistent policies and activities damage the cultural and intellectual property rights of indigenous peoples."[30]

Te Papa Tongarewa – the grand and magnificent Museum of New Zealand – has coordinated a formal repatriation programme of Māori and Moriori ancestral remains to New Zealand, and to *iwi* communities of origin.[31] Brian Hole reflects upon the origins of this institution:

> The old Dominion Museum in Wellington had from time to time been described as a 'colonialist' and 'monocultural' institution. In 1998 it reopened its doors in a new location on Wellington's waterfront with a redefined mission of representing a bicultural New Zealand, involving Maori staff and cultural participation, and housing a functioning *marae* (community meeting place), the only museum in the country to do so. The museum also contains an 'ancestral remains vault', or *wahi tapu*, which is the only place in New Zealand specifically designed to hold unprovenanced Maori ancestral remains. Because of this, Te Papa is the receiving museum for the majority of remains returned to New

---

[29] *Mataatua Declaration on Cultural and Intellectual Property Rights of Indigenous Peoples* 1993.

[30] *Mataatua Declaration on Cultural and Intellectual Property Rights of Indigenous Peoples* 1993.

[31] Te Papa Tongarewa, "Repatriation," *Museum of New Zealand*, http://www.tepapa.govt.nz/AboutUs/Repatriation/Pages/overview.aspx (accessed 21 February 2011).

> Zealand, and has a proactive programme of researching and requesting repatriation from overseas institutions.[32]

The museum has developed an extensive Kōiwi Tangata Policy, last updated in October 2010.[33] The policy notes:

> The Museum of New Zealand Te Papa Tongarewa (hereinafter referred to as Te Papa) regards the kōiwi tangata in its guardianship as tūpuna to be cared for in a consistent and culturally appropriate manner until such time as the kōiwi tangata are returned to their place of provenance or to an appropriate final resting place.[34]

The purpose of the policy is "to provide guidelines that ensure the kōiwi tangata Māori and Moriori in the guardianship of Te Papa will be managed and cared for in a consistent and culturally appropriate manner." The Museum prepared a report on its efforts at repatriating Māori remains in April 2010.[35]

The New Zealand Government has elaborated six principles for the Karanga Aotearoa Repatriation Programme. First, the government's role is mainly one of facilitation – it does not claim ownership of *kōiwi/koimi tangata*. Second, repatriation is by mutual agreement only. Third, the programme does not cover Māori or Moriori remains in war graves. Fourth, *kōiwi/koimi tangata* must be identified as originating in New Zealand or the Chatham Islands. Māori and Moriori are able to be involved in the repatriation of *kōiwi/koimi tangata* and to determine the final resting place. No payment will be made for *kōiwi/koimi tangata*.

In 2008, the Trustees of the British Museum considered a request by the New Zealand Te Papa Tongarewa under section 47 of the Human Tissue Act

---

[32] Brian Hole, "Playthings for the Foe: The Repatriation of Human Remains in New Zealand," *Public Archaeology* 6.1 (2007): 18.

[33] Te Papa Tongarewa, "Kōiwi Tangata Policy," *Museum of New Zealand* (1 October 2010), http://www.tepapa.govt.nz/SiteCollectionDocuments/AboutTePapa/Repatriation/Draft%20Koiwi%20Tangata%20Policy%20dated%201%20October%202010.pdf (accessed 21 February 2011).

[34] Te Papa Tongarewa, "Kōiwi Tangata Policy."

[35] Nicola Kira Smith, "Waiuku Kōiwi Tangata Report," *Museum of New Zealand* (April 2010), http://www.tepapa.govt.nz/SiteCollectionDocuments/AboutTePapa/Repatriation/Waiuku%20Kōiwi%20Tangata%20Report.pdf (accessed 21 February 2011).

2004 (UK) for the repatriation of seven preserved tattooed heads and nine human bone fragments.[36]

The British Museum has developed a policy on human remains.[37] It is informed by an array of ideologies – including what John Henry Merryman would call 'cultural property internationalism', which he defines as follows:

> the proposition that everyone has an interest in the preservation and enjoyment of cultural property, wherever it is situated, from whatever cultural or geographical source it derives.[38]

The policy is a somewhat bizarre document – reflecting the defensive posture of the organization to requests for the repatriation of human remains, and its somewhat recalcitrant presumption that it should be the custodian of human remains in the collection.

The British Museum has three main justifications for its retention of human remains. First, the Museum emphasizes that "human remains in the Collection are a record of the varied ways that different societies have conceived of death and disposed of the remains of the dead." It notes:

> human remains in various contexts and forms constitute an important part of the Collection: from Lindow Man, an ancient inhabitant of Britain, who may have been ritually murdered and his body then deposited in a bog in Cheshire, to the ancient mummies from Egypt, consistently voted among the Museum's most popular exhibits.

Moreover, the Museum argues that

> the study of human remains provides one of the most direct and insightful sources of information on different cultural approaches to death, burial practices and belief systems, including ideas about the afterlife.

---

[36] The British Museum, "Request for Repatriation of Human Remains to New Zealand," *The British Museum* (April 2008), http://www.britishmuseum.org/the_museum/news_and_press_releases/statements/human_remains/repatriation_to_new_zealand.aspx (accessed 21 February 2011).

[37] The British Museum, "Policy on Human Remains," *The British Museum* (6 October 2006), http://www.britishmuseum.org/PDF/Human%20Remains%206%20 Oct%202006.pdf (accessed 21 February 2011). Unless otherwise indicated, quotations that follow are from this document.

[38] John Henry Merryman, "Cultural Property Internationalism," *International Journal of Cultural Property* 12 (2005): 11.

It contends that the "collection should be protected, because it provides an opportunity to look at the diversity of human ideas about death and the human body across cultures of vastly different times and places."

Second, the Museum stresses that "human remains in the Collection help advance important research in fields such as the history of disease, changing epidemiological patterns, forensics and genetics." It emphasizes that "challenging theories about human evolution are being developed from the study of human remains in museum collections such as, for example, the likelihood that there is no genetic basis for modern concepts of race."

Third, the Museum argues that "human remains, which have been physically modified by a person working within a cultural context, or which form part of an archaeological record, illuminate other objects in the Collection." In its view, "these remains may be very important and sometimes irreplaceable records and symbols of one or more of the world's cultures." The underlying philosophy of the document is that human remains are part of world heritage – civilization, as it were – and that global institutions, such as the British Museum, are the proper custodians of such collections. Further,

> the Trustees consider that the public interest is strongly in favour of the retention in the Collection of human remains that have been modified for a purpose other than mortuary disposal (e.g., made into a Tibetan Buddhist thighbone trumpet) and will not accept claims for transfer in respect of them.

The British Museum's 'Public Interest Test' provides for the following:

> Having taken account of all the principles set out in part 5 of this Policy, the Trustees will then decide whether they believe that: either, (in the case of human remains less than 100 years old) the significance of the direct and close genealogical link with the human remains demonstrated by the claimants outweighs the public benefit to the world community of retaining the human remains in the Collection; or (in the case of human remains more than 100 years old) the significance of the Cultural Continuity and the Cultural Importance of the human remains demonstrated by the community making the claim outweighs the public benefit to the world community of retaining the human remains in the Collection.

The British Museum commented on their response to the application from Te Papa Tongarewa:

> The museum policy starts from a presumption of retention which can be outweighed in certain circumstances. They concluded that in the case of the seven preserved tattooed heads it was not clear whether or not a process of mortuary disposal had been interrupted or disturbed; and that it was not clear that the importance of the remains to an original community outweighed the significance and importance of the remains as sources of information about human history.
>
> On the other hand, they concluded that it was very probable that the fragments of human bone had been part of a process of mortuary disposal, and that the importance of the remains to the claimants outweighed any likely public benefit of retaining the remains in the collection. Therefore they agreed that the nine human bone fragments should be transferred to Te Papa Tongarewa.

The rather recalcitrant position of the British Museum in seeking to maintain ownership of the seven preserved tattooed Māori heads seems quite bizarre. It is certainly not evident that such remains should be retained, given "the significance and importance of the remains as sources of information about human history."

In 2009, Te Papa Tongarewa had somewhat more success in negotiating the repatriation of thirty-three ancestral remains from five museums and institutions in the European Union – including the National Museum of Wales in Cardiff, the Gothenburg Natural History Museum and the Museum of World Culture in Sweden, the Hunterian Museum in Glasgow University, Scotland, and Trinity College Dublin, Republic of Ireland.[39] Michelle Hippolite, the acting executive chief of the Karanga Aotearoa unit, in the repatriation unit at Te Papa, commented: "This is both a time for sad reflection on the turbulent journeys these ancestors experienced and, at the same time, a cause for joy and hope as they are returned," and observed: "On behalf of the Karanga Aotearoa Repatriation Team, I thank the institutions involved for their positive decisions to repatriate and for their support in the repatriation planning."[40]

---

[39] Te Papa Tongarewa, "Te Papa to undertake second largest repatriation of Māori human remains from the United Kingdom and Europe," *Museum of New Zealand* (13 November 2009), http://www.tepapa.govt.nz/AboutUs/Media/MediaReleases/2009/Pages/Secondlargestrepatriation.aspx (accessed 21 February 2011).

[40] "Te Papa to undertake second largest repatriation of Māori human remains from the United Kingdom and Europe."

Catherine Morin–Desailly, the Senator for the Seine–Maritime in France, was responsible for the development of a bill to allow for the repatriation of the remains. The Act aims to authorize the restitution by France of Māori heads to New Zealand and pertains to the management of collections.[41] It is designed to modify a disposition in French law which prevents museums from 'alienating' their assets, including human remains. In addition to dealing specifically with *Toi Moko*, the Act also creates a committee to assess restitution requests on a case-by-case basis.

Catherine Morin–Desailly observed, in an interview about the legislative regime:

> First, these are really special human remains. They are the result of barbarian methods, I mean, in some cases, slaves were tattooed and then decapitated to feed the global mummified heads traffic / trade at the time. And this was considered art work! So our repatriation bill doesn't apply to remains like Egyptian mummies for instance. Second criterion remains can be repatriated if they are to be buried in their home country. It's not a deal from a museum to another. Third, remains can be repatriated if they are not used or useful for scientific research. And fourth, we can repatriate human remains if it's requested by a contemporary people. In the case of the Maoris, it wasn't a long time ago, imagine, it could almost be your great great great grand parents![42]

She scoffed at the argument of some French cultural institutions that such ancient remains were part of the French cultural heritage or patrimony – "A part of French heritage? Well I don't think I as a French person own these Māori human remains" – and concluded: "We western people end up asking

---

[41] *ADOPTED TEXT NO. 455: TO AUTHORIZE THE RESTITUTION OF MAORI HEADS TO NEW ZEALAND AND CONCERNING MANAGEMENT OF COLLECTIONS*, NATIONAL ASSEMBLY, CONSTITUTION OF OCTOBER 4, 1958, THIRTEENTH PARLIAMENT, 2009–2010 ORDINARY SESSION, MAY 4, 2010 (FRANCE).

[42] Caroline Lafarguie, "French Parliament Passes Bill To Repatriate Maori Heads," *Radio Australia* (6 May 2010), http://www.radioaustralia.net.au/international/radio/onairhighlights/french-parliament-passes-bill-to-repatriate-maori-heads (accessed 21 February 2011).

ourselves the same big question, Can we do and accept anything in the name of art and culture?"[43]

Te Papa in Wellington welcomed the French Parliament's *Toi Moko* repatriation-act decision.[44] Michelle Hippolite, Te Papa's Acting Chief Executive and *Kaihautū*, observed: "The passing of this Act is a significant move by the French Government, and a major acknowledgment to indigenous people."[45] She said: "this Act now opens the door for Te Papa, on behalf of Māori, to approach and formally negotiate repatriation with Museums in France."[46] The Museum has been coordinating the repatriation of Māori remains from all over the world for fifty years now. Hippolite commented upon the decision:

> The Museum of Natural History of Rouen in France will be the first to hand back its unique Maori head to the Te Papa Museum, probably before the end of 2010. But the Quai Branly Museum, the largest collection of items from Oceania in France, simply refused to comment on the repatriation bill. The direction of the Museum is not keen on giving back its 7 Toi Moko, which haven't been shown to the public for decades. The director of the heritage department at the Quai Branly Museum, Yves Le Fur, said last year that these 7 Maori heads, in quotes, 'are part of New Zealand's heritage, but also part of French heritage'.[47]

As part of her work, Hippolite seeks to identify any repatriated Māori remains and send them to the appropriate *iwi* or tribe. She notes: "Having them here in New Zealand enables us to consider where these toui moukou may have been tattooed and that gives us some sense of where this person or people have come from."[48]

Hippolite laments that sometimes there are difficulties in identifying the provenance of the Māori remains:

---

[43] "French Parliament Passes Bill To Repatriate Maori Heads."

[44] Te Papa Tongarewa, "Te Papa Welcomes French Government Repatriation Act Decision," *Museum of New Zealand* (5 May 2010), http://www.tepapa.govt.nz /Site CollectionDocuments/Media/Media.Release_TE.PAPA.WELCOMES.FRENCH .GOVERNMENT.REPATRIATION.ACT.DECISION.pdf (accessed 21 February 2011).

[45] "Te Papa Welcomes French Government Repatriation Act Decision."

[46] "Te Papa Welcomes French Government Repatriation Act Decision."

[47] "Te Papa Welcomes French Government Repatriation Act Decision."

[48] "Te Papa Welcomes French Government Repatriation Act Decision."

> Having them here at Aotearoa makes the spiritual connexions become real. Because the toui moukou do then return to their Aotearoa to the place where these people were born. But our experience is that very small per centage has been repatriated to iwi. Sometimes, iwi think about whether or not they should receive these remains, because sometimes we aren't able to say that they are people from the tribal group. Where we do find that person enable to talk about the repatriation of the human remains, it takes us as short as 6 months for us to organise for the repatriation. So many Maori heads remain at the Te Papa museum until their story can be understood.[49]

As a result, the Te Papa museum operates as a steward or custodian of human remains, while the identification process proceeds.

In addition to France, cultural institutions and museums in Scotland, Wales, England, the Netherlands, Australia, Ireland, and Sweden have repatriated remains to Te Papa Tongarewa. Te Papa estimates that there are approximately five hundred Māori human remains which are still kept in overseas museums, such as Germany and Russia. Brian Hole contends that the repatriation of ancient remains could serve an important purpose in modernizing and re-invigorating cultural institutions:

> When many archaeologists and museum personnel around the world view repatriation and reburial of human remains to be a threat to their institutions and the very knowledge they stand for, they are in fact missing a critical and exciting opportunity. By repatriating remains they are taking an important step towards redressing serious issues of cultural ownership and breaking down the barriers that have been constructed over centuries between themselves and the very cultures they claim to represent and interpret. By using repatriation as a first step, museums can form partnerships with representative groups of those cultures (such as *tangata whenua* in New Zealand), and ultimately reinvigorate and maintain the relevance of their collections.[50]

He concludes: "Unless museums begin to reflect the increasingly global world in which cultures are mixing and learning to work together on a scale

---

[49] "Te Papa Welcomes French Government Repatriation Act Decision."

[50] Hole, "Playthings for the Foe: The Repatriation of Human Remains in New Zealand," 25.

never seen before, they run the risk of losing their relevance in the present, becoming little more than showcases of outdated world-views in themselves."[51]

Conclusion

This essay has explored an array of jurisdictional approaches to the repatriation of Indigenous ancestral remains – looking at Australia, and New Zealand, in particular. Arguably, there needs to be further legislative work in such jurisdictions in implementing the obligations of the United Nations Declaration on the Rights of Indigenous Peoples 2007.

At the behest of the United Nations Permanent Forum on Indigenous Issues, the United Nations Assembly adopted the United Nations Declaration on the Rights of Indigenous Peoples 2007.[52] Notably, settler countries such as the USA, Canada, New Zealand, and Australia initially voted against the international instrument; however, all of these have belatedly endorsed the Declaration.

A number of provisions are pertinent to the protection of Indigenous ancient remains, cultural property, and traditional knowledge. Article 11 (1) of the Declaration states:

> Indigenous peoples have the right to practise and revitalize their cultural traditions and customs. This includes the right to maintain, protect and develop the past, present and future manifestations of their cultures, such as archaeological and historical sites, artefacts, designs, ceremonies, technologies and visual and performing arts and literature.

Article 11 (2) emphasizes that

> States shall provide redress through effective mechanisms, which may include restitution, developed in conjunction with indigenous peoples, with respect to their cultural, intellectual, religious and spiritual property taken without their free, prior and informed consent or in violation of their laws, traditions and customs.

Moreover, Article 12 (1) of the Declaration emphasizes that

> Indigenous peoples have the right to manifest, practice, develop and teach their spiritual and religious traditions, customs and ceremonies; the right to maintain, protect, and have access in privacy to their reli-

---

[51] "Playthings for the Foe: The Repatriation of Human Remains in New Zealand."

[52] *United Nations Declaration on the Rights of Indigenous Peoples*, 61st sess., UN Doc. A/61/L.67 (2007).

> gious and cultural sites; the right to the use and control of ceremonial objects; and the right to the repatriation of human remains.

Article 12 (2) states:

> States shall seek to enable the access and/or repatriation of ceremonial objects and human remains in their possession through fair, transparent and effective mechanisms developed in conjunction with indigenous peoples concerned

– while Article 31 (1) provides for the following:

> Indigenous peoples have the right to maintain, control, protect and develop their cultural heritage, traditional knowledge and traditional cultural expressions, as well as the manifestations of their sciences, technologies and cultures, including human and genetic resources, seeds, medicines, knowledge of the properties of fauna and flora, oral traditions, literatures, designs, sports and traditional games and visual and performing arts

– adding:

> They also have the right to maintain, control, protect and develop their intellectual property over such cultural heritage, traditional knowledge, and traditional cultural expressions.

Furthermore, Article 31 (2) emphasized the need for state protection: "In conjunction with indigenous peoples, States shall take effective measures to recognize and protect the exercise of these rights."

It is particularly important to develop international standards in this field, given the development of large-scale biology projects, like the Genographic Project.[53] Spencer Wells has led the five-year research project under the auspices of the National Geographic Society. He and his collaborator Theodor Schurr have defended the programme thus:

> The many Genographic participants, including more than 30,000 members of indigenous and traditional groups from around the world, *do* want to understand more about this aspect of their history. Their participation reflects the fact that all Indigenous peoples do not hold an identical view of genetic research and demonstrates both the desire and the ability of these individuals to construct a more nuanced under-

[53] Matthew Rimmer, "The Genographic Project: Traditional Knowledge and Population Genetics," *Australian Indigenous Law Review* 11.2 (2007): 33–55.

> standing of their history, one that takes into account both traditional beliefs and scientific evidence.[54]

The Genographic Project defends itself with respect to research relating to ancestral remains:

> The ancient DNA center will only work on remains that have been cleared for analysis by appropriate indigenous and traditional communities with oversight of a burial ground as well as relevant government agencies and other authorities.[55]

Moreover, the Genographic Project has argued that its consent procedures conform to the Native American Graves Protection and Repatriation Act 1990 (US). Nonetheless, further work needs to be done on the legal, ethical, and social implications of the genetic testing of ancient remains by the Australian Centre for Ancient Remains on behalf of the Genographic Project. There needs to be further discussion as to whether the Genographic Project meets the standards of the United Nations Declaration on the Rights of Indigenous Peoples 2007.[56]

Works Cited

Anon. "Aboriginal Remains Repatriation," *Creative Spirits*, http://www.creativespirits.info/aboriginalculture/people/aboriginal-remains.html (accessed 20 February 2011).

The British Museum. "Policy on Human Remains," *The British Museum* (6 October 2006):http://www.britishmuseum.org/PDF/Human%20Remains%206%20Oct%202006.pdf (accessed 21 February 2011).

——. "Request for Repatriation of Human Remains to New Zealand," *The British Museum* (April 2008): http://www.britishmuseum.org/the_museum/news_and_press_releases/statements/human_remains/repatriation_to_new_zealand.aspx (accessed 21 February 2011).

[54] Spencer Wells & Theodor Schurr, "Response to Decoding Implications of the Genographic Project," *International Journal of Cultural Property* 16 (2009): 185.

[55] The Genographic Project, "Frequently Asked Questions: Ethics and Privacy", https://genographic.nationalgeographic.com/genographic/lan/en/faqs_privacy.html (accessed 9 March 2011).

[56] *United Nations Declaration on the Rights of Indigenous Peoples.*

Chapman, Anna. "Australian Racial Hatred Law: Some Comments on Reasonableness and Adjudicative Method in Complaints Brought by Indigenous People," *Monash University Law Review* 30.1 (2004): 27–48.

Davies, Chris, & Kate Galloway. "The Story of Seventeen Tasmanians: the Tasmanian Aboriginal Centre and Repatriation from the Natural History Museum," *Newcastle Law Review* 11.1 (2008–2009).

Falko, Kathleen. "Grave Concerns: The Repatriation of Tasmanian Indigenous Remains and the Nature of Cultural Property Rights" (Canberra: ANU College of Law, Second Semester, 2007).

The Genographic Project. "Frequently Asked Questions: Ethics and Privacy," https://genographic.nationalgeographic.com/genographic/lan/en/faqs_privacy.html (accessed 9 March 2011).

Hole, Brian. "Playthings for the Foe: The Repatriation of Human Remains in New Zealand," *Public Archaeology* 6.1 (2007): 5–27.

Lafarguie, Caroline. "French Parliament Passes Bill To Repatriate Maori Heads," *Radio Australia* (6 May 2010), http://www.radioaustralia.net.au/pacbeat/stories/201005/s2891653.htm (accessed 21 February 2011).

Langton, Marcia et al., ed. *Honour Among Nations?: Treaties and Agreements with Indigenous People* (Melbourne: Melbourne UP, 2004): 273–87.

McGlade, Hannah. "Race Discrimination in Australia: A Challenge for Treaty Settlement," in *Honour Among Nations?: Treaties and Agreements with Indigenous People*, ed. Marcia Langton et al. (Melbourne: Melbourne UP, 2004): 273–87.

——. "Race Vilification Before the Human Rights and Equal Opportunity Commission," *Indigenous Law Bulletin* 5.7 (2001): 8.

——. "The Repatriation of Yagan: A Story of Manufacturing Dissent," *Law Text Culture* 4.1 (1998): 244–55.

Merryman, John Henry. "Cultural Property Internationalism," *International Journal of Cultural Property* 12 (2005): 11–39.

Nason, James. "Beyond Repatriation: Cultural Policy and Practice for the Twenty-First Century," in *Borrowed Power: Essays on Cultural Appropriation*, ed. Bruce Ziff & Pratima Rao (New Brunswick NJ: Rutgers UP, 1997): 291–312.

National Museum of Australia. "Repatriation," *National Museum of Australia*, http://www.nma.gov.au/collections/repatriation/ (accessed 20 February 2011).

Paradise, Jordan, & Lori Andrews. "Tales from the Crypt: Scientific, Ethical, and Legal Considerations for Biohistorical Analysis of Deceased Historical Figures," *Temple Journal of Science, Technology and Environmental Law* 26.2 (Fall 2007): 223–99.

Rimmer, Matthew. "The Genographic Project: Traditional Knowledge and Population Genetics," *Australian Indigenous Law Review* 11.2 (2007): 33–55.

Roach, Archie. "Travellin' Bones," on *Journey* (Liberation Music LBCD92595, 2007).

Selby, Hugh, & Ian Freckleton, ed. *Appealing to the Future: Michael Kirby and his Legacy* (Sydney: Thomson Reuters, 2009).

Smith, Nicola Kira. "Waiuku Kōiwi Tangata Report," *Museum of New Zealand* (April 2010): http://www.tepapa.govt.nz/SiteCollectionDocuments/AboutTePapa/Repatriation/Waiuku%20Kōiwi%20Tangata%20Report.pdf (accessed 21 February 2011).

Tasmanian Aboriginal Centre. "Aborigines are able to collect human remains now scientists at Natural History Museum have finished testing them" (27 April 2007): http://www.eniar.org/pdf/PR_%20TAC.pdf (accessed 20 February 2011).

Te Papa Tongarewa. "Kōiwi Tangata Policy," *Museum of New Zealand* (1 October 2010): http://www.tepapa.govt.nz/SiteCollectionDocuments/AboutTePapa/Repatriation/Draft%20Koiwi%20Tangata%20Policy%20dated%201%20October%202010.pdf (accessed 21 February 2011).

——. "Repatriation," *Museum of New Zealand*, http://www.tepapa.govt.nz/AboutUs/Repatriation/Pages/overview.aspx (accessed 21 February 2011).

——. "Te Papa to undertake second largest repatriation of Māori human remains from the United Kingdom and Europe," *Museum of New Zealand* (13 November 2009): http://www.tepapa.govt.nz/AboutUs/Media/MediaReleases/2009/Pages/Secondlargestrepatriation.aspx (accessed 21 February 2011).

——. "Te Papa Welcomes French Government Repatriation Act Decision," *Museum of New Zealand* (5 May 2010): http://www.tepapa.govt.nz/SiteCollection Documents/Media/Media.Release_TE.PAPA.WELCOMES.FRENCH.GOVERNMENT.REPATRIATION.ACT.DECISION.pdf (accessed 21 February 2011).

Wells, Spencer, & Theodor Schurr. "Response to Decoding Implications of the Genographic Project," *International Journal of Cultural Property* 16 (2009): 182–87.

Ziff, Bruce, & Pratima Rao, ed. *Borrowed Power: Essays on Cultural Appropriation* (New Brunswick NJ: Rutgers UP, 1997).

Case Law

Bropho v. *HREOC* FCA 1510 (2002).

Bropho v. *HREOC* FCAFC 16 (2004).

Bropho v. *HREOC & Anor* HCATrans 9 (4 February 2005).

Danish Muslim Organisations v. *Jyllands-Posten* ('The Danish Cartoons case') (Denmark: City Court of Aarhus, 2006).

Jamiat-ul-ulama of Transvaal v. *Johncom Media Investment Ltd* 1127/06 (Unreported, High Court of South Africa, Witwatersrand Local Division, 2006).

Muslim World League v. *Charlie–Hebdo and Philippe Val* (France, 2006).

Legislation

ADOPTED TEXT NO. 455: TO AUTHORIZE THE RESTITUTION OF MAORI HEADS TO NEW ZEALAND AND CONCERNING MANAGEMENT OF COLLECTIONS, NATIONAL ASSEMBLY, CONSTITUTION

OF OCTOBER 4, 1958, THIRTEENTH PARLIAMENT, 2009–2010 ORDINARY SESSION, MAY 4, 2010 (FRANCE)

Human Tissue Act 2004 (UK)

Native American Graves Protection and Repatriation Act 1990 (USA).

Racial Discrimination Act 1975 (Cth)

International Treaties

Mataatua Declaration on Cultural and Intellectual Property Rights of Indigenous Peoples 1993.

United Nations Declaration on the Rights of Indigenous Peoples, 61st sess., UN Doc. A/61/L.67 (2007).

⌘

Material Legacies

Indigenous Remains and Contested Values in UK Museum Collections

Lisa O'Sullivan

On 23 November 2006, the journal *Nature* reported that the London Natural History Museum (NHM) had agreed to the return of seventeen Tasmanian ancestral remains held at the Museum. For the reporter, this was a case where "indigenous claims trump scientific value." These remains, he argued, were for scientists a "crucial illustration on a page in human history," representing as they did "specimens of the now dead race." However, "for modern Australians descended from the Aborigines, they are stolen property." The report describes the return as a "loss" and "blow" for science and includes statements from scientists critical of the Museum's decision. Finally, there is the suggestion that in agreeing to return remains to the Tasmanian Aboriginal Centre (TAC), the NHM acted against newly established UK Government guidelines on repatriation, having ignored the requirement for a strong connection to be established between contemporary claimants and the remains.[1]

Nature's brief report condenses many of the issues at stake in debates around repatriation, and packages them as a dichotomy in which specific Indigenous sensibilities and belief-systems are pitched against a scientific world-view encompassing, and able to speak for, all of humanity. The rhetorical position taken in *Nature* is not merely journalistic shorthand, but has permeated much of the discussion surrounding repatriation within scientific and

---

[1] Jim Giles, "Aboriginal remains head for home," *Nature* 444.7118 (23 November 2006): 411.

museum communities.[2] In such representations, human remains are either specimens or ancestors, incommensurable positions, one of which must ultimately triumph over the other. In some contexts, such as in North America, this may see repatriation claims interpreted within the broad nexus of emerging 'anti-science' cultural forces, such as anti-Darwinism, creationism, and extreme cultural relativism.[3] However, human remains came into European museums for a myriad of reasons, and under the rubric of a variety of epistemological practices. Acquiring, exchanging, and gifting Indigenous remains facilitated the entry of colonial enthusiasts into prestigious networks and societies. They were valuable as commodities of influence, status, and reciprocal exchange.[4] Fascination with the 'new' world and its peoples, the search for the macabre, the romance of illicit collection, ethnographic study, and straightforward curiosity are just some of the reasons that have contributed to the collections now held throughout Europe. As a result, human remains can be found not only in the research collections of natural history and medical museums, but across a broad range of anthropological, social, and even art museums.

While the language of science remains the dominant rhetorical discourse, other appeals to a universal position also appear in arguments for retention of remains. These arguments rely on the vision of museums as universal institutions, based on Enlightenment values and caring for and representing global culture. Simultaneously, claimant groups have pointed an alternate frame of analysis in universal human-rights laws, as well as freedom of worship, the rights of Indigenous peoples and trade in illicitly acquired material.[5] Requests

---

[2] There is frequent concern expressed within the scientific community about the reporting of science in the media, where complexities are, perhaps inevitably, reduced to bite-sized language. This is distinct from questions about the nature of science itself under scrutiny here. See Sharon Dunwoody, "Science journalism," in *Handbook of Public Communication of Science and Technology*, ed. Massimiano Bucchi & Brian Trench (London: Routledge, 2008): 15–26.

[3] For an overview of the issues at stake in US debates about repatriation, see David Hurst Thomas, *Skull Wars: Kennewick Man, Archaeology and the Battle for Native American Identity* (New York: Basic Books, 2000): xxxvii–xli.

[4] Cressida Fforde, *Collecting the Dead: Archaeology and the Reburial Issue* (London: Duckworth, 2004): 68–75.

[5] The TAC referenced the *International Covenant on Civil and Political Rights*; the United Nations Draft Declaration on the Rights of Indigenous Peoples and the Vermillion accord in their arguments to the NHM. See Paul Turnbull, "Scientific theft of

for the return of human remains from museums are thus simultaneously part of global movements, such as indigenous self-determination and human rights, and highly contingent on local contexts, histories, and institutional and legal frameworks. This essay examines how these issues have been expressed in the contemporary British context, and suggests that more nuanced explorations of the nature of scientific and museological practice provide a method by which to escape the cul-de-sacs created by relying on the language of universalism.

The TAC Claim In Its British Context

The successful claim for repatriation by the TAC was widely commented on in Australia and the UK as a landmark case.[6] The claim acts as a snapshot of a period when opinions and approaches to repatriation were changing rapidly in Britain.[7] Factors which made this particular case notable included legal action taken by the claimants against the NHM, relating to testing being undertaken on remains before their return; and the NHM's statements about the basis for return, including the illegality of the remains' acquisition and recognition of Indigenous perspectives on cultural affiliation.[8] Indigenous Australian groups, including the TAC, had been pressing for repatriation from the UK since the mid-1980s. While some museums had returned material, these were primarily

---

remains in colonial Australia," *Australian Indigenous Law Review* 7/11.1 (2007): 92–104, note 56.

[6] Detailed analyses of specific aspects of the case can be found in Michael Mansell, "The War of the Dead," *Indigenous Law Bulletin* 46/6.29 (2007): 19–21, and Paul Turnbull, "Scientific theft of remains in colonial Australia."

[7] Since 2008, for instance, the NHM has a Human Remains Unit, which works directly with claimant groups, including proactively visiting communities in Australia whose remains are still retained at the Museum.

[8] The TAC's legal action related to the NHM condition that the remains in question would be retained until a variety of scientific tests were completed. Effectively, the institution appeared to be asserting the NHM's right to retain intellectual control of the remains, and the use to which they were put, until such time as they ceded physical control to the TAC. The TAC objected to continued testing and successfully obtained an interim injunction against the NHM. Legal action was halted after mediation in early 2007 and the return of the remains to the TAC without further testing. See "Human remains to be returned," NHM Press release (17 November 2006): http://www.nhm.ac.uk/about-us/news/2006/november/news_10019.html (accessed 20 October 2010).

institutions with small holdings, whose focus was not in areas which involved the study of human remains.[9]

The success of the Tasmanian claim is a testament to the persistence of Indigenous activists, and simultaneously highlights significant changes in attitude, working practices, and legislation in British museums. Much of this change, without which the success of claims such as the TAC's was unlikely if not impossible, was generated through dialogue with claimants and activists. However, other concerns relating to specifically British human remains also changed the legislative and cultural landscape in which repatriation claims were considered.

In the late-1990s and early-2000s, a combination of medical scandals, new legislation, and controversial exhibitions saw public opinion and professional approaches to human remains become sensitized in the UK. This sensitivity was largely in reaction to a scandal centred on Alder Hey Hospital in Liverpool, where in 1999 details emerged of the systematic retention and use of remains, often those of children and babies, without consent. Subsequent public outcry influenced the development of a new Human Tissue Act, passed into law in 2004 in England and Wales.[10] Informed consent is the centrepiece of the *Act*, which deals with the use of the human body in a variety of mostly clinical and research settings. However, the Act also includes the storage and public display of human remains, bringing museums and museum collections into its legislative framework.

The scope of the Human Tissue Act, hence the requirement for consent, is limited to remains less than one hundred years old, a cut-off point which excludes many – but not all – colonial-era scientific and ethnographic collections of Indigenous remains. The hundred-year point was considered "a sensible and pragmatic cut-off point, being one that means that there is unlikely to be a living relative with a memory of the individual concerned."[11] This time-scale broadly reflects European ideas and sensitivities about what constitutes direct 'family', often incompatible with those of Indigenous communi-

[9] Cressida Fforde & Lyndon Ormond Parker, "Repatriation Developments in the UK," *Indigenous Law Bulletin* 10/5.6 (2001): 9–13.

[10] *Human Tissue Act 2004* (London: HMO Stationery Office, 2004), www.legislation.gov.uk/ukpga/2004/30/enacted (accessed 20 October 2010).

[11] Lord Warner speaking in a Lords Debate on the Human Tissue Bill (Hansard, 16 September 2004): www.publications.parliament.uk/pa/ld200304/ldhansrd/vo040916/text/40916-58.htm (accessed 20 October 2010).

ties, many of whom see no distinction between recent and ancestral family.[12] The inclusion of public displays of remains in the Act related to ethical concerns about the source of bodies displayed in exhibitions such as Gunther von Hagens' *Bodyworlds*, as much, if not more, than existing museum collections.[13] However, the Act included a clause allowing a number of named national institutions, including the NHM, to de-accession human remains in their collections. This clause was directly related to questions of repatriation to Indigenous communities, removing legal obstacles to the return of materials to claimants.

The other prompt to re-evaluate human remains in museum collections was explicitly political. In 2000, a joint statement by British and Australian Prime Ministers committed both governments to increasing their efforts in the repatriation of Indigenous Australian material.[14] A Working Group on human

---

[12] Jim Berg, one of the first Indigenous Australians to successfully regain control and oversee reburial of a collection of remains makes direct links between his work ensuring the return of recently deceased Aboriginal people to their home country for burial, and his determination to ensure the return of ancestral skeletal remains. See "Jim's story," Part One, in Shannon Faulkhead & Jim Berg, *Power and the Passion: Our Ancestors Return Home* (Melbourne: Koorie Heritage Trust, 2010): 24–32. See also Tom Griffiths, *Hunters and Collectors*: *The Antiquarian Imagination in Australia* (Melbourne: Cambridge UP, 1996): 96.

[13] The *Bodyworlds* exhibition, on display in London March 2002–February 2003, was repeatedly referred to in Parliamentary discussions relating to the passing of the Human Tissue Act Bill. See Standing Committee G, Human Tissue Bill (3 February 2004), http://www.publications.parliament.uk/pa/cm200304/cmstand/g/st040203/am/40203s01.htm (accessed 20 October 2010).

Human Tissue Bill (15 January 2004): http://www.publications.parliament.uk/pa/cm200304/cmhansrd/vo040115/debtext/40115-13.htm#40115-13_head0 (accessed 20 October 2010).

Human Tissue Bill (22 July 2004): http://www.publications.parliament.uk/pa/ld200304/ldhansrd/vo040722/text/40722-09.htm#40722-09_head5 (accessed 20 October 2010).

There have been ongoing concerns raised about the source of bodies used in *Bodyworlds* and the similar *Bodies ... The Exhibition*. See Helen MacDonald, "Under Your Skin," *New Scientist* 190.2550 (6–12 May 2006): 51, and Annette Tuffs, "Von Hagens faces investigation over use of bodies without consent," *British Medical Journal* 327.7423 (6 November 2003): 1068.

[14] John Howard, "Joint statement with Tony Blair on Aboriginal remains," media release (4 July 2000): http://parlinfo.aph.gov.au/parlInfo/search/display/display.w3p

remains was established by the UK Government in 2001, headed by Professor Norman Palmer, a cultural-property expert, and composed of members of the museum, legal, and anthropological professions. Potential claimant communities were included in the process of consultation for the Working Group, a broadening of scope that reflected earlier changes in North America and Australia, in which the scientific and museum communities were no longer seen as the sole groups with the authority to determine the meaning and future of remains held in collections.[15] The Working Group's report was published in November 2003, and set forth recommendations, later consolidated into *Guidance for the Care of Human Remains in Museums*, which stressed the need to consider the multiple, and potentially competing, notions of 'value' attached to remains held in museum collections.[16] Neil Chalmers, then director of the NHM, formed part of the Working Group, and made a dissenting minority report on the Group's findings.[17] He argued that, while he was broadly sympathetic to the concerns of claimant communities, the recommendations of the Report were overly skewed towards these concerns, and were "tantamount to mandatory repatriation" rather than recognizing the "benefits that these collections provide to humanity."[18]

As the misconception that they are a 'dead race', as described in *Nature*, continues to have a strong after-life outside Australia, issues of identity are particularly acute for Indigenous Tasmanians. The genocidal nature of British colonization in Tasmania was deployed by a number of respondents to the Working Group's call for consultation, who argued that contemporary Indigenous Tasmanians had no mandate to make claims, as "true" Tasmanians have no descendants.[19] One commentator went as far as to suggest that in demand-

;query=Id%3A%22media%2Fpressrel%2FFC026%22.

[15] For more on changes to UK policies relating to human remains, see Fforde, *Collecting the Dead*, 135–36.

[16] Department for Culture Media and Sport, *Report of the Working Group on Human Remains* (London: HMO Stationery Office, 2003), and Department for Culture Media and Sport, *Guidance for the Care of Human Remains in Museums* (London: HMO Stationery Office, 2005).

[17] Neil Chalmers, "Statement of dissent," in *Report of the Working Group on Human Remains:* 220–29.

[18] Neil Chalmers, letter to *Nature* 427.6972 (22 January 2004): 287.

[19] *Report of the Working Group on Human Remains*, 38. The assertion that Indigenous people making claims are politically motivated and not 'authentic' has been made

ing the return of remains, contemporary Tasmanians were themselves practising "a form of racism if not genocide," because they were attempting to remove "particular genotypes from the possibility of scientific investigation."[20] For nineteenth-century anthropologists, Tasmanians were low on the scale of humanity, an example of remnant populations akin to prehistoric groups, destined to die out when exposed to 'civilized' European culture.[21] As such, Tasmanian remains were highly sought-after, even as British colonists in Tasmania clashed violently with living Tasmanians. The 'extermination' of Indigenous Tasmanians was extensively documented and strenuous efforts were made by European collectors to ensure that the remains of the last 'full-blood' individuals were not buried, but instead returned to Britain for analysis.[22] Remnant populations of Indigenous Tasmanians, who survived primarily outside Tasmania, were discounted as not Tasmanian, on the basis of being of mixed descent, or redefined as if they were from mainland populations.[23] Hence the criticism of the NHM from scientists reported in *Nature*, as the recognition of the TAC as legitimate claimants meant acknowledging contemporary Tasmanian identity, still clearly not universally accepted in the scientific community.

---

in arguments against repatriation in numerous settings. See Cressida Fforde, "Collection, repatriation and identity," in *The Dead and Their Possessions: Repatriation in Principle, Policy and Practice*, ed. Cressida Fforde, Jane Hubert & Paul Turnbull (London: Routledge, 2002): 36–38. Requests for repatriation go hand in hand with the increased agency of Indigenous groups, and a community needs a certain level of resource, organization, and above all acceptance as a legitimate claimant, before their voices are included in consultations and debates. Hence, silence about artefacts and remains from any particular culture cannot necessarily be read as a lack of interest.

[20] *Report of the Working Group on Human Remains*, 52.

[21] George W. Stocking, Jr, "Epilogue: the extinction of Palaeolithic man," in *Victorian Anthropology* (New York: Free Press, 1987): 274–83.

[22] On the collection of skeletal material in colonial Tasmania, see Helen MacDonald, "The Bone Collectors," and "Death and Dissection, 1869," in *Human Remains: Dissection and Its Histories* (New Haven CT & London: Yale UP, 2006).

[23] For the history of one Tasmanian community, see Rebe Taylor, *Unearthed: The Aboriginal Tasmanians of Kangaroo Island* (Adelaide, SA: Wakefield, 2008). Complicating questions of identity, the TAC are recognized by the Australian government as able to speak for Indigenous Tasmanians. However, the Lia Pootah community of Tasmania also consider themselves Aboriginal and dispute the ability of the TAC to speak on their behalf.

Repatriation and the Nature of Science

Before 2006, the NHM had been perceived in the UK as one of the museums more fundamentally opposed to repatriation, a perception based in part on statements from its then director, Neil Chalmers.[24] In addition to his minority report in the Working Group *Report*, Chalmers frequently made public statements relating to repatriation, in which he argued for the retention of collections on the basis of their service in the good of 'humanity'. In a letter to *Nature* in 2004, for instance, Chalmers argued for a "balance between the concerns of claimant communities on the one hand, and on the other, the loss to humanity resulting from wholesale return on request."[25]

The gesture towards an inherent dichotomy between science as a universal humanitarian project and specific Indigenous beliefs, culture, or spirituality that pervade discussions about repatriation is generally deployed in arguments for retention of collections. These arguments endorse the still-dominant vision of scientific practice found in most science policies, textbooks, public discussion, and often accounts by scientists themselves. Here, 'science' stands for rationality, objectivity, and humanistic research. Science is intrinsically progressive, and its self-correcting internal processes mean that the social and institutional settings of scientific practice are relevant only in contextual terms, rather than being considered as potential influences on the knowledge subsequently produced.

How, then, does this model of science meet the challenge of repatriation claims, particularly in terms of unethical practices of the past? There is a tendency, particularly when dealing with now highly sensitive areas, such as racial science, for those aspects of past practice that are now considered unethical or immoral to be relegated to the fringes of definitions of 'science'. This view could be described as a 'contamination' model, in which science is rendered 'impure' to one degree or another by the world-views of its practitioners. Put broadly, the argument acknowledges that in the past scientists

---

[24] This perception was also based on reports by Indigenous activists relating to the attitude displayed by some NHM staff during talks in 2003. For example, Major Sumners, a Ngarrindjeri elder, reports a senior NHM staff member asking another Indigenous elder if he would donate his remains when he died, "so we can research you." See Tim Elliott, "Jealous keepers of the sacred bones," *The Age* (13 March 2010): http://www.theage.com.au/national/jealous-keepers-of-the-sacred-bones-20100312 -q48m.html (accessed 20 October 2010).

[25] Neil Chalmers, letter to *Nature*, 287.

engaged in practices, and developed theories, which were, by contemporary standards, unethical, racist, and discriminatory. However, it then points to the improved ethical standards and humanistic values and goals of contemporary scientists. In other words, 'good' science can be done on specimens or data collected in the name of 'bad' science in the past.[26] In the case of human remains, retention of material collected in the name of racial difference is justified by its subsequent use in research looking at human unity, and by its potential to "explode many myths about race and ethnicity."[27] References to the future relevance of collections are frequently presented in pro-retention arguments as overriding the methods by which they were collected, or the intentions of the collectors. Hence, remains collected in pursuit of metric measurements are now seen as potential subjects for genetic, molecular, and isotope analysis, to name only a few techniques.[28] Just as knowledge of DNA has opened up a myriad of new techniques and analytical possibilities, so, too, might now still-to-be-developed technologies generate knowledge, the impact of which cannot be known.

Analyses that rely on the future possibilities inherent in a collection ignore the ways in which the history of collecting can shape, and more often limit, the interpretative usefulness of remains. Take the Murray Black collection, one of Australia's largest collection of Indigenous skeletal remains. The collection has now been repatriated, but for many decades was considered the

---

[26] Ongoing controversies about using scientific data acquired through experiments in concentration camps during World War Two show the limits of this argument. Bernard Dixon, "Citations of shame," *New Scientist* 1445 (28 February 1985): 31.

[27] Robert Foley, Director of evolutionary anthropology at Cambridge University's Leverhulme Centre for Human Evolutionary Studies, cited in David Derbyshire, "Folly to give back ancient bones, say scientists," *Telegraph UK* (16 May 2003): http://www.telegraph.co.uk/news/uknews/1430184/Folly-to-give-back-ancient-bones-say-scientists.html (accessed 20 October 2010).

[28] Potential techniques able to be applied to human remains include morphometrics, 3-D imaging, x-rays, bone and dental histology, trace-element bone chemistry, microscopy, ancient DNA, laser ablation, and protein-chemistry pathogen analysis. Robert Foley, "Report on the scientific significance of two cremation bundles from Tasmania held in the British Museum" (3 November 2005): 4, http://www.britishmuseum.org/the_museum/news_and_press_releases/statements/human_remains/repatriation_to_tasmania.aspx (accessed 20 October 2010).

"mainstay of bio-anthropological research in Australia."[29] Black was a pastoralist and amateur archaeologist, whose donations to the Australian Institute of Anatomy and Anatomy Department of the University of Melbourne between 1929 and 1950 formed the largest skeletal collection in the country. Even at the time of collecting, relatively little active research took place on the donated remains. However, despite some reservations about the thoroughness of Black's practices, scientists at both institutions encouraged him to continue to collect, arguing in 1938 that "At some future date the aboriginal material will probably be much more highly esteemed than it is at present, and we are custodians for the future."[30] Despite this appeal to the future, and the material's centrality to Australian archaeology, the collection, as Sarah Robertson has demonstrated, has serious biases, including site selection, decisions about retention, lack of contextual information, and subsequent curatorial decisions, which in combination mean that it cannot be considered representative of a population, and is thus of limited use in archaeological study.[31]

A focus on the future may be framed not only in terms of the potential new data collections may yield but also with regard to future generations of Indigenous people likely to enter professional life as scientists. By increased immersion in the world view of archaeology, genetics or biosciences, they will, it is presumed, develop new (and, by implication, more rational and universal) attitudes to the use of remains as scientific specimens. As one British archaeologist argues,

> new laboratory techniques and new research questions demand that this material remains available for future scientists to investigate, and it is very satisfying to see that some of the new generation of investigators are from tribal groups.[32]

However, the increased involvement of Indigenous scholars in the sciences – and non-Indigenous engagement with relevant communities in shaping re-

---

[29] Sarah Robertson, "Sources of Bias in the Murray Black Collection: Implications for Palaeopathological Analysis," *Australian Aboriginal Studies* 1 (2007): 116. See *The Power and Passion* for details of the collection's return to Indigenous communities.

[30] Charles MacKay, Acting Director, Australian Institute of Anatomy, letter to George Murray Black, 14 January 1938, Murray Black Correspondence, National Museum of Australia file 02/637.

[31] Robertson, "Sources of bias in the Murray Black collection."

[32] Don Brothwell, "Bring out your dead: people, pots and politics," *Antiquity* 78.300 (2004): 418.

search projects – suggests that the identity, viewpoint, and approach to the science of individuals involved, is challenged and contextualized through this process, rather than the development of a wholesale commitment to a universal vision of science.[33]

Commentators more favourable to repatriation point out that recourse to arguments about global benefits of collections tends to elide significant historical, ethical, legal, and socio-political aspects to the collection and retention of human remains. As the archaeologist Laurajane Smith argues, reducing debates about repatriation to questions of science and religion

> misses the point. The issues are political ones that revolve around the politics of identity and recognition – in which the disciplinary and individual identity of archaeologists and other scientists are as much at stake as those of Indigenous peoples.[34]

However, even when calls are made to consider points of ethics, legality, and politics in decision-making processes, these are often treated as separate categories from that of science. The argument then becomes one about the competing importance of such values. What is largely missing is a consideration of the nature of science itself, or engagement with the extensive literature examining the construction of science as a specific world-view, fundamentally questioning its claims to objectivity, neutrality, and universalism.

Since the Second World War, science has been analysed as an historical, socio-cultural, and political category by historians, sociologists, and philosophers of science.[35] Thomas Kuhn, in his landmark book *The Structure of*

---

[33] See Chris Wilson on the relationship between his professional and his Indigenous (*Ngarrindjeri*) identity. Christopher L. Wilson, "Indigenous Research and Archaeology: Transformative Practices in/with/for the Ngarrindjeri Community," *Journal of the World Archaeological Congress* 3.3 (2007): 320–34. In the US context, see the discussions generated by Robert McGee's article "Aboriginalism and the Problems of Indigenous Archaeology," *American Antiquity* 73.4 (2008): 579–97. Responses from Dale R. Croes, Stephen W. Silliman, Michael Wilcox, Chip Colwell–Chanthaphonh et al. and Robert McGhee can be found in *American Antiquity* 75.2 (2010): 211–43.

[34] Laurajane Smith, "The repatriation of human remains: problem or opportunity?" *Antiquity* 78.300 (2004): 406.

[35] For a summary of approaches to understanding the nature of scientific practice, see Sandra Harding, "A role for postcolonial histories of science in theories of knowledge? Conceptual shifts," in *Is Science Multicultural? Post-Colonialisms, Feminisms and Epistemologies* (Bloomington & Indianapolis: Indiana UP, 1998).

Scientific Revolutions (1962), demonstrated that scientific practice does not always operate in a linear, additive manner, with new data bringing scientific explanations ever closer to reality. Instead, science periodically undergoes crises, which see the underlying structures, or paradigms, of practice and theory overturned, changing the entire conceptual framework in which scientific investigation takes place. A classic example is the shift from a geocentric to a heliocentric model of the universe.[36]

Feminist, Indigenous, and postcolonial scholars have expanded on this analysis, arguing from different perspectives that the practice of science cannot be understood outside its cultural, historical, institutional, and ideological contexts. Such arguments do not merely suggest that the a-priori beliefs – whether conscious or unconscious – of individual scientists may shape the questions they ask, or the conclusions they draw from their data. Such 'constructivist' theories suggest that scientific knowledge cannot be considered as a "transcendent mirror of reality," but "both embeds and is embedded in social practices, identities, norms, conventions, discourses, instruments and institutions – in short, in all the building blocks of what we term of the *social*."[37] The feminist theorist Donna Haraway points out that this is not necessarily an argument for relativism, and stresses the need to retain a "no-nonsense commitment to faithful accounts of the real world" able to be produced through scientific practice.[38] The solution for Haraway is to reject the idea of a "gaze from nowhere": i.e. to acknowledge as fantasy the concept of a viewpoint which is able to transcend its own position and become universal.[39]

Collections of Indigenous human remains in museum collections are the product of an historical process by which the vast numbers of specimens were brought together, particularly in Imperial centres, through networks of acquisition and collecting on a vast scale. As such, they are embedded in histories of empire and colonialism which cannot be separated from their scientific history. As the historians Paolo Palladino and Michael Worboys have pointed

---

[36] Thomas Kuhn, *The Structure of Scientific Revolutions* (Chicago & London: U of Chicago P, 1962).

[37] Sheila Jasanoff, "The idiom of coproduction," in *States of Knowledge: The Co-production of Science and Social Order*, ed. Sheila Jasanoff (Abingdon & New York: Routledge, 2004): 3.

[38] Donna Haraway, "Situated Knowledges: The Science Question in Feminism and the Privilege of the Partial Perspective," *Feminist Studies* 14.3 (Fall 1988): 579.

[39] Haraway, "Situated Knowledges," 581.

out, "for most of humanity, the history of science and imperialism *is* the history of science."[40] This is not to argue that science played a uniform role in colonial projects across the globe: rather, that science and technology were an integral part of the apparatus of colonialism, not only in the practical technologies of empire-building but also in establishing the criteria through which Indigenous cultures were condemned as backward or 'savage' and in justifying colonial projects as enlightened and rational – the 'civilizing' mission.

Science and technology studies informed by postcolonial theory have demonstrated the complexities of local encounters between knowledge systems, the "messy translations and transactions" that take place in contact zones.[41] In their summary and analysis of work in the field of science and postcolonialism, Warwick Anderson and Vincanne Adams bring together case studies from multiple localities to argue for the situated and heterogeneous nature of scientific projects, even those which identify themselves as global, which clearly mark science as "constitutively social and local, and always particular and political."[42] As Anderson and Adams point out, the contemporary interactions between scientists and Indigenous peoples highlight the mobility, hybridity, and heterogeneity of knowledge systems. The work of Helen Verran and David Turnbull in the Australian context, looking at the relationships between specific Indigenous knowledge systems and science, exemplifies such approaches.[43]

Such interactions point towards another factor influencing debates about repatriation in the UK context, where, beyond the specific context of repatriation claims, there has been only a limited interface between Indigenous communities and scientific and museological institutions. As such, discussions relating to Indigenous issues continue to be distinguished, as one British

---

[40] Paolo Palladino & Michael Worboys, "Science and Imperialism," *Isis* 84.1 (1993): 102.

[41] Warwick Anderson & Vincanne Adams, "Pramoedya's chickens: postcolonial studies of technoscience," in *The Handbook of Science and Technology Studies*, ed. Edward J. Hackett, Olga Amsterdamska, Michael Lynch & Judy Wajcman (Cambridge MA: MIT Press, 2007): 184.

[42] Anderson & Adams, "Pramoedya's chickens," 181.

[43] Helen Verran, "Science and the Dreaming," *Issues* 82 (2008): 23–26, Helen Watson–Verran & David Turnbull, "Science and Other Indigenous Knowledge Systems," in *Handbook of Science and Technology Studies*, ed. Sheila Jasanoff, Gerald Markel, James Peterson & Trevor Pinch (Thousand Oaks CA: Sage, 1995): 115–39.

anthropologist has put it, by "extreme political and social distance from indigenous peoples."[44] For some commentators, this distance is not necessarily negative, as it allows a degree of objectivity and disengagement from the social, political, and emotional legacies of colonialism, impossible to avoid for those nations directly engaged with their own Indigenous populations.[45] This argument reflects a distinct displacement of historical responsibility for colonialism onto settler communities, rather than former Imperial powers; and a significant gap between what could be described as 'New-' and 'Old-World' understandings of the implications of repatriation for the communities involved, whether museological, indigenous, or scientific.

In Australia, Canada, New Zealand, and the USA, debates over repatriation began gaining traction in the 1980s. Repatriation has since become established as part of the mainstream working practices of museums and research institutions and in some cases enshrined in legislation.[46] Scientists at museums wishing to work with Indigenous material in these countries are normally required to approach the appropriate communities for permission. Such negotiations shape the type of questions asked and approaches taken. Concurrently, much of the rhetoric of repatriation as a threat to the very practice of science itself has dissipated.[47] At the same time, scientists and scientific in-

---

[44] Laura Peers, "Repatriation: A Gain for Science?" *Anthropology Today* 20.6 (2004): 3.

[45] For comments of this nature made at a Musée du Quai Branly international conference on repatriation, see Michael Pickering, "Lost in Translation," *borderlands* 7.2 (2008): 9.

[46] Museums Australia sets out guidelines for working with Indigenous cultural heritage in *Continuous Cultures, Ongoing Responsibilities* (2005), 136.154.202.120 /dbdoc/ccor_final_feb_05.pdf (accessed 20 October 2010). — In the USA, the Native American Graves Protection and Repatriation Act (NAGPRA) was passed in 1990. See http://www.nps.gov/nagpra/MANDATES /25USC3001etseq.htm (accessed 20 October 2010). — In Canada, there are provincial and municipal rather than federal laws relating to Indigenous heritage. See Joe Watkins, "Through Wary Eyes: Indigenous Perspectives on Archaeology," *Annual Review of Anthropology* 34 (2005): 429–49. — In New Zealand, repatriation is supported by the government, and administered through the National Museum of New Zealand, Te Papa Tongarewa. See "Repatriation," National Museum of New Zealand, Te Papa Tongarewa: http://www.tepapa.govt .nz/AboutUs /Repatriation/Pages/Overview.aspx (accessed 20 October 2010).

[47] There continue to be debates about the impact of repatriation in particular scientific disciplines, which often flare up in response to discoveries seen of signal valuable

stitutions have in many cases become highly self reflexive about their own practice.[48] Collaborative and more self-consciously situated and contingent models of scientific practice such as those analysed by Verran and Turnbull appear to be emerging, in which common ground can be found between Indigenous and scientific goals.

Museums, Universalism, and 'New Museology'

The rhetoric of universalism and universal good is not restricted to science but also remains central to the self-identification of many museums. In 2002, eighteen major museums signed a declaration that reiterated the 'universal' role and nature of museums.[49] Signatories were exclusively European and North American museums, many with an art focus. While opening with a statement condemning the illegal trade in artefacts, the Declaration was widely interpreted as a defence against claims for repatriation and the restitution of

---

to science, particularly ancient remains, which from a non-Indigenous perspective are more difficult to understand as connected to contemporary communities. In the USA, many of these issues have crystallized in disputes over the fate of Kennewick Man or the Ancient One. See David H. Thomas, *Skull Wars*.

[48] The prominent Australian archaeologist Colin Pardoe is a good example of a scientist who has documented his changing views about repatriation and its impact on his vision of science and professional practice. See Pardoe, "Arches of Radii, Corridors of Power: Reflections on Current Archaeological Practice," in *Power, Knowledge and Aborigines*, ed. Bain Attwood & Jon Arnold, special edition of the *Journal of Australian Studies* (Pandora, Victoria: Latrobe UP, 1992): 132–41, and "Sharing the Past: Aboriginal Influence on Archaeological Practice, a Case Study from New South Wales," *Aboriginal History* 14.2 (1990): 208–23.

[49] The Declaration is reproduced in Neil G.W. Curtis, "Universal museums, museum objects and repatriation: The tangled stories of things," *Museum Management and Curatorship* 21.2 (2006): 117–27, and in Geoffrey Lewis, "The 'Universal Museum': a case of special pleading?" in *Art and Cultural Heritage: Law, Policy and Practice*, ed. Barbara T. Hoffman (Cambridge & New York: Cambridge UP, 2006): 381–82. Both reproduce a version of the document that includes the British Museum as a signatory, bringing the number of museums to nineteen. However, as Curtis notes (119), while the director, Neil MacGregor, has indicated that the British Museum supports the declaration and made his own statements on the 'Universal Museum', it is not clear whether it is in fact an official signatory.

objects.[50] The Declaration argued that objects become "part of the heritage of the nations which house them," that the space of the museum itself becomes "a valid and valuable context of objects that were long ago displaced from their original source," and that "museums serve not just the citizens of one nation but the people of every nation."[51]

Like the concept of universal science, the 'universal museum' is an ideal based on Enlightenment values. The goal was to amass world cultures within the aegis of a single institution, so that the space of the museum itself became a universalizing cultural context. Museums are now global in reach, but in their modern form developed from a specifically European tradition beginning with cabinets of curiosities and private collections, many of which became public institutions with research and education at their heart in the eighteenth and nineteenth centuries.[52] The Enlightenment values on which such museums self-consciously base their legacy has been subject to critique as disguising their role in the determination and representation of identity and social and political purposes to which museum collections and exhibitions can be put.[53] As critics have pointed out, museums identifying themselves as universal tend to be situated in Imperial centres of power, and their development was entangled with, and relied on, the processes of European expansion, particularly in the acquisition of artefacts on a global scale.

'New museology', emerging in the 1980s, began to question the idea that museums could in fact act as neutral and universal spaces. Instead, scholars highlighted the potential multiplicity of meanings attached to artefacts, and the underlying philosophies guiding museum displays, whether these are explicit or unspoken. Museums, although often treated as naturalized spaces, act as repositories and reifications of national and cultural legacies, places where

---

[50] See: Curtis, "Universal museums, museum objects and repatriation"; Magnus Fiskesjö, "Global Repatriation and 'Universal' Museums," *Anthropology News* (March 2010): 11–12; and Mark O'Neill, "Enlightenment museums: universal or merely global?" *Museum and Society* 2.3 (2004): 190–202.

[51] "Declaration on the importance and value of universal museums," reproduced in Geoffrey Lewis, "The 'Universal Museum'," 381–82.

[52] Tony Bennett, "The formation of the Museum," in *The Birth of the Museum: History, Theory, Politics* (London & New York: Routledge, 1995): 17–58.

[53] The first anthology to bring together such approaches was *The New Museology*, ed. Peter Vergo (London: Reaktion, 1989).

origin stories can be told and reworked.[54] The stories told in them have traditionally been represented as authoritative versions of history, hence the ferocity of debates surrounding contentious interpretations, or reinterpretations, of objects and histories.[55] In practice, value judgments are made at every stage of museum acquisition, interpretation, and display. As Sharon MacDonald has noted, these judgments tend to be glossed over in exhibitions, which are "presented to the public rather as scientific facts: as unequivocal statements rather than as the outcome of particular processes and contexts."[56] Just as the presentation of scientific practice in the media disguises messier realities, museum exhibitions are rarely explicit about the processes of negotiation and interpretation involved in their production. What makes an artefact of value changes over time and can be based on issues of aesthetics, rarity, and cultural importance (whether to producers or collectors). Further significant disjunctions can exist between the meanings and value attached to objects by their producers and by collectors and curators.[57]

Another of the concerns of new museology relates to questions of authority and expertise in the interpretation and representation of cultures. Historically, Indigenous people have been dissociated from their own material culture, and more commonly seen as resources for interpretation than than active participants in the process of representation.[58] One significant change in museum practice, often related to negotiations about repatriation, has been the redefinition of what constitutes a bona fide researcher or interest group, or specialist knowledge relating to an object or culture. Repatriation claims can also lead to ongoing relationships in which Indigenous groups offer museums

---

[54] Flora E.S. Kaplan, "Introduction: Displays of power," *in Museums and the Making of "Ourselves": The Role of Objects in National Identity*, ed. Flora E.S. Kaplan (London & New York: Leicester UP, 1994): 3.

[55] See Steven C. Dubin, *Displays of Power: Controversy in the American Museum from the Enola Gay to Sensation* (New York & London: New York UP, 1999).

[56] Sharon MacDonald, *The Politics of Display: Museums, Science, Culture* (London & New York: Routledge, 1998): 2.

[57] See, for example, Sally Price's analysis of the interpretation of artefacts at the Musée du Quai Branly. Sally Price, *Paris Primitive: Jacques Chirac Museum on the Quai Branly* (Chicago & London: U of Chicago P, 2007).

[58] Moira G. Simpson, "Charting the boundaries: Indigenous models and parallel practices in the development of the post-Museum," in *Museum Revolutions: How Museums Change and Are Changed*, ed. Sheila Watson, Suzanne MacLeod & Simon Knell (London: Routledge, 2007): 238.

new objects or experiences, whether tangible and intangible expressions of their culture.

Conclusion: The Challenge of Repatriation

Human remains in museum collections are a material legacy raising a number of often unsettling issues about the practice of science and museology; engaging with Indigenous communities often demands considerable self-reflection for institutions. Questions about identity – professional, institutional, and Indigenous – are inevitably central to claims. The majority of remains in museum collections are there precisely because of the way identity was understood through the lens of European anthropology and science, particularly racial science; and factors used to determine the legitimacy of groups making claims remain deeply embedded in scientific notions of identity, whether relying on DNA analysis or on affiliation as determined by physical and cultural anthropologists. As such, repatriation needs to be considered in relation to a number of overlapping disciplinary concerns within museology – the treatment and representation of human remains; the treatment and representation of Indigenous peoples; questions of what constitutes authority and expertise; and the responsibilities contemporary institutions and communities bear for wrongdoings in the past that may form part of their legacy.

In recent decades, Indigenous activism, particularly in the New-World context, has resulted in greater emphasis being placed on self-representation, including, but not confined to, physical and intellectual control of remains and material culture, and the authority accorded to particular identities and worldviews.[59] Such negotiations are, perhaps inevitably, less fraught in museums whose central focus is not scientific research. For these, one of the key questions is the relation of contemporary scientific practice to the past. Are past wrongs a legacy to be addressed in current practice, or does the research potential of collections outweigh any historical injustices associated with their collection?

The UK government guide to the treatment of historical human remains in museums focuses on the processes of assessing claims for the return of human remains, and recommends that each is assessed on a case-by-case basis.[60] These guidelines leave considerable room for local interpretation, and there

---

[59] Kylie Message, *New Museums and the Making of Culture* (Oxford & New York: Berg, 2006): 1.

[60] *Guidance for the Care of Human Remains in Museums*, 23.

has been a wide range of responses across the museum sector in the UK, influenced by local politics, the academic expertise and background of curatorial teams, and individual political commitments. For many UK museums, the need to respond to a repatriation claim was the trigger to considering their treatment of human remains on a systematic policy level for the first time. Very few institutions have the legal ability, even if they have the will, to de-accession material from their collections without following strict procedures.

At the same time, claimant communities have been asked to legitimate their interest in, and speak dispassionately about, remains they regard as stolen ancestors. The identities of curators and museum professionals may be challenged by the notion that their 'care' of collections and the meanings they attach to them are considered not only inaccurate but offensive to the very communities represented. Competing notions of 'expertise' are highlighted when, for example, institutions seek expert opinions on questions of 'value' attached to remains from academics relying on historical records, as opposed to Indigenous knowledge relating to remains, which may be oral and/or secret and sacred, making it challenging for community representatives asked to share these with museum personnel. Perhaps most crucial to working through these complexities is the interaction between individuals from claimant and museum communities. Such conversations may be largely unrecorded, and are often specifically framed as confidential. The reasons for a return or rejection of a claim are various, and often not in the public domain. As such, the nuances of individual case studies do not always lend themselves easily to broader theorization. This underlines the need to recognize the use of arguments about universal systems of knowledge and the nature of Enlightenment ideals as rhetorical positions, arguments and positions which have only limited usefulness in practical processes and decision-making on local levels.

For European museums, repatriation is perhaps the moment at which the Empire most acutely strikes back, when the colonized speak, and repressed narratives of violence and displacement undermine the edifices of enlightened culture on which their institutions have been built. Recognizing the inevitably situated and historically contingent nature of science and museums is perhaps one way in which discussions of repatriation can be redirected away from reductionist dichotomies of competing value-systems, and instead acknowledge the inherently messy and political nature of the processes which create new knowledge.

WORKS CITED

Anderson, Warwick, & Vincanne Adams. "Pramoedya's chickens: postcolonial studies of technoscience," in *The Handbook of Science and Technology Studies*, ed. Edward J. Hackett, Olga Amsterdamska, Michael Lynch & Judy Wajcman (Cambridge MA: MIT Press, 2007): 181–204.

Attwood, Bain, & John Arnold, ed. *Power, Knowledge and Aborigines* (Pandora, Victoria: La Trobe UP, 1992).

Bennett, Tony. *The Birth of the Museum: History, Theory, Politics* (London & New York: Routledge, 1995).

Brothwell, Don. "Bring out your dead: people, pots and politics," *Antiquity* 78.300 (2004): 414–18.

Chalmers, Neil. Letter to *Nature* 427.6972 (22 January 2004): 287.

Colwell–Chanthaphonh, Chip et al. "The premise and promise of indigenous archaeology," *American Antiquity* 75.2 (2010): 228–38.

Croes, Dale R. "Courage and thoughtful scholarship = indigenous archaeology partnerships," *American Antiquity* 75.2 (2010): 211–16.

Curtis, Neil G.W. "Universal museums, museum objects and repatriation: The tangled stories of things," *Museum Management and Curatorship* 21.2 (2006): 117–27.

Department for Culture, Media & Sport. *Guidance for the Care of Human Remains in Museums* (London: HMO Stationery Office, 2005).

——. *Report of the Working Group on Human Remains* (London: HMO Stationery Office, 2003).

Derbyshire, David. "Folly to give back ancient bones, say scientists," *Telegraph* (UK; 16 May 2003): http://www.telegraph.co.uk/news/uknews/1430184/Folly-to-give-back-ancient-bones-say-scientists.html (accessed 20 October 2010).

Dixon, Bernard. "Citations of shame," *New Scientist* 1445 (28 February 1985): 31.

Dubin, Steven C. *Displays of Power: Controversy in the American Museum from the Enola Gay to Sensation*; with a new afterword (New York & London: New York UP, 1999).

Dunwoody, Sharon. "Science journalism," in *Handbook of Public Communication of Science and Technology*, ed. Massimiano Bucchi & Brian V. Trench (London: Routledge, 2008): 15–26.

Elliott, Tim. "Jealous keepers of the sacred bones," *The Age* (Australia; 13 March 2010): http://www.theage.com.au/national/jealous-keepers-of-the-sacred-bones-20100312-q48m.html (accessed 20 October 2010).

Faulkhead, Shannon, & Jim Berg. *Power and the Passion: Our Ancestors Return Home* (Melbourne: Koorie Heritage Trust, 2010).

Fforde, Cressida. *Collecting the Dead: Archaeology and the Reburial Issue* (London: Duckworth, 2004).

——, & Lyndon Ormond Parker. "Repatriation Developments in the UK," *Indigenous Law Bulletin* 10/5.6 (2001): 9–13

——, Jane Hubert & Paul Turnbull, ed. *The Dead and Their Possessions: Repatriation in Principle, Policy and Practice* (London: Routledge, 2002).

Fiskesjö, Magnus. "Global Repatriation and 'Universal' Museums," *Anthropology News* (March 2010): 11–12.

Foley, Robert. "Report on the scientific significance of two cremation bundles from Tasmania held in the British Museum" (3 November 2005), http://www.britishmuseum.org/the_museum/news_and_press_releases/statements/human_remains/repatriation_to_tasmania.aspx (accessed 20 October 2010).

Giles, Jim. "Aboriginal remains head for home," *Nature* 444.7118 (23 November 2006): 411.

Griffiths, Tom. *Hunters and Collectors: The Antiquarian Imagination in Australia* (Cambridge & Melbourne: Cambridge UP, 1996).

Haraway, Donna. "Situated Knowledges: The Science Question in Feminism and the Privilege of the Partial Perspective," *Feminist Studies* 14.3 (Fall 1988): 575–99.

Harding, Sandra. *Is Science Multicultural? Post-Colonialisms, Feminisms and Epistemologies* (Bloomington & Indianapolis: Indiana UP, 1998).

Howard, John. "Joint statement with Tony Blair on Aboriginal remains," Australian Government media release (4 July 2000): http://parlinfo.aph.gov.au/parlInfo/search/display/display.w3p;query=Id%3A%22media%2Fpressrel%2FFC026%22 (accessed 20 October 2010).

Human Tissue Act 2004. (London: HMO Stationery Office, 2004), www.legislation.gov.uk/ukpga/2004/30/enacted (accessed 20 October 2010).

Human Tissue Bill. (Hansard Commons Debates, 15 January 2004): http://www.publications.parliament.uk/pa/cm200304/cmhansrd/vo040115/debtext/40115-13.htm#40115-13_head0 (accessed 20 October 2010).

Human Tissue Bill. (Hansard Lords Debates, 16 September 2004): http://www.publications.parliament.uk/pa/ld200304/ldhansrd/vo040916/text/40916-58.htm (accessed 20 October 2010).

Human Tissue Bill. (Hansard Lords Debates, 22 July 2004): http://www.publications.parliament.uk/pa/ld200304/ldhansrd/vo040722/text/40722-09.htm#40722-09_head5 (accessed 20 October 2010).

Jasanoff, Sheila, ed. *States of Knowledge: The Coproduction of Science and Social Order* (Abingdon & New York: Routledge, 2004).

Kaplan, Flora E.S., ed. *Museums and the Making of "Ourselves": The Role of Objects in National Identity* (London & New York: Leicester UP, 1994).

Knell, Simon, Sheila Watson & Suzanne MacLeod, ed. *Museum Revolutions: How Museums Change and Are Changed* (London: Routledge, 2007).

Kuhn, Thomas. *The Structure of Scientific Revolutions* (Chicago & London: U of Chicago P, 1962).

Lewis, Geoffrey. "The 'Universal Museum': a case of special pleading?" in *Art and Cultural Heritage: Law, Policy and Practice*, ed. Barbara T. Hoffman (Cambridge & New York: Cambridge UP, 2006): 379–85.

MacDonald, Helen. *Human Remains: Dissection and Its Histories* (New Haven CT & London: Yale UP, 2006).

MacDonald, Sharon. *The Politics of Display: Museums, Science, Culture* (London & New York: Routledge, 1998).

——. "Under Your Skin," *New Scientist* 190.2550 (6–12 May 2006): 51.

McGhee, Robert. "Aboriginalism and the Problems of Indigenous Archaeology," *American Antiquity* 73.4 (2008): 579–97.

Mansell, Michael. "The War of the Dead," *Indigenous Law Bulletin* 46/6.29 (2007): 19–21.

Message, Kylie. *New Museums and the Making of Culture* (Oxford & New York: Berg, 2006).

Museums Australia. *Continuous Cultures, Ongoing Responsibilities* (2005), 136.154 .202.120/dbdoc/ccor_final_feb_05.pdf (accessed 20 October 2010).

National Museum of New Zealand, Te Papa Tongarewa, *Repatriation*, http://www .tepapa.govt.nz/AboutUs/Repatriation/Pages/Overview.asp (accessed 20 October 2010).

Native American Graves Protection and Repatriation Act, USA, http://www.nps .gov/nagpra/MANDATES/25USC3001etseq.htm (accessed 20 October 2010).

Natural History Museum, London. "Human remains to be returned," press release (17 November 2006), http://www.nhm.ac.uk/about-us/news/2006/november /news_10019.html (accessed 20 October 2010).

O'Neill, Mark. "Enlightenment museums: universal or merely global?" *Museum and Society* 2.3 (2004): 190–202.

Palladino, Paolo, & Michael Worboys. "Science and Imperialism," *Isis* 84.1 (1993): 91–102.

Pardoe, Colin. "Arches of Radii, Corridors of Power: Reflections on Current Archaeological Practice," in *Power, Knowledge and Aborigines* (special edition of the *Journal of Australian Studies*), ed. Bain Attwood & Jon Arnold (Pandora, Victoria: La Trobe UP, 1992): 132–41.

——. "Sharing the Past: Aboriginal Influence on Archaeological Practice, a Case Study from New South Wales," *Aboriginal History* 14.2 (1990): 208–23.

Peers, Laura. "Repatriation: A Gain for Science?" *Anthropology Today* 20.6 (2004): 3–4.

Pickering, Michael. "Lost in Translation," *borderlands* 7.2 (2008): 1–18.

Price, Sally. *Paris Primitive: Jacques Chirac Museum on the Quai Branly* (Chicago & London: U of Chicago P, 2007).

Robertson, Sarah. "Sources of Bias in the Murray Black Collection: Implications for Palaeopathological Analysis," *Australian Aboriginal Studies* 1 (2007): 116–30.

Silliman, Stephen W. "The value and diversity of indigenous archaeology: a response to McGhee," *American Antiquity* 75.2 (2010): 217–20.

Smith, Laurajane. "The repatriation of human remains: problem or opportunity?" *Antiquity* 78.300 (2004): 404–13.

Standing Committee G. *Human Tissue Bill* (London: UK Parliament, 3 February 2004): http://www.publications.parliament.uk/pa/cm200304/cmstand/g/st040203/am/40203s01.htm (accessed 20 October 2010).

Stocking, George W., Jr. *Victorian Anthropology* (New York: Free Press, 1987).

Taylor, Rebe. *Unearthed: The Aboriginal Tasmanians of Kangaroo Island* (Adelaide, SA: Wakefield, 2008).

Thomas, David Hurst. *Skull Wars: Kennewick Man, Archaeology, and the Battle for Native American Identity* (New York: Basic Books, 2000).

Tuffs, Annette. "Von Hagens faces investigation over use of bodies without consent," *British Medical Journal* 327.7423 (8 November 2003): 1068.

Turnbull, Paul. "Scientific theft of remains in colonial Australia – postscript," *Australian Indigenous Law Review* 38/11.2 (2007): 72–75.

Turnbull, Paul. "Scientific theft of remains in colonial Australia," *Australian Indigenous Law Review* 7/11.1 (2007): 92–104.

Vergo, Peter, ed. *The New Museology* (London: Reaktion, 1989).

Verran, Helen. "Science and the Dreaming," *Issues* 82 (2008): 23–26.

Watkins, Joe. "Through Wary Eyes: Indigenous Perspectives on Archaeology," *Annual Review of Anthropology* 34 (2005): 429–49.

Watson–Verran, Helen, & David Turnbull. "Science and Other Indigenous Knowledge Systems," in *Handbook of Science and Technology Studies*, ed. Sheila Jasanoff, Gerald Markel, James Peterson & Trevor Pinch (Thousand Oaks CA: Sage, 1995): 115–39.

Wilcox, Michael. "Saving indigenous peoples from ourselves: separate but equal archaeology is not scientific archaeology," *American Antiquity* 75.2 (2010): 221–27.

Wilson, Christopher L. "Indigenous Research and Archaeology: Transformative Practices in/with/for the Ngarrindjeri Community," *Journal of the World Archaeological Congress* 3.3 (2007): 320–34.

⌘

Aboriginal Claims

DNA Ancestry Testing and Changing Concepts of Indigeneity

NATASHA GOLBECK AND WENDY D. ROTH

IN THE LAST FIFTY YEARS, THE NUMBER OF NORTH AMERICANS claiming Aboriginal heritage has increased dramatically. This is not a product of high birth-rates, but is because previously non-Aboriginal-identified people are coming to identify as Aboriginal.[1] Although stereotypes of Aboriginal people as noble, spiritual, and connected to nature are not new,[2] evidence suggests that non-Aboriginal identified people are increasingly viewing Aboriginal heritage as desirable and are seeking out evidence of some Aboriginal ancestry.[3]

The dramatic increase in claims to Aboriginal heritage by previously non-Aboriginal-identified people raises questions about how these individuals and others view the legitimacy of the new claims. As Canada and the USA are colonial nations, North American Aboriginal identity is deeply political and has historically been contested and managed through identity legislation[4] as

---

[1] Jack Hitt, "The Newest Indians," *New York Times Magazine* (21 August 2005): 36–41; Joane Nagel, *American Indian Ethnic Renewal: Red Power and the Resurgence of Identity and Culture* (New York: Oxford UP, 1996); Jeffrey S. Passal, "The Growing American Indian Population, 1960–1990: Beyond Demography," *Population Research and Policy Review* 16.1–2 (1997): 11–31.

[2] See, for example, Phillip J. Deloria, *Indians in Unexpected Places* (Lawrence: UP of Kansas, 2006).

[3] Hitt, "The Newest Indians."

[4] Identity legislation refers to the state policies that regulate the status and classification of Aboriginal subjects during and post-colonization. In Canada, identity legislation

well as coercion and violence.[5] In the contemporary context, the legitimacy of claims to Aboriginality has important practical and political implications in terms of state benefits, sovereignty, tribal registration, and social relations between Aboriginal and non-Aboriginal people.

The emergence of the genetic-ancestry-testing industry provides a new avenue for people to claim Aboriginal heritage and identity. Today more than thirty companies offer to test individuals' DNA and send them a report stating, for example, what proportion of their ancestry is Native American or trace the Native American tribes from which they descend. With this information, test-takers can also trace new lines of ancestry to previously unknown distant relatives, often claiming new ethnic and racial identities in the process.

Genetic-ancestry testing also shapes notions of legitimacy in terms of making claims to a particular identity or heritage. Many test-takers see DNA ancestry tests as a way to offer 'proof' of their Native American ancestry. The tests' scientific nature appears to offer consumers a sense of objectivity. Although the tests and the direct-to-consumer testing industry have themselves been widely criticized, test-takers view them as more reliable than traditional methods of demonstrating ancestry, which they see as flawed by the self-serving interests of colonial identity legislation. As a result, more people than ever before are accessing and claiming this newly desirable identity.

In this essay, we ask what Indigenous identities mean in the genomic age when people who previously did not identify as Aboriginal can claim Aboriginal group membership on a genetic level. We examine how people who take genetic-ancestry tests conceive of Aboriginality and the legitimacy of those who claim it. When Aboriginal identities are increasingly desirable and accessible, the right to legitimate claims-making also becomes more competi-

usually refers to the Indian Act, and in the USA it refers to policies such as the Dawes Rolls, a master list commissioned by the US government to record the names of the Aboriginal people officially enrolled in the Five Civilized Tribes. Identity legislation also refers to regulations restricting tribal membership to those with more than a certain 'blood quantum' level, or proportion of Aboriginal ancestry. The purpose of identity legislation with regard to Aboriginal populations has generally been to limit the number of people who can claim Aboriginal status with the federal government.

[5] Eva Marie Garroutte, *Real Indians: Identity and Survival of Native America* (Berkeley: U of California P, 2003); Bonita Lawrence, *"Real" Indians and Others: Mixed-Blood Urban Native Peoples and Indigenous Nationhood* (Vancouver: U of British Columbia P, 2004); Alexandra Harmon, *Indians in the Making: Ethnic Relations and Indian Identities Around Puget Sound* (Berkeley: U of California P, 2000).

tive. If every person in North America were considered legitimately Aboriginal, the identity claim would no longer be meaningful. This creates a need to measure the legitimacy of mixed-ancestry people's claims to Aboriginal identity, even among those who make putative claims themselves.

We draw on data from a qualitative study with 111 people who have taken DNA ancestry tests. Many of those individuals identified as part-Native American before taking a test, while others did not identify racially or ethnically as Aboriginal but believed they had some Native American heritage which genetic-ancestry testing could help them confirm. Several test-takers also discovered Native American heritage that they did not expect.

Test-takers claiming Aboriginal ancestry in our study distinguish three discursive levels of authentic Aboriginality.[6] The first level is made up of 'Wannabes', whom test-takers see as the least legitimate. Having heard rumours in their family of a distant Aboriginal ancestor, wannabes call upon this connection to give them access to Indigenous identity and/or heritage, but have little to offer by way of proof. The second level, where many test-takers place themselves, consists of 'Lost Descendants' – mixed-blood people whose ties to their Aboriginal heritage are seen as wrongfully severed due to discrimination, identity legislation, and other historical acts. Here, the use of genetic testing is viewed as a crucial way to offer proof of one's long-lost connections to an Aboriginal group. The third level consists of 'Real Indians', whom test-takers see as the most legitimate Aboriginal people. These people typically have legal status as tribal members, or they may simply have been raised in an Aboriginal community, or with regular practice of Aboriginal customs and traditions. Their families have maintained a continuous connection to their Aboriginal roots. Most importantly, those in this group are accepted as Aboriginal by other tribal members. Test-takers criticize the standards required for such social acceptance. They view the enrolment criteria for tribal membership, based on proportions of demonstrable Native American blood quantum, as too narrow. Since DNA test-takers believe that all people are mixed-race and racial purity is a myth, they tend to see tribal membership restrictions as archaic and rigid in their conception of race. Further-

[6] In naming these levels of authenticity, we draw on the works of Bonita Lawrence, *"Real" Indians and Others*, and Eva Marie Garroutte, *Real Indians*. These authors use the term 'real' critically, emphasizing the socially constructed notion of racial authenticity.

more, they challenge social standards of acceptance that fail to recognize the way that Aboriginal descendants were excluded historically.

DNA ancestry test-takers are engaging in 'boundary work' – drawing conceptual distinctions between groups of people to create new systems of classification or new relations between groups.[7] In doing so, they try to redefine who is included and who is excluded from the category 'Aboriginal'. Their efforts attempt to create new symbolic boundaries that use genetic 'evidence' of Aboriginal ancestry as defining characteristics. While test-takers draw symbolic boundaries to differentiate themselves from Wannabes, claiming greater legitimacy and therefore status, they attempt to break down the boundaries that distinguish them from Real Indians by arguing that new standards are needed for who is included and excluded within that group. Through these discourses, genetic genealogy communities are challenging conventional means of legitimating Aboriginality and opening doors for more people to have access to Aboriginal identity.

Direct-to-Consumer DNA Ancestry Testing

Direct-to-consumer DNA ancestry testing offers customers a genetic profile indicating where their ancestors likely came from. For a fee of usually a few hundred dollars, consumers send companies a sample of their DNA, usually cells from the inside of their cheek on a swab, and receive back a package containing information on test-takers' possible heritage in a global context. Close to half a million people worldwide have taken these tests.[8] Among the most common DNA ancestry tests are mitochondrial (mtDNA) tests, Y-chromosome (Y-DNA) tests, and autosomal or genome-wide tests.

---

[7] Michèle Lamont, *Money, Morals & Manners: The Culture of the French and the American Upper-Middle Class* (Chicago: U of Chicago P, 1992); Michèle Lamont, *The Dignity of Working Men: Morality and the Boundaries of Race, Class, and Immigration* (New York: Russell Sage Foundation, 2000); Michèle Lamont & Virág Molnár, "The Study of Boundaries in the Social Sciences," *Annual Review of Sociology* 28 (2002): 167–95; Charles Tilly, "Social Boundary Mechanisms," *Philosophy of the Social Sciences* 34 (2004): 211–36.

[8] Deborah A. Bolnick, Duana Fullwiley, Troy Duster, Richard S. Cooper, Joan H. Fukimura, Jonathan Kahn, Jay S. Kaufman, Jonathan Marks, Ann Morning, Alondra Nelson, Pilar Ossorio, Jenny Reardon & Kimberly TallBear, "The Science and Business of Genetic Ancestry Testing," *Science* 318 (2007): 399–400.

An mtDNA test traces the ancestry of a person's direct maternal line – in other words, their mother's mother's mother's heritage and so on. Y-chromosome DNA tests trace a person's direct paternal line - their father's father's father's ancestry and so on. Y-DNA is passed down from fathers to sons, so only males can take Y-DNA tests while both males and females can take mtDNA tests. These tests look at how genes have changed from random mutations and environmental selection pressures as groups of people have migrated over tens of thousands of years. By identifying 'markers' or places on a DNA chain and matching them to population databases, the tests identify the migration routes taken by that ancestral line and the associated historical population group that took that route, known as a 'haplogroup'.[9] The tests also typically provide information about other people in the database whose markers provide a genetic match to the test-taker, with various degrees of removal, and information about where matches are located around the world.[10] People often use this test to find relatives or common ancestors.

Autosomal or genome-wide testing maps a person's entire genome. Rather than direct maternal or paternal lines, it looks at genetic information given to a person by all their ancestors. Some companies offer these results as percentages of ancestry (e.g., fifty-five percent European, three percent Native American, etc.).[11] Many companies also offer tests that identify from which Native American tribe the test-taker is descended. Native American tribal-specific tests measure either the Y-DNA or mtDNA of test-takers. By comparing a test-taker's DNA markers with those of Native American tribes sampled by the testing company, test results offer consumers information about which (if any) tribe they match. Academics, journalists, scientists, and tribal councils have engaged in charged debates about the usefulness of these tests

[9] Henry T. Greeley, "Genetic Genealogy: Genetics Meets the Marketplace," in *Revisiting Race in a Genomic Age*, ed. Barbara A. Koenig, Sandra Soo-Jin Lee & Sarah S. Richardson (New Brunswick NJ: Rutgers UP, 2008): 215–34; Charmaine D. Royal, John Novembre, Stephanie M. Fullerton, David B. Goldstein, Jeffrey C. Long, Michael J. Bamshad & Andrew G. Clark, "Inferring Genetic Ancestry: Opportunities, Challenges, and Implications," *American Journal of Human Genetics* 86.5 (May 2010): 661–73.

[10] Alondra Nelson, "The Factness of Diaspora: The Social Sources of Genetic Genealogy," in *Revisiting Race in a Genomic Age*, 253–68.

[11] Henry T. Greeley, "Genetic Genealogy."

for tribal enrolment policies, financial reparations including casino money, or the right to speak and identify as a Native American person.[12]

Critics of DNA ancestry testing argue that test-takers may misinterpret their results, both in terms of the amount of ethnic heritage they have and by drawing from it a biological understanding of race.[13] For example, if a non-black consumer takes an mtDNA test and the results report that their maternal line originated in Africa 20,000 years ago, they may begin to think of themselves as having African genes. In the Native American context, DNA testing companies claim that they can prove a person has DNA markers found "only in Native Americans," but which are in fact found in other populations, leading to "false positives."[14] Given the desirability of Aboriginal heritage to many contemporary North Americans, critics are concerned that companies exaggerate the certainty of Native American heritage in order to generate higher customer satisfaction.

Methodology

Data from this essay draw from a qualitative study of 111 DNA test-takers, mostly in the USA and Canada. Interviews focused on respondents' experiences taking the tests, their racial identity before the test, and any ways in which the test has changed their identity and behaviour, their racial attitudes, and their conceptions about race. To generate a wide population of DNA test-takers to use as a sampling frame for in-depth interviews, we created an online survey which individuals could complete anonymously. At the end, survey respondents were asked if they would be interested in participating in an in-depth interview over the telephone. Only those who said they would were asked for contact information. Three DNA testing companies agreed to send out notices about the online survey to their customers. Other survey respon-

---

[12] Kimberly TallBear & Deborah Bolnick, "'Native American DNA' Tests: What are the risks to tribes?" *The Native Voice* (3–17 December 2004): D2.

[13] Lynne B. Jorde & Stephen P. Wooding, "Genetic variation, classification and 'race'," *Nature Genetics* 36 (2004): S28–S33; John Dupré, "What Genes Are and Why There Are No Genes for Race," in *Revisiting Race in a Genomic Age*, 39–55; Kimberly TallBear, "Narratives of Race and Indigeneity in the Genographic Project," *Journal of Law, Medicine & Ethics* 35.3 (Fall 2007): 412–24; Kimberly TallBear, "Native-American-DNA.com: In Search of Native American Race and Tribe," in *Revisiting Race in a Genomic Age*, 235–52.

[14] TallBear & Bolnick, "Native American DNA Tests," 4.

dents were informed of it through online genetic genealogy communities, by posting notices on listservs and discussion forums for people who have taken the tests, and snowball sampling from other respondents.

We conducted qualitative interviews with nineteen people who identified as part Native American before the test, although none identified as only Native American and none were registered tribal members or Status Indians prior to taking the test. In addition, several respondents who previously identified as White (N=8), Black (N=5), Hispanic (N=5), or Asian (N=1) discovered Native American heritage from taking the test, including several who came to identify as partly or entirely Native American. We focus primarily on these 32 cases in this chapter.

Buried Indigeneity

Aboriginal identities are considered attractive to many test-takers, more so than other types of identities they might uncover. In part this stems from stereotypes of Native Americans as spiritual, righteous, and living in harmony with their environment – an image which takes on a new cachet as environmental consciousness is re-branded as desirable. Several respondents report having an interest in Native Americans that stems from the television shows, movies, and books they used to read as children. For instance, James, a 35-year-old computer technician who identified his race before the tests as white with some Native American, describes his disappointment at not uncovering any Native American heritage:

> I was a little surprised, and disappointed, of course, because I wanted to find that Native American and point to it. And, you know, I guess growing up in America and watching shows, […] I guess being a Native American is probably something that a lot of people want to have in their family, if you're American. And, because it's kind of neat to have some type of Aboriginal in your DNA.

Respondents report that people having family rumours of an ancestor who was a Native American princess is practically a contemporary American cliché – just about every other person doing genealogy seems to be searching for a link to his or her Aboriginal heritage.

Although our respondents have lived largely outside of an Aboriginal group – their cultural upbringing and appearance is typically non-Aboriginal – many share such family stories of a Native American ancestor to emphasize their connection to aspects of Aboriginality that they deem to be positive.

Sometimes test-takers express their connection as a kind of 'blood memory'– the idea that a sense of collective identity can be felt in the body even if it is not known in the mind.[15] This sense of blood memory often mirrors stereotypes about Aboriginal people. Respondents report that they have always felt close to nature or spirituality, and attribute these characteristics to a presumed Aboriginal heritage. In other cases, uncovering Aboriginal ancestry can lead people to search back through their past and uncover anything related to Native American customs which can now be interpreted as evidence that they felt that connection all along. In other words, these respondents feel that they always knew they were Native American, even when they had no evidence to prove it.

As part of their sense that there is Aboriginality buried within them, many of these respondents feel that they were never able to identify with the racial group they grew up as an ostensible part of, and report always having felt 'different' from their peers. Louise, a 68-year-old arborist from Oregon, grew up in a family who identified as white but had some knowledge of Native American heritage. She says:

> I always knew I was different, and people who knew me said, "You're not white. I don't know what you are but you're not white. You don't look white." Well, I look white, but there was something about me.

White respondents in particular report not identifying with white North American culture and attribute this feeling to their hidden racial background. Such discourses, together with genetic 'evidence' of Native American ancestry, give respondents a way to distance themselves from the racial group in which they were raised. Joe, a mixed-race landscaper who was raised in the US South, claims that one of the benefits of DNA ancestry testing is that "it might give [someone who hated identifying in their own racial group] enough data to identify with somebody else that they feel more comfortable with." This sense that DNA testing can give test-takers the proof they need to be flexible about their racial identity is indicative of the high number of people looking to do so. As a result, many test-takers use the DNA results as a way of giving themselves greater legitimacy and differentiating themselves from the masses of other Americans claiming similar stories. In the following sections, we illustrate each of the three layers of legitimacy that test-takers

---

[15] Lawrence, *"Real" Indians and Others*, 200.

describe in making sense of their own claims to Aboriginality and the boundaries they draw or attempt to break down between these groups.

Wannabes

The category of Aboriginal claimants whom test-takers identify as Wannabes is closely in line with Bonita Lawrence's conception of white or "*virtually* white people"; these

> resurrect an extremely distant Native ancestor whose existence has no other tangible implications for an otherwise white family than to 'boundary cross' in the name of that ancestor and have access to otherwise forbidden Native spaces, usually for some form of personal gratification.[16]

The test-takers who identify as part Native American, either before taking the tests or since, are highly aware of the influx of people claiming Aboriginal heritage, and the effect this has on the legitimacy of their own claims. Respondents quickly acknowledge the vast numbers of claimants, partly in order to pre-emptively differentiate themselves from such Wannabes. Kevin, a 40-year-old event planner from Arizona, found he had Aboriginal ancestry after being raised thinking that his family's descent was purely French. After finding links to the tribe he is related to, his Aboriginal legitimacy was strongly criticized by genealogical researchers. He says, "I know it's almost a joke among Indians but it's like everybody in the US had a grandmother who was a Cherokee princess." Kevin was concerned that he would be treated as a 'Wannabe', lumped together with everyone else making the same claim without anything to back it up.

Janet, a 64-year-old teacher from New Jersey, who identified as white with vague knowledge of Aboriginal heritage before taking a DNA ancestry test, discusses how Aboriginality has become trendy:

> Well, [Native Americans] are in vogue now. It's all the thing to be Native American, if you can claim, 'Well, I'm part Native American' because they get the passion and the glory days now of everything with the sweat lodges and medicine healing and all that, their Native spirit thing. [...] They may have lived as an American-Caucasian all their life and [Native Americans] had to go through the discrimination, and the reservation, and all the other stuff. They don't know about the

---

16 *"Real" Indians and Others*, 4.

> language or mythology, but they can say they're Native American […] I guess it's kind of like an ego-status thing.

Janet expresses frustration at the illegitimacy of Wannabe claims, since they can take the good – identifying with a trendy racial group – without the bad – experiencing racism and the legacy of colonialism.

The distinction respondents make between legitimate and illegitimate claims often rests on the ability to 'prove' one's Aboriginal heritage. Lisa is a 25-year-old woman from Kentucky who previously identified as white Hispanic, but found that she had Mayan and Apache ancestry through a DNA test, and as a result she now identifies as part-Native American. Lisa states:

> There's some people that will – I don't want to say fake, but they'll just present themselves the wrong way, like they'll present themselves like they're all this when they really don't have, you know, any documentation or proof. […] There's always people trying to make themselves look like something they're not.

Those who claim an Aboriginal identity without such 'proof' risk being stigmatized by others. People who take DNA ancestry tests and are seeking such proof are particularly aware of this risk, and some respondents are therefore hesitant to make such claims if their test results did not provide the proof they were looking for. James, the 35-year-old computer technician who was disappointed not to uncover Aboriginal ancestry, expresses this concern. Even though his family's stories made him proud of his putative Native American connections, the lack of genetic evidence led him to curtail his claims:

> When I was a child, I was told that we have Cherokee in us. And, now that I'm older and I hear other people's stories, it seems like every person I talk [to] always has some Indian in them, they say. Even if they look totally White, like I do. As a child, I used to read Indian books in the library […] and think it was really neat, and I always thought, 'hey, that's something that I know I am, that I can kind of point to'. Because everyone wants to identify some way. So I knew I was White, and I knew I had some Indian in me. So, that's why I've been looking around for it. Since I have uncles and aunts that look Indian, it's not like it's totally foreign. It's kind of there, I can see some evidence of it. So that's probably influenced me liking Native American type ancestry.
>
> But I also don't want to be a poser, you know. I don't want to say, 'oh, I have Native American,' when I really have nothing in me.

> Because I already look like a White guy, so, it's kind of silly of me to say I have Native American if I really don't. And so far we have none.

Now James identifies himself just as white, to avoid the label of a poser, or Wannabe. The stigma of a Wannabe is such that those without the 'goods' to back up their claims see themselves as better-off avoiding such claims altogether.

Lost Descendants

Identity legislation geared toward limiting the number of Aboriginals recognized by the federal government,[17] and ancestors' efforts to 'pass' as non-Aboriginal to avoid discrimination, have cut many descendants of Aboriginal people of from connections to their origins. Respondents in this group see themselves as joining up the wrongfully severed ties of the past by trying to locate evidence of Aboriginality. In their efforts to do so, they often become frustrated with traditional genealogical research, which presents the official classifications motivated by such identity legislation. For example, their ancestors are often excluded from the Dawes Rolls or other Native American registries that listed thousands of part-Aboriginal people as white, black, or mulatto. Respondents also report that their elder family members are secretive about known Aboriginal heritage, either because they are ashamed of it or to improve their family's life-chances. These test-takers see genetic genealogy as a way past these barriers to knowledge about their heritage, as if science had objectively outsmarted the irrational wrongs of history.

Racial categorization, according to many of these test-takers, is political and has changed frequently to suit the state's interest in having fewer Aboriginal people. Julie, a 57-year-old retired nurse from New Jersey who identifies as white but heard rumours in her family of an Aboriginal ancestor, discusses an example of how identity legislation can erase textual records of indigeneity. She describes the Racial Integrity Act, a law that registered all people of colour into one racial category:

> In Virginia, during the time of eugenics in the country – I think it was 1924 – that one of the fellows who was in position in the government or whatever, I don't know what his title was, but he's decided that there are no Indians in the States. You could be black or you could be white, and that was it. There were no Indians in Virginia.

---

[17] Garroutte, *Real Indians*, 48.

By grouping Aboriginals, blacks, and anyone with non-white ancestry into one 'coloured' category, and expunging an 'Indian' designation from their birth certificates and other records,[18] the details of Aboriginal ancestry are lost to descendants trying to trace their heritage through historical documentation from the period. Lost Descendants cite such examples to illustrate how state policies and practices, influenced by political motivations, cut them off from their 'rightful' heritage.

For people with partial or mixed Aboriginal heritage whose Aboriginal identity was lost through such legislation, DNA ancestry testing is a way to re-claim proof of their heritage. Respondents also describe DNA testing as a way to go over the heads of elder family members determined to conceal their Indigenous heritage. Julie explains:

> The reason I got the test in the first place was because my godmother, who's my mother's oldest sister, she will be one hundred and three this year, and she supposedly knows the family history, but she won't talk about it. And I had heard from my mother that my grandfather's mother was Indian, and, at that time, a hundred years ago, it was a thing that was shame[ful], and people didn't talk about that. And so she may have known what was going on, or who people were that were part of the family, but she's not talking about it. She's going to take that to the grave with her. So, I figured, okay, well let me see what's coming up here, and uh [*laughs*] surprise!

Despite her godmother's concealing information about her heritage, Julie is able to identify as a Lost Descendant due to the knowledge she gained from her DNA ancestry test.

The reliance of Lost Descendants on genetic 'proof' for legitimacy is particularly important to them as a way of differentiating themselves from Wannabes. Kevin explains:

> everybody wants to be an Indian and then I come along and have DNA proof of what I am and they can't deny that. It struck me that they were jealous of that.

Kevin is demonstrating his awareness that some people are sceptical of Aboriginal identity claimants, especially if their appearance or upbringing does not

---

[18] Ben A. Franklin, "Legacy of Racism," *Washington Spectator* (1 November 2004), http://www.washingtonspectator.org/articles/20041101fyi.cfm (accessed 15 October 2010).

help them assert their claims. These respondents themselves view Wannabes' claims with suspicion. Offering the 'proof' they see DNA tests as providing is a way, in their eyes, in which to distinguish themselves from Wannabe claimants and assert a higher level of legitimacy.

The interest of Lost Descendants in searching for their roots also emanates from their belief that discrimination is a thing of the past. They believe there is no longer a need to hide one's racial origins, as their ancestors might have done. As people who are visually and culturally non-Aboriginal, test-takers seek the positive associations with an Aboriginal identity, without any of the costs.

Real Indians

Respondents see members of registered tribes and those who have grown up with Aboriginal communities and traditions as the most legitimate claimants to Aboriginality. Tribal membership in the USA and Canada is usually granted based on a demonstrated 'blood quantum' – often one-half or sometimes one-fourth, but the amount typically depends on the tribe. These restrictive criteria are in place at least partly because the federal governments of both countries govern Aboriginal groups differently from non-Aboriginal groups, including providing some financial benefits such as casino income, tax exemptions, and postsecondary scholarships.[19] Tribes and bands tend to be protective of their membership because it is tied to resources as well as cultural meaning.

Our respondents see the criteria for tribal membership as too restrictive. Although tribal membership is not necessarily desirable for test-takers who identify as part-Aboriginal, some see their exclusion from tribal classifications as undermining their access to their Aboriginal heritage. Respondents report that such restrictive classification systems are archaic and based on an erroneous belief that it is possible to be racially 'pure'. However, respondents acknowledge that tribal membership is the gold standard of legitimate Aboriginality. Caitlin is a 26-year-old filing clerk who came to identify as part-Aboriginal in addition to white Mexican-American after taking a DNA test. As she puts it, "in most cases when [someone] asks you if you're any sort of Native American, they usually don't consider you Native American unless

[19] Steven Light & Kathryn Rand, *Indian Gaming and Tribal Sovereignty: The Casino Compromise* (Lawrence: UP of Kansas, 2007); Richard H. Bartlett, "Indians and Taxation in Canada," *American Indian Law Review* 7.2 (1979): 185–243.

you're officially enrolled in a tribe." This idea that there is an external measure of legitimate Aboriginality and that they do not meet the criteria is upsetting to some respondents. Roberto, a 31-year-old white Hispanic college instructor, explains:

> Well, I know one person that actually tried to get tribal affiliation but she was unable to do it and I was very saddened for her because she had the DNA plus she even had a document, she could measure ancestry to 1850 and she actually had the mitochondrial DNA to prove that she was Native American. [...] Well she goes and she tries to get recognized, but they won't recognize her because she doesn't have the percentile that she needs to be Native American.

Test-takers who feel excluded from tribal membership or social acceptance within Aboriginal communities tend to resent what they see as overly restrictive and obsolete criteria. Since most test-takers believe that everyone is racially mixed (and, indeed, racial purity is a myth), respondents consider blood quantum regulations to be outdated. Winona, a woman of black and Native American ancestry, demonstrates this feeling, describing racial categorization as a function of "the Western colonizing mind." She describes her reaction when a Choctaw activist challenged her claim to Aboriginality because she is mixed:

> I told him, "do you understand the history of my family? And if you don't understand the history of my family and who I am, then don't comment on it or tell me who I should be because that's not your job or your right" – who somebody wants to be, like the government told you who you could be and what blood quantum you are. This is not South Africa, nor is this apartheid, and I will not give into that.

Winona's statement shows her powerful reaction to the legitimacy of her Aboriginality being questioned, because racial groups in the USA are highly mixed as a consequence of slavery and colonization. She feels that the larger, federally recognized Aboriginal tribes who look to blood quantum for legitimacy are a product of procrustean state policy, and she applauds the smaller tribes that accept DNA results as proof of membership. Other DNA test-takers echo her feeling that those who have 'legitimacy' as Aboriginal people are a product of political, subjective, and oppressive historical actions, and that the 'objective' science of DNA ancestry testing can undo the wrongs of this history.

Test-takers' criticism of tribal membership criteria is meant to redraw the boundaries of Aboriginal group membership, and change the criteria on which it is based. To further this effort, they also engage in boundary work to emphasize their similarities to Real Indians and weaken the boundary that separates them. They point to tribal members who themselves have mixed origins or experience living outside of Aboriginal communities. And they often try to minimize the importance of appearance as an arbiter of group membership by pointing to its inability to determine who is authentically Aboriginal, with examples of Real Indians who have a non-Aboriginal appearance. Peter, a retired biologist from rural Indiana who claims that he looks "like an Irish boy," joined a non-federally recognized tribe for mixed-blood people. His brother got a DNA test, which Peter used as evidence of his own genetic line. He describes his reaction to the Chief of that tribe:

> [When my brother] got the DNA results, it confirmed what he thought and at that time he started talking to Dallas King, who is our Chief. And Dallas was raised on a reservation, and he's 7/8ths or something Cherokee. [He's a] reservation Indian, [a] speaker, [he] understands the language. First time I saw him, I was thunderstruck: he's a blonde-haired, blue-eyed Irish man as far as I could tell, because he had a blonde-haired, blue-eyed grandfather. You meet his siblings and his mother, they are dusky skinned Native Americans, they're Cherokees, and here he is – the speaker, the Chief – and he is blue-eyed, blonde-haired, and his mission is to move in both worlds and to give a place for those who are mixed and willing to learn, a place where they can learn something about the culture.

Peter's story communicates his view that the non-Aboriginal appearance of mixed-blood people can be deceiving. By rejecting the idea that a person has to look like an Aboriginal person to really be Aboriginal, he legitimizes his own claim to Indigenous ancestry. In effect, this story allows Peter to redefine Aboriginality as something that should include himself.

Conclusion

DNA ancestry test-takers attempt to create a new standard for Aboriginal legitimacy. Their understanding of Aboriginal authenticity relies on the notions that race is something that can be biologically measured, that the proportion of Aboriginal ancestry one has is a less important measure of authenticity than having 'proof' of any Native American heritage, and that registered tribes' blood quantum rules for membership are archaic, since they are based on an erroneous concept of 'pure' racial heritage. These test-takers put forth genetic evidence as the new standard for legitimate Aboriginality to differentiate themselves from the innumerable other people who claim Aboriginal heritage. At the same time, they attempt to diminish the boundaries between themselves and those who have been legally and socially accepted as Aboriginal. The boundaries they try to draw and tear down are intended to redefine the meaning of indigeneity.

As more people take DNA ancestry tests to find proof of an Aboriginal ancestor, both they and the general public may conceive of Aboriginality differently. The rapid increase in people claiming Aboriginal identity could challenge state benefits for Aboriginal people, such as tax exemptions, casino money, or scholarships. If every other person identifies as Indigenous, critics may argue, why should the state accord them rights and reparations?

The trend towards genetic genealogy testing to prove Native American heritage could also result in Aboriginality becoming less tribe-centred and potentially less meaningful. Many DNA tests cannot demonstrate tribal affiliation,[20] so, if genetic testing becomes a popular basis for Aboriginal identification, Native American identity may increasingly be pan-Indian. Further, if claims to Aboriginality are based primarily on genetic proof and not recent ancestry or cultural affiliation, Native American representations may become centred in commercial markets rather than in individual tribal cultures. Aboriginal people and tribes therefore have a stake in the field of DNA ancestry testing and may desire participation in managing the industry's growth.

DNA ancestry testing is a growing phenomenon, and many of the people who take these tests seek or hope to find evidence of a long-lost indigenous past. As more people rely on DNA ancestry testing for 'proof' of their heritage, existing Aboriginal groups will face increasing challenges to defend their membership. Is Aboriginality something that can exist on a genetic level, separate from the social experience of living – and being seen – as a member of an Aboriginal group, with all the negative as well as positive repercussions that entails? The different definitions maintained by Aboriginal communities and those who claim new Aboriginal identities are likely to force that conversation to take place.

---

[20] TallBear & Bolnick, "Native American DNA Tests."

Works Cited

Bartlett, Richard. "Indians and Taxation in Canada," *American Indian Law Review* 7.2 (1979): 185–243.

Bolnick, Jonathan, Deborah A., Duana Fullwiley, Troy Duster, Richard S. Cooper, Joan H. Fukimura, Kahn, Jay S. Kaufman, Jonathan Marks, Ann Morning, Alondra Nelson, Pilar Ossorio, Jenny Reardon & Kimberly TallBear. "The Science and Business of Genetic Ancestry Testing," *Science* 318 (2007): 399–400.

Deloria, Phillip J. *Indians in Unexpected Places* (Lawrence: UP of Kansas, 2006).

Dupré, John. "What Genes Are and Why There Are No Genes for Race," in *Revisiting Race in a Genomic Age*, ed. Barbara A. Koenig, Sandra Soo-Jin Lee & Sarah S. Richardson (New Brunswick NJ: Rutgers UP, 2008): 39–55.

Franklin, Ben A. "Legacy of Racism," *Washington Spectator* (1 November 2004), www.washingtonspectator.org/articles/20041101fyi.cfm (accessed 15 October 2010).

Garroutte, Eva Marie. *Real Indians: Identity and Survival of Native America* (Berkeley: U of California P, 2003).

Greeley, Henry T. "Genetic Genealogy: Genetics Meets the Marketplace," in *Revisiting Race in a Genomic Age*, ed. Barbara A. Koenig, Sandra Soo-Jin Lee & Sarah S. Richardson (New Brunswick NJ: Rutgers UP, 2008): 215–34.

Harmon, Alexandra. *Indians in the Making: Ethnic Relations and Indian Identities Around Puget Sound* (Berkeley: U of California P, 2000).

Hitt, Jack. "The Newest Indians," *New York Times Magazine* (21 August 2005): 36–41.

Jorde, Lynn B., & Steven P. Wooding. "Genetic variation, classification and 'race'," *Nature Genetics* 36 (2004): S28–S33.

Lamont, Michèle. *The Dignity of Working Men: Morality and the Boundaries of Race, Class, and Immigration* (New York: Russell Sage Foundation, 2000).

——. *Money, Morals & Manners: The Culture of the French and the American Upper-Middle Class* (Chicago: U of Chicago P, 1992).

——, & Virág Molnár. "The Study of Boundaries in the Social Sciences," *Annual Review of Sociology* 28 (2002): 167–95.

Lawrence, Bonita. *"Real" Indians and Others: Mixed-Blood Urban Native Peoples and Indigenous Nationhood* (Vancouver: U of British Columbia P, 2004).

Light, Stephen, & Kathryn Rand. *Indian Gaming and Tribal Sovereignty: The Casino Compromise* (Lawrence: UP of Kansas, 2007).

Nagel, Joane. *American Indian Ethnic Renewal: Red Power and the Resurgence of Identity and Culture* (New York: Oxford UP, 1996).

Nelson, Alondra. "The Factness of Diaspora: The Social Sources of Genetic Genealogy," in *Revisiting Race in a Genomic Age*, ed. Barbara A. Koenig, Sandra Soo-Jin Lee & Sarah S. Richardson (New Brunswick NJ: Rutgers UP, 2008): 253–68.

Passal, Jeffrey S. "The Growing American Indian Population, 1960–1990: Beyond Demography," *Population Research and Policy Review* 16.1–2 (1997): 11–31.

Royal, Charmaine D., John Novembre, Stephanie M. Fullerton, David B. Goldstein, Jeffrey C. Long, Michael J. Bamshad & Andrew G. Clark. "Inferring Genetic Ancestry: Opportunities, Challenges, and Implications," *American Journal of Human Genetics* 86.5 (May 2010): 661–73.

TallBear, Kimberly. "Narratives of Race and Indigeneity in the Genographic Project," *Journal of Law, Medicine & Ethics* 35.3 (Fall 2007): 412–24.

——. "Native-American-DNA.com: In Search of Native American Race and Tribe" in *Revisiting Race in a Genomic Age*, ed. Barbara Koenig, Sandra Soo-Jin Lee & Sarah Richardson (New Brunswick NJ: Rutgers UP, 2008): 235–52.

——, & Deborah Bolnick. "'Native American DNA' Tests: What are the risks to tribes?" *The Native Voice* (3–17 December 2004): D2.

Tilly, Charles. "Social Boundary Mechanisms," *Philosophy of the Social Sciences* 34 (2004): 211–36.

⌘

Notes on Contributors and Editors

IAN ANDERSON is the foundation Chair of Indigenous Health at the University of Melbourne and the Director of Murrup Barak – Melbourne Institute for Indigenous Development at the University of Melbourne. He is also the Research Director for the Lowitja Institute, Australia's National Institute for Aboriginal and Torres Strait Islander Health Research, incorporating the Cooperative Research Centre for Aboriginal and Torres Strait Islander Health. Ian also chairs the National Aboriginal and Torres Strait Islander Health Equality Council, the peak Indigenous health advisory group for the Australian Government. Ian Anderson's background is in medicine and social sciences and he has worked in Aboriginal Health for more than twenty-five years as a health worker, educator, general practitioner, policy-maker, and academic. Professor Anderson has been a full-time research academic since 1998. He completed a medical degree at the University of Melbourne in 1989 and has a PhD in Sociology and Anthropology from La Trobe University. His family are Palawa Trowerna from the Pyemairrenner mob in Tasmania, which includes Trawlwoolway and Plairmairrenner and related clans.

RENATE BARTL holds an MA in American Cultural History, Ethnology, and Political Science from the University of Munich. She wrote her master's thesis (1987) on the relations between Native Americans and African Americans in the USA. Her present research project is on "American Identity: Groups of Mixed African-American – Native American – European Ancestry in the Eastern USA." She has taught Native American Studies in the Department of American Cultural History, University of Munich, and First Nation Studies for the Institute for Canadian Studies, University of Augsburg. She currently teaches the Virtual Canadian Studies (VCS) online seminars on Canadian First Nations for the Association of Canadian Studies in German-Speaking Countries. She administrates the American Indian Workshop (AIW) general items and is responsible for the AIW webpage (www.amer

ican-indian-workshop.org) and mailserver. Her major publications are "Tri-racial groups," in *The Encyclopedia of New York State*, ed. Peter Eisenstadt (2005): 1578; "The Importance of the 'Indian Church' for Native American Survival in the Eastern United States," *Acta Americana* 8.2 (2000): 37–53; "Native American Tribes and Their African Slaves" in *Slave Cultures and the Cultures of Slavery*, ed. Stephan Palmié (1995): 162–75; and, with Barbara Göbel & Hanns J. Prem, "Los Calendarios Aztecas de Sahagún," *Estudios de Cultura Náhuatl* 19 (1990): 13–82.

SUSANNE BERTHIER–FOGLAR is Professor of American Studies and Native American Studies at the University of Grenoble. She is the project leader of a French research project on Indigenous Peoples Facing Global Change (LabexITEM, Work Package 1, Project CGPM). In 2008–11 she was in charge of a (French) regionally funded research project on the Indigenous discourse on human genomics (Cluster 14, Rhône–Alpes: The Discourse of Indigenous Refusal of Human Genome Studies in Anglo-Saxon Settler Colonies). Her research and publications are about Native American history and ethnic identities in the Southwest from the first contact with Europeans to contemporary Indigenous activism. Her recent books include *Les Indiens Pueblo du Nouveau-Mexique* (2010), *La France en Amérique: Mémoire d'une conquête* (ed. 2009), and, with François Bertrandy, *La montagne, pouvoirs et conflits de l'Antiquité au XXIe siècle* (2011). Recent articles include: "A Patient/Hospital Relationship in 1863–65: Mainstream Doctors and Navajo Patients in the Bosque Redondo Camp," in *The Patient*, ed. Kimberly R. Myers & Harold Schweizer (2010): 79–95; "The 1889 World Exhibition in Paris: The French, the Age of Machines, and the Wild West," *Nineteenth–Century Contexts* 31.2 (2009): 129–42; "The Zia Sun Symbol and the Pueblo's New Tribalism," *Perspectives* 7 (Cahiers du CICLaS, 2006): 9–25; and "Gastronomy and Conquest in the Mexican-American War: Food in the Diary of Susan Magoffin," *Diálogos Latinoamericanos* (October 2005): 1–27.

SHEILA COLLINGWOOD–WHITTICK is a senior lecturer in the English section of the Department of Foreign Languages at Stendhal University, Grenoble 3, where she has taught for the past sixteen years. From 2002 to 2004, she also taught postcolonial studies at the University of Geneva. Over the last thirty years her field of research has been, broadly, that of postcolonial literatures and she has published widely on fictional and autobiographical writings from several former British settler colonies. For the past twelve years her scholarship has focused increasingly on Indigenous and non-Indigenous Australian fiction. During that time she has edited a collection of essays entitled *The Pain of Unbelonging: Alienation and Identity in Australasian Literature* (2007), and has published several essays on the tortuous relationship between history and fiction in contemporary Australian literature. The scope of her research has also widened to encompass non-literary issues and, latterly, her work has been devoted to the ongoing impact of the trauma of colonization on the lives, culture, and environment of Australia's Indigenous peoples. Her most recent publication is a chapter in *Indigenous Rights in the Age of the U.N. Declaration*, ed. Elvira Pulitano (2012), and forthcoming publications include book chapters on reconciliation discourse and non-Indigenous cultural appropriation strategies. She is a member of the editorial committee of the international journal *Middle Ground: Journal of Literary and Cultural Encounters*.

JAROSŁAW DERLICKI is Assistant Professor in the Institute of Archaeology and Ethnology, the Polish Academy of Sciences, Warsaw. He is also editor of *Polish Ethnography*. In 2007–2008, he was awarded the scholarship of the Foundation for Polish Science for young scientists. He is a specialist in ethnicity studies and has worked on the cultural borderlands in Moldova as well as on the Yakut peoples. His research on the Yakut has been awarded several grants (Grant Committee for Scientific Research, S. Batory Foundation Grants). Among his publications are *First Nations* (co-ed. W. Lipinski, 2002) and numerous articles on the Yakut, the latest being "I am the son of Oliero: Yukaghir identity and land issues," *Alternative: An International Journal of Indigenous Peoples* 6.3 (2010): 272–82.

SÉVERINE GAUTHIER–LABOUROT defended her doctoral dissertation on the Cherokees' fight for the preservation of their identity and sovereignty (1838–2008) at the Sorbonne (University Paris IV) in 2010. She conducted field work on ethnicity and racial issues as well as on politics from 2003 to 2010 in Norman, Oklahoma, where she was a visiting scholar at the University of Oklahoma. She has taught English in French secondary schools and has been recruited as a junior lecturer at the University of Reims Champagne–Ardenne. Her contribution to this work is her first publication. She is also a published author of juvenile literature.

EMMA KOWAL is a postdoctoral research fellow in anthropology at the University of Melbourne supported by a National Health and Medical Research Council Aboriginal and Torres Strait Islander Training Fellowship. She is a cultural anthropologist who has previously worked as a doctor and public health researcher in Indigenous health settings. Her research interests include race, racism and anti-racism, whiteness and indigeneity, settler colonialism, and contestations between biological and cultural difference. She is currently pursuing these interests through ethnographic research with genetic researchers who work in Indigenous communities in Australia. Her forthcoming monograph 'Caught in the Gap: The Cultural Politics of White Anti-Racism' is based on ethnography of progressive white professionals who work in Indigenous health in the north of Australia. She is the co-editor, with Tess Lea and Gillian Cowlishaw, of *Moving Anthropology: Critical Indigenous Studies* (2006), and her research has been published in the *American Anthropologist*, *Social Science and Medicine*, *Cultural Studies*, the *Asia Pacific Journal of Anthropology*, the *Health Sociology Review*, and in national and international medical journals.

FRANK KRESSING wrote his doctoral dissertation at Ulm University, Germany, on Western and traditional health-care systems in rural Bolivia. He is a lecturer at Ulm University and participates in the research project "Evolution and Classification in Biology, Linguistics, and the History of Sciences" funded by the Federal German Ministry of Education and Research. His chief research interests cover Indigenous land rights, Indigenous knowledge and health-care systems, action anthropology (advocacy), medical anthropology, and the anthropology of religions. The areas of his field-work include North and South America, South and Central Asia, and the Balkans. Among his recent publications are "Computergestützte Netzwerkanalyse in Biologie, Sprach- und Geschichtswissenschaften," in *Mit Leben rechnen: Zur Geschichte des Wissentransfers zwischen Computer- und Biowissenschaften* (with Matthis Krischel & Heiner Fangerau, 2009), "Religion als Bestandteil von Ethnizitätskonstruktionen," *Augsburger Volkskundliche Nachrichten* (2009), "The Increase of Shamans in Contemporary Ladakh," *Asian Folklore Studies* 62.1 (2003): 1–23, and "Das 'Human Genome Diversity Project': Rassismus im neuen Gewand oder erkenntnistheoretische Grundlagenforschung zum Wohle der Menschheit?" *INFOE-Magazin* 1 (1994): 16–21.

ULIA GOSART (POPOVA–GOSART) is a descendant of the Udmurt Indigenous peoples, an ethnic group that historically inhabited the Western parts of the Ural region of Russia. For a number of years Ulia has been involved in activities related to the protection of the rights of Indigenous peoples. She worked as a volunteer for an umbrella Indigenous peoples' organization from

Russia – LIENIP (www.indigenous.ru) – from 2004 to 2009. As a researcher she investigates issues related to the protection of the traditional knowledge of Indigenous peoples and has published on the subject. Currently, Ulia is a doctoral candidate in the Department of Information Studies, University of California at Los Angeles, and a principal investigator for a grant funded initiative between the UCLA academic community, the Hopi Preservation office, and the World Intellectual Property Organization, focused on the protection of the cultural and intellectual property of Indigenous peoples.

LISA O'SULLIVAN completed her doctorate at Queen Mary University of London in 2006. Her thesis examined the construction of 'nostalgia' as a clinical and political category in nineteenth century France. From 1991 until 1997 she worked with the Australian Science Archives Project, University of Melbourne, in the preservation and communication of Australia's scientific, technological and medical heritage. Between 2005 and 2009 she was Chair of the Human Remains Subject Specialist Network, a network of UK museum professionals responsible for human remains collections. Since 2003, she has been Senior Curator of medicine at the Science Museum, London, where she curates the Wellcome medical collections. As part of this role she is responsible for issues relating to human remains and culturally sensitive objects. During the year 2010–11 she was a postdoctoral research fellow at the University of Sydney, investigating the material cultures of anthropological and anatomical collecting within the context of scientific studies of race.

MATTHEW RIMMER is an Australian Research Council Future Fellow working on Intellectual Property and Climate Change. He is an associate professor at the ANU College of Law, and an associate director of the Australian Centre for Intellectual Property in Agriculture (ACIPA). He holds a BA (Hons) and a University Medal in literature, and a LLB (Hons) from Australian National University. Rimmer received a PhD in law from the University of New South Wales for his dissertation on "The Pirate Bazaar: The Social Life of Copyright Law" (2001). He is a member of the ANU Climate Change Institute, and a director of the Australian Digital Alliance. Rimmer has published widely on copyright law and information technology, patent law and biotechnology, access to medicines, clean technologies, and traditional knowledge. He is the author of three research monographs, *Digital Copyright and the Consumer Revolution: Hands off my iPod* (2007), *Intellectual Property and Biotechnology: Biological Inventions* (2008), and *Intellectual Property and Climate Change: Inventing Clean Technologies* (2011). He also edited the thematic issue of *Law in Context* entitled "Patent Law and Biological Inventions" (2006), and co-edited a collection on access to medicines entitled *Incentives for Global Public Health: Patent Law and Access to Essential Medicines* (with Kim Rubenstein & Thomas Pogge, 2010). Rimmer was also a chief investigator in an Australian Research Council Discovery Project, "Gene Patents In Australia: Options For Reform" (2003–2005), and an Australian Research Council Linkage Grant, "The Protection of Botanical Inventions" (2003). He is currently a chief investigator in an Australian Research Council Discovery Project, "Promoting Plant Innovation in Australia" (2009–2011) and the holder of an Australian Research Council Future Fellowship for work on "Intellectual Property and Climate Change" (2011–14).

WENDY D. ROTH is an Assistant Professor of Sociology at the University of British Columbia. Her research focuses on racial classifications and boundaries, and how social processes such as intermarriage, immigration, and genetic technology challenge those categories. Her current project on how genetic ancestry testing affects racial and ethnic identities and definitions of race is funded by grants from the Canada Foundation for Innovation and the Social Sciences and Humanities Research Council of Canada. She is the winner of the 2007 Dissertation Award from the American Sociological Association. She received her PhD in Sociology and Social Policy from Harvard University in 2006. Her chief publications are: "Racial Mismatch: The Divergence Between Form and Function in Data for Monitoring Racial Discrimination of Hispanics," *Social Science Quarterly* (2010), "'Latino Before the World': The Transnational Extension of Panethnicity," *Ethnic and Racial Studies* (2009), and "The End of the One-Drop Rule? Labeling of Multiracial Children in Black Intermarriages," *Sociological Forum* (2005).

MARIE-CLAUDE STRIGLER is an Associate Professor at the University Paris III – Sorbonne Nouvelle, and a member of APSAM (Political and social anthropology of North America), a Paris XII research group. After her doctoral dissertation on the economic policy of the Navajo tribal government, she published several books on various aspects of Navajo culture and economy, and a history of American Indians. She has also written a number of articles, both in French and in English, about the current economic and political evolution of Native American nations. She is a member of the network of experts for GITPA / IWGIA (International Work Group for Indigenous Affairs). Her main published works are: *Moi Sam Begay, homme-médecine navajo* (2010); *Histoire des Indiens des États-Unis – l'autre FarWest* (2007, co-ed. with

Christian Gros), *Être indien dans les Amériques* (2006); *La Nation Navajo, tradition et développement* (2000); *La Médecine navajo* (2001); and *Parlons navajo, mythes, langue et culture* (2002).

SANDRINE TOLAZZI is an Associate Professor in the Department of Foreign Languages of the University of Grenoble, where she teaches Canadian and Australian civilization; she is a member of the research group on Modes of Representation in English Studies (CEMRA). She is also the Vice-President of Grenoble's Centre for Canadian Studies. After a comparative analysis of multiculturalism policies in Canada and Australia, whose aim was to underline the necessity of developing a sense of national identity while managing cultural diversity in liberal societies (doctoral dissertation), she went on to conduct research related to questions of citizenship, group identity, and the integration of ethnic minorities in these two countries. More recently, she has focused on Indigenous peoples, for whom these questions can also be raised, though in a very different context owing to the history of colonization and exploitation that they have known. She sees the reluctance of many Indigenous populations to have their genes studied through the prism of this history and thus tries to point out the difficulties of reconciling what some present as biomapping and others see as biocolonizing.

YU-UEH TSAI received her PhD in Sociology from the National Taiwan University in 2006. She is Assistant Research Fellow in the Institute of Sociology, Academia Sinica, Taiwan, and also Adjunct Assistant Professor in the Department of Sociology, National Taiwan University. She was a Postdoctoral Researcher on the Science Studies Program, University of California, San Diego (September 2007–August 2008) and Visiting Research Fellow at the Department of Social Medicine, Harvard Medical School (July 2004–August 2005). She works in the areas of medical sociology, cultural psychiatry, race/ethnicity, Aboriginal health, and sociology through documentary films. She is currently researching the complex relationships between biomedicine, identity-politics, and modernity involved in the geneticization of Aboriginal health and identity in Taiwan and has published, among other works, *Mental Disorder of the Tao Aboriginal Minority in Taiwan: Modernity, Social Change, and the Origin of Social Suffering* (2009, in Chinese). This book explores the difficult issue of the rising rate of mental disorders among the Taos over the past three decades, which has been typically attributed to their genetic background, demonstrating that it must be explained mainly in terms of socio-economic and cultural influences that are closely related to their social position as an ethnic minority.

SHEILA VAN HOLST PELLEKAAN, RN., BA (Hons), MA (Hons), Dip. Ed. Sydney College of Advanced Education, PhD University of Sydney, is currently a Visiting Senior Research Fellow in the School of Biotechnology and Biomolecular Sciences, University of New South Wales, Australia. Her research includes ethically approved DNA studies with medical, anthropological and Indigenous colleagues and continued liaison with participants and communities regarding the reporting of results and future research. Her early career was in nursing and medical research prior to academic studies in anthropology and archaeology. Her focus on Aboriginal Australian studies sprang from an awareness that a trusting relationship between researchers and researched required development and her research in genetics arose from this background at a time when Indigenous concerns were influencing the establishment of guidelines for researchers. While holding a position as Senior Lecturer in the Faculty of Nursing at the University of Sydney, she devoted a great deal of effort both to improving opportunities for Aboriginal and Torres Strait Islander students and staff and to the inclusion of relevant content into courses. She has published two articles on genetic research in the historical context.

GERALD VIZENOR is Distinguished Professor of American Studies at the University of New Mexico, and Professor Emeritus of American Studies at the University of California, Berkeley. He is a citizen of the White Earth Nation and has published more than thirty books, including, more recently, *Native Liberty: Natural Reason and Cultural Survivance*, *Survivance: Narratives of Native Presence*, *Native Stories*, *Father Meme*, *Fugitive Poses: Native American Indian Scenes of Absence and Presence*, *Manifest Manners: Narratives of Postindian Survivance*, *Hiroshima Bugi: Atomu 57*, and *Shrouds of White Earth*. Vizenor was the Principal Writer of the Constitution of the White Earth Nation in Minnesota. He received the American Book Award for *Griever: An American Monkey King in China*, the Western Literature Association Distinguished Achievement Award, and the Lifetime Literary Achievement Award from the Native Writer's Circle of the Americas. He is the series editor with Diane Glancy of "Native Stories: A Series of American Narratives." He is also series editor with Deborah Madsen of "Native Traces."

NATASHA GOLBECK completed her MA in Sociology at the University of British Columbia in Summer 2011. Her research was on violence and the social meaning of space in Vancouver's Stanley Park. For this research she received the Joseph–Armand Bombardier scholarship from the Social Sciences and Humanities Research Council. Natasha works in the non-profit sector in Vancouver; she has worked in administration and governance for a sex-worker support agency for the past several years, and is currently running a writing centre and publishing house for people facing mental illness and addiction issues.

ANDREA ZITTLAU works currently as a research assistant and lecturer in the department of North American Studies at the University of Rostock, Germany. Additionally, she coordinates the Graduate School "Cultural Encounters and Discourses of Scholarship," also at the University of Rostock. She has written about museums and the representation of Native Americans in exhibitions, and on issues of tourism. Her latest project is concerned with aspects of medical history and disability studies.

⌘

Index